Basic Electrical Engineering and Instrumentation for Engineers
SECOND EDITION

E.C. Bell
R.W. Whitehead

Department of Electrical and Electronic Engineering University of Bradford

COLLINS
8 Grafton Street, London W1

Collins Professional and Technical Books
William Collins Sons & Co. Ltd
8 Grafton Street, London W1X 3LA

First published in Great Britain by
Granada Publishing, 1977 (ISBN 0-258-97051-0)
Reprinted 1978, 1979
Second edition 1981 (ISBN 0-246-11477-0)
Reprinted 1982
Reprinted by Collins Professional and Technical Books, 1985

Distributed in the United States of America
by Sheridan House, Inc.

Copyright © 1981, E. C. Bell and R. W. Whitehead

ISBN 0-00-383181-7

Printed and bound in Great Britain by
Mackays of Chatham Ltd, Kent

All rights reserved. No part of this publication may
be reproduced, stored in a retrieval system or transmitted,
in any form, or by any means, electronic, mechanical, photocopying,
recording or otherwise, without the prior permission of the
publishers.

Preface

It is our experience when teaching electrical engineering to students in other disciplines that the text books available provide a depth of treatment of certain sections of the syllabus far in excess of that required, whilst not providing an overall coverage of the basic principles of the subject. As the interpretation of a syllabus is very much a matter of personal interests, no doubt the same criticism will be levelled at this text, but we hope we have produced a book that is not only particularly suitable for service courses, but which may be useful as a first reader for electrical engineers.
In line with current trends, the chapter on Digital Systems has been extended and up-dated, and a new chapter on microprocessors included.
Our grateful thanks are extended to Mr. R. Hartley for his care with the illustrations and to Mrs. M. Mawson and Mrs. M. Bell for their patience and perseverance with the typing.

E.C.B.
R.W.W.

Contents

1   FUNDAMENTALS OF ELECTRIC, ELECTROMAGNETIC AND ELECTROSTATIC SYSTEMS   1

    1.1 Physically analogous systems   1
    1.2 Electric circuit   3
    1.3 Electromagnetic system equations   4
    1.4 Electrostatic system equations   17
    1.5 Energy stored in inductors and capacitors   21
    Examples   24

2   ELECTRIC CIRCUITS   26

    2.1 Power and energy   26
    2.2 Solution of circuits   27
    2.3 Solution of simple RL and RC circuits under transient conditions   28
    2.4 Solution of steady-state d.c. circuits   35
    Examples   48

3   STEADY-STATE ALTERNATING CURRENTS   49

    3.1   Differential equation for RLC circuit   49
    3.2   Definition of quantities   50
    3.3   Sinusoidal response of RLC circuit   52
    3.4   Phasor diagrams   52
    3.5   Root mean square value   54
    3.6   The use of complex numbers or 'j' notation   55
    3.7   Power in a.c. circuits   58
    3.8   Frequency response of circuits   60
    3.9   Three-phase circuits   62
          Examples   78

4   DEVICES   80

    4.1   Semiconductors   80
    4.2   Doped semiconductors   84
    4.3   The p-n junction diode   86
    4.4   The diode in a circuit   95
    4.5   The Zener diode   106
    4.6   The junction transistor   108
    4.7   The transistor in a circuit   115
    4.8   The field effect transistor (FET)   123
    4.9   Switching devices   129
          Examples   138

5   AMPLIFIERS   141

    5.1   Amplifier gain   141
    5.2   Input and output impedance   147
    5.3   Determination of amplifier input and output resistance   152
    5.4   Logarithmic gain units   The decibel   155
    5.5   Amplifier bandwidth   161
    5.6   The application of feedback to amplifiers   172
    5.7   Transistor amplifiers   183
    5.8   Operational amplifiers   192
          Examples   196

6   POWER SUPPLIES   198

    6.1   The half wave rectifier with resistive load   199
    6.2   The full wave rectifier with resistive load   201
    6.3   Polyphase rectifiers   205
    6.4   Ripple reduction   208
    6.5   Voltage multiplying rectifiers   216
    6.6   Power supply regulation   217
    6.7   Controlled power rectification   223
          Examples   224

7   TRANSFORMERS   226

    7.1   Transformer action   226
    7.2   The equivalent circuit of a transformer   228
    7.3   Transformer tests   233
    7.4   Phasor diagram of a transformer on load   234
    7.5   The voltage equation of a transformer   234
    7.6   Voltage regulation   235
    7.7   Efficiency of a transformer   237
    7.8   Auto-transformers   238
    7.9   Three-phase transformers   239
          Examples   243

8   DIGITAL SYSTEMS   245

    8.1   Analogue and digital systems   245
    8.2   Encoding the amplitude samples - number systems and codes   255
    8.3   Mathematical treatment of coded signals - Boolean algebra   269
    8.4   Electronic gates   277
    8.5   The realisation of switching functions using simple logic gates   290
    8.6   The Karnaugh map   295
    8.7   Realisations using NAND or NOR gates   307
    8.8   The use of the American military specification gate symbols   316
    8.9   Sequential networks   319
    8.10 Registers and counters   328
    8.11 The multivibrator   335
          Examples   341

9   MICROPROCESSORS   343

    9.1   Digital computers   343
    9.2   The power of a C.P.U.   347
    9.3   Memory organisation   348
    9.4   The Motorola 6800 microprocessor   355
    9.5   Machine code programming   364
    9.6   Entering the programme   375
    9.7   Assembly language programme   377
    9.8   Input/Output programming   382
    9.9   The use of microprocessors in digital system design   389

10 ELECTRICAL TRANSDUCERS 391

- 10.1 Static properties of transducers 393
- 10.2 Dynamic properties of transducers 396
- 10.3 Variable resistance transducers 403
- 10.4 Variable reactance transducers 419
- 10.5 Electrodynamic transducers 429
- 10.6 Piezoelectric transducers 431
- 10.7 Thermocouples 431
- Examples 433

11 SIGNAL PROCESSING 435

- 11.1 Noise in systems 435
- 11.2 Analogue digital conversions 440
- 11.3 Real signals 448
- 11.4 Modulation systems 457
- Examples 468

12 ELECTRICAL MACHINES 469

- 12.1 The d.c. generator 470
- 12.2 D.C. motors 479
- 12.3 Three-phase rotating fields 486
- 12.4 Three-phase alternators 487
- 12.5 Synchronous motors 489
- 12.6 Induction motors 490
- 12.7 Single-phase induction motors 502
- 12.8 Plain series motor 504
- Examples 507

13 MEASUREMENTS 509

- 13.1 The basic types of ammeters and voltmeters 509
- 13.2 Bridge methods to measure resistance, inductance and capacitance 516
- 13.3 Measurement of small values of resistance 519
- 13.4 Measurement of large values of resistance 520
- 13.5 Digital instruments 521
- Examples 525

Appendix 1   Heating, Lighting and Tariffs   516
Appendix 2   The measurement of flux density   533
Appendix 3   A note on the manipulation of complex numbers   535
Appendix 4   537

CHAPTER ONE

Fundamentals of Electric, Electromagnetic and Electrostatic Systems

1.1 <u>Physically Analogous Systems</u>

Shown below are three basically similar systems involving the flow of liquid, of heat and of electricity.

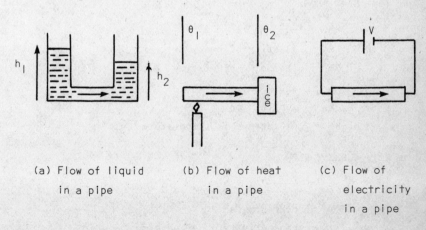

    (a) Flow of liquid   (b) Flow of heat   (c) Flow of
          in a pipe   in a pipe   electricity
                                                                                     in a pipe

Figure 1.1

In each system there is a relationship between the flow and the cause of flow. Let the quantity that flows be termed the FLUX and the quantity that causes the flow be termed the POTENTIAL DIFFERENCE (p.d.), then in each case, for a fixed value of the p.d., the FLUX is proportional to the AREA of the pipe and inversely proportional to the LENGTH of the pipe. The FLUX is also dependent on the material. In case (a) there would be greater flow if the liquid were water rather than syrup, and in cases (b) and (c) there would be greater flow if the pipe were metallic rather than plastic. In general therefore

$$\text{FLUX} = \frac{\text{MEDIUM FACTOR} \cdot \text{AREA}}{\text{LENGTH}} \cdot \text{p.d.} \qquad 1.1$$

Case (a), LIQUID FLOW

$$Q = \frac{\frac{1}{\eta} \cdot A}{\ell} (h_1 - h_2) \qquad 1.2$$

where
Q — quantity of liquid m^3/sec
η — viscosity of liquid Ns/m^2
A — area of pipe m^2
ℓ — length of pipe m
$h_1 - h_2$ — difference in head N

Case (b), HEAT FLOW

$$H = \frac{k\,A}{\ell}(\theta_1 - \theta_2) \qquad 1.3$$

H — quantity of heat W
k — thermal conductivity $W/m\,^\circ C$
A — area m^2
ℓ — length m
$\theta_1 - \theta_2$ — difference in temperature $^\circ C$

Case (c), CURRENT FLOW

$$I = \frac{\sigma \cdot A}{\ell} \cdot V \qquad 1.4$$

I — current A
σ — conductivity $\text{siemens}/m$
A — area m^2

ℓ - length   m
V - potential difference or voltage   V

1.2 Electric Circuit

If the term $\frac{\sigma A}{\ell}$ of equation 1.4 is written as the conductance G then

$$I = GV \qquad 1.5$$

The reciprocal of conductance is called the resistance R, and

$$R = \frac{1}{G} = \frac{\ell}{\sigma A} \qquad 1.6$$

Therefore $\quad I = \frac{V}{R} \qquad 1.7$

which is recognisable as OHM's LAW.

Equation 1.4 may be re-written as

$$\frac{I}{A} = \frac{\sigma V}{\ell}$$

or $\quad J = \sigma E \qquad 1.8$

where J - current density A/m^2
and   E - potential gradient   V/m

It is more usual, however, to define the resistance of a conductor in terms of the resistivity, rather than the conductivity, of the material.

$$R = \frac{\rho \ell}{A} \qquad 1.9$$

where ρ is the resistivity, and $\rho = \frac{1}{\sigma}$.

Example

The resistivity of copper at $20°C$ is 1.76 ohm-m 10^{-8}. The resistance of a 10 metre length of copper wire whose cross sectional area is 1 mm^2 is

$$R = \frac{1.76 \cdot 10^{-8} \cdot 10}{1 \cdot 10^{-6}} = 0.176 \text{ ohm}$$

The resistivity of nichrome at $20°C$ is 108 ohm-m 10^{-8}. The resistance of an equivalent length of nichrome wire is,

4 Basic Electrical Engineering and Instrumentation for Engineers

$$R = \frac{108 \cdot 10^{-8} \cdot 10}{1 \cdot 10^{-6}} = 10.8 \text{ ohm}$$

Copper is the material that is used in the manufacture of electric cables. Nichrome is used to make heating elements.

1.3 Electromagnetic System Equations

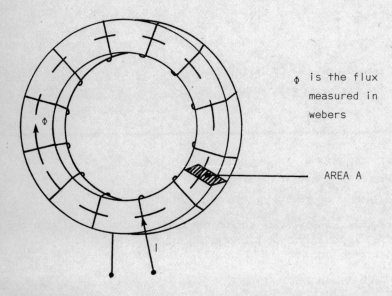

ϕ is the flux measured in webers

AREA A

Figure 1.2

Consider a toroidal coil of N turns carrying a current I amperes. A MAGNETIC FLUX ϕ is produced as shown by the dotted line in figure 1.2. In this case the p.d. which causes the flux to flow is the MAGNETIC POTENTIAL DIFFERENCE, which is termed the MAGNETO-MOTIVE-FORCE, F and is equal to the number of turns on the coil multiplied by the current in the coil. F = I N ampere turns. The medium factor of electromagnetic systems is the PERMEABILITY, (μ). Therefore substituting in equation 1.1

$$\phi = \frac{\mu A}{\ell} \cdot F \qquad \qquad 1.10$$

where ϕ - magnetic flux   webers
        μ - permeability   henry/m

A - cross sectional area of the toroid m^2

ℓ - length of the flux path, i.e. mean circumference of the toroid m

F - m.m.f. At

The term $\frac{\mu A}{\ell}$ is written as the PERMEANCE Λ, thus

$$\phi = \Lambda F \qquad 1.11$$

The reciprocal of permeance is termed the RELUCTANCE, S,

$$S = \frac{\ell}{\mu A} \qquad 1.12$$

and

$$\phi = \frac{F}{S} \qquad 1.13$$

which might be called the "electromagnetic ohms law". Equation 1.10 may be re-written as

$$\frac{\phi}{A} = \mu \frac{F}{\ell}$$

or $\qquad B = \mu H \qquad\qquad 1.14$

where $\quad B$ - flux density Wb/m^2 or Tesla

and $\qquad H$ - magnetic intensity At/m

This relationship is particularly important when dealing with iron cored electromagnetic systems, and will be referred to again in Section 1.3.5.

1.3.1 Direction of magnetic flux

The direction of the magnetic flux ϕ produced by the current I in the toroidal coil of figure 1.2 was not given an arbitrary direction, it was obtained from the "screw rule", as shown in figure 1.3. The magnetic field produced by a current is in the direction given by turning a right hand screw thread.

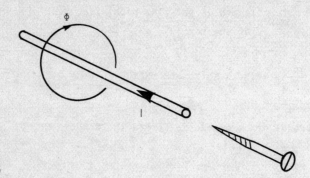

Figure 1.3

1.3.2 Faraday's Law

One of the most fundamental and important laws of electromagnetic systems is FARADAY'S LAW. It states that the voltage induced in a magnetic circuit is equal to the rate of change of flux linkage in that circuit. In equation form it is written as,

$$e = N \frac{d\phi}{dt} \qquad 1.15$$

The generation of all domestic and industrial electric power utilises this effect.

In many texts Faraday's Law is written as,

$$e = -N \frac{d\phi}{dt} \qquad 1.16$$

This equation, however, inherently incorporates LENZ'S LAW which states that the direction of the induced voltage is such as to produce a current which will in turn produce a flux which opposes the initial flux change. Throughout this text equation 1.15 will be used to obtain the magnitude of the induced voltage, and when the direction of this voltage is important LENZ'S LAW will be incorporated.

The rate of change of flux linkage can be achieved by (a) having a static coil linked by a time varying field, or (b) by having a rate of change of space displacement between the coil and the field, i.e. by moving the coil relative to the field, figures 1.4(a) and 1.4(b).

Fundamentals of Electric, Electromagnetic and Electrostatic Systems 7

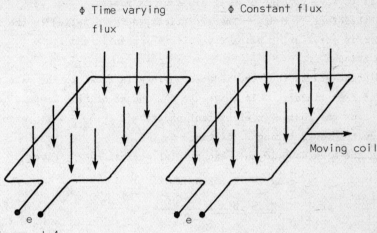

Figure 1.4

The latter case is the technique used in all rotating electric generators.

An expression for the force on a current carrying conductor in a magnetic field can now be derived by using Faraday's Law. This force expression is fundamental to the operation of all electric motors. Figure 1.5 shows a conductor of length ℓ metres carrying a current I amperes, lying perpendicular to a magnetic field of flux density B tesla.

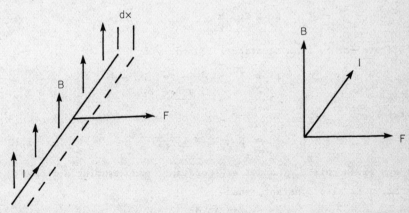

Figure 1.5

Assume that the force produced electromagnetically on the conductor causes it to move through a small distance dx metres. The work done is therefore $\int F\,dx$. The electrical energy supplied to the conductor is $\int e\,I\,dt$, e being the voltage across it and I the current through it.

Faraday's Law states that the induced voltage is equal to the rate of change of flux linkage. In this case the change of flux linkage is $B.\ell.dx$, thus the change in electrical energy $= I \int \frac{B\,\ell\,dx}{dt}\,.\,dt$ when the conductor moves through a distance dx.

Equating the work done to the change in electrical energy gives

$$\int F\,dx = I \int B\,\ell\,dx$$

or $\qquad \underline{F = B\,\ell\,I} \quad \text{newtons} \qquad\qquad 1.17$

1.3.3 Self Inductance

A coil is said to possess self inductance if, when a current is passed through it, a magnetic flux is produced which links the coil. When dealing with electric circuits it is convenient if the voltage across individual circuit elements can be written as a function of the current flowing through them, and to facilitate this for electromagnetic coil systems, Faraday's Law is presented in an alternative form. From equation 1.15

$$e = N \frac{d\phi}{dt}$$

Substitute for ϕ, from equations 1.13 and 1.12

$$\phi = \frac{F}{S} = \frac{N\,i}{\ell/\mu A} = N\,i\,\frac{\mu A}{\ell} \qquad\qquad 1.18$$

therefore

$$e = N\frac{d}{dt}(N\,i\,\frac{\mu A}{\ell}) \qquad\qquad 1.19$$

For any given coil N, A and ℓ are constant, and assuming a linear system, μ is also constant, thus

$$e = N^2\,\frac{\mu A}{\ell}\,\frac{di}{dt} \qquad\qquad 1.20$$

The term $N^2 . \frac{\mu A}{\ell}$ or $N^2 \Lambda$ is called the SELF INDUCTANCE of the coil, and is given the symbol L, and the unit, henry. The alternative form of Faraday's Law is therefore

$$e = L \frac{di}{dt} \qquad 1.21$$

The self inductance L is defined as

$$L = N^2 \frac{\mu A}{\ell} \text{ henry} \qquad 1.22$$

or by substituting for ϕ from equation 1.18

$$L = \frac{N \phi}{i} \text{ henry} \qquad 1.23$$

1.3.4 Mutual Inductance

Two coils are said to possess Mutual Inductance if a current in one produces a magnetic flux which links the other. The mutual inductance M between two coils is

$$M = \frac{N_2 \phi}{I_1} \text{ henry} \qquad 1.24$$

where
N_2 is the number of turns on coil 2
I_1 is the current through coil 1
ϕ is the flux linking coils 1 and 2

The mutual inductive effect is utilised in all electrical machines and transformers.

1.3.5 Magnetic circuits with Iron Cores

Almost all practical applications of electromagnetic systems involve iron (ferrous) cores of some form or other, as it is much easier to establish magnetic flux in an iron cored coil than it is in an air cored coil. Consider two identical toroidal cores, one wound on a wooden* former (air cored) and the other on a cast steel former (iron cored). If the current in the coil is increased in steps and the flux density B in the core measured at each step, then a plot of

---

*Any non-ferrous material used as a former for a coil may be considered as having the magnetic properties of air.

B against H can be obtained. The measurement of B is discussed in Appendix 2. From the B-H or "magnetisation" curves shown in figure 1.6, it can be seen that for the same magnetic intensity H_1 the flux density produced in the cast steel core is about 600 times that produced in the air core. It is also seen from figure 1.6 that the ratio of B-H (i.e. permeability) is not constant in the case of cast steel. This is true for all ferrous cored coils, where at large values of flux density the core is said to "saturate" and a large increase in m.m.f. will produce only a small change in flux density.

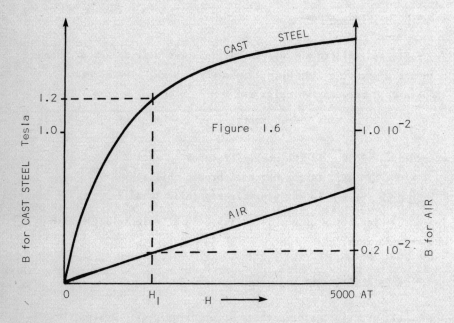

Figure 1.6

To take account of this the permeability is re-written as

$$\mu = \mu_o \, \mu_r \qquad 1.25$$

where μ is the permeability and is the ratio of B to H for any point on the magnetisation curve

μ_o is the permeability of free space, and is constant and equal to $4\pi \times 10^{-7}$

μ_r is the relative permeability, which is variable depending on the operating point on the magnetisation curve

The relative permeability for cast steel varies from about 400 to 900. The relative permeability can be greatly increased ($\approx$100,000) by using nickel iron alloy cores.

1.3.6 Hysteresis

Consider a toroidal coil wound on an iron core which is completely demagnetised. Let the current in the coil be taken through a cyclic change from zero to + I to zero to - I to zero to + I. If the flux density in the core is measured as the current and hence the magnetic intensity is changed then a curve of B versus H can be plotted, figure 1.7. This curve is termed the "hysteresis loop" for the core material and information regarding the core material and its practical uses can be obtained directly from it.

It can be seen from figure 1.7 that when the current is reduced to zero the core retains some magnetism. This means that all the electrical energy supplied to the coil to magnetise the core is not returned when the coil current is reduced to zero. There is therefore a nett loss in energy which is termed "hysteresis loss", and which manifests itself as heat in the core. An analogy may be drawn to the heating of car tyres which are subjected to a cyclic variation of deformation. The portion of the tyre in contact with the road is under compression which is gradually released as it begins to break contact with the road. The rubber does not regain its initial no load condition because of "hysteresis", and the "hysteresis loss" is manifested as heat.

The hysteresis loss can be shown to be proportional to the area of the hysteresis loop in the following way.

The electrical energy supplied to the toroidal coil is

$$w_{elec} = \int v \, i \, dt$$

By Faraday's Law the voltage induced in the coil is

$$e = N \frac{d\phi}{dt}$$

If the coil is assumed to have no resistance then the induced voltage e is equal to the applied voltage v, thus the energy supplied is

$$w_{elec} = \int i N \frac{d\phi}{dt} dt = \int i N \, d\phi$$

Now $\phi = BA$ and $H = \frac{iN}{\ell}$

Therefore

$$w_{elec} = A\ell \int H \, dB$$

or the electrical energy supplied per unit volume of core is

$$w_{elec} = \int H \, dB$$

Referring to figure 1.7, as the magnetic intensity is increased from zero to AQ (by increasing the current in the coil), the electrical energy supplied to the coil is given by the area FGPRF. The energy returned when the magnetic intensity is reduced to zero, is given by the area PCRP. Thus the nett loss in energy is area FGPCA. A similar argument may be used as the magnetising force is increased in the negative direction so that the total loss in energy for one complete cycle of magnetisation is the area of the hysteresis loop.

1.3.7 Eddy Current Loss

In Section 1.3.2, figure 1.4(a), it was shown that a time varying flux which links a coil will induce an e.m.f. in that coil, the magnitude of which is given by Faraday's Law, equation 1.15. If the ends of that coil are connected together to form a closed circuit, the induced voltage will cause a current to flow.

Consider now an iron core in which there exists a time varying flux as shown in figure 1.8. A typical path PQRS in the core is equivalent to a closed coil; therefore, because iron is a conductor, a current will flow. There are of course many such paths and currents in the core. These currents are termed EDDY CURRENTS and because of the ohmic resistance of the core there will be an energy loss associated with these currents, which is called the EDDY CURRENT LOSS and is dissipated as heat in the core.

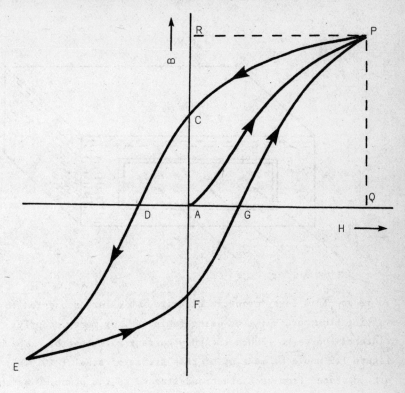

Figure 1.7   HYSTERESIS LOOP

- AC REMANENT FLUX DENSITY i.e the flux density remaining in the iron core when the magnetic intensity has been reduced to zero.

- AD COERCIVE FORCE i.e. the magnetising force to reduce the residual or remanent flux density to zero

Figure 1.8

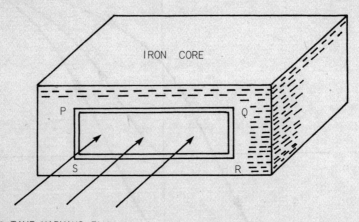

Φ TIME VARYING FLUX

To reduce this loss, magnetic circuits, which are subjected to a time varying flux, are built up using laminated, or more recently, sintered* cores to reduce the eddy current path length. The core of figure 1.8 would be made up of thin strips of steel ($\approx 0.014"$) which are insulated from each other and stacked in the plane PQ as shown. The phenomenon of "eddy currents" is put to good use in several heating applications which go under the name of "Induction Heating". The metallic workpiece to be heated is placed inside a coil which is supplied with alternating current (discussed in Chapter 3). Eddy currents are induced in the workpiece which manifest themselves as heat. The depth of penetration of the heat is dependent on the depth of penetration of the eddy currents which in turn is dependent on the frequency of the alternating current supplied to the coil. Very high frequencies are used to produce surface heating, and this type of induction furnace is extensively used to surface harden castings.

* Sintered cores are made by bonding ferrous dust with epoxy resin

1.3.8 Some applications of Iron Cored Magnetic Systems

It was shown in section 1.3.6 that the area of the hysteresis loop was proportional to the hysteresis loss. It is essential therefore that any magnetic system which is to be subjected to a continuous cyclic variation of flux density (an alternating current system) must have a magnetic core whose hysteresis loop has a small area. All electrical machines, transformers, electromagnetic lifting magnets, relays, solenoids, etc. operating off the a.c. mains, would have a silicon iron, low loss core, loop 1 in figure 1.9. This type of core is termed a "soft" magnetic core.

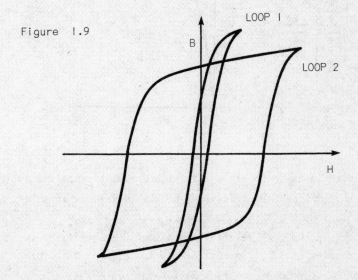

Figure 1.9

Permanent magnets, however, require a high remanent flux density, and as there is no cyclic variation of flux density, there will be no hysteresis loss and thus the loop area is unimportant. A "hard" magnetic material, loop 2, in figure 1.9 would be the type used for a permanent magnet.

Example

Consider the two "C" cores shown in figure 1.10. The magnetic length ℓ of each core is 50 cm, and the cross sectional area is 10 cm². A coil of 1000 turns is wound on the cores and they are clamped together with non-magnetic spacers to form an air gap of g mm in each limb.

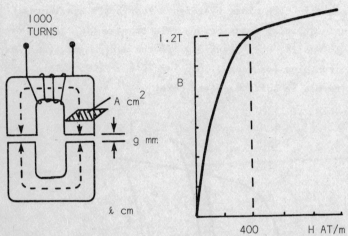

Figure 1.10

If the flux density in the core has not to exceed 1.2 T, calculate the current in the coil and the inductance of the coil for a range of spacers 0.5, 1, 2, 5 mm.

From the B-H curve of the core material the magnetic intensity H_c to produce a flux density of 1.2 T in the core is 400 AT/m. Therefore the m.m.f. required to produce the flux in the core is

$$F_c = 400 (50 + 50) 10^{-2} = 400 \text{ AT}$$

The magnetic intensity required to produce a flux density of 1.2 T in the air gap is

$$H_g = \frac{B}{\mu_o} = \frac{1.2}{4\pi \, 10^{-7}}$$

Therefore the m.m.f. required to produce the flux in the air gap, for a gap of 0.5 mm, is

$$F_g = \frac{1.2}{4\pi \, 10^{-7}} (2 \times 0.5) 10^{-3} = 0.95 \cdot 10^3$$

Fundamentals of Electric, Electromagnetic and Electrostatic Systems 17

The total m.m.f. required to produce the flux in the core and the air gap is,

$$F = Fc + Fg$$
$$F = 400 + 950 = 1350 \text{ AT}$$

The current in the coil is

$$I = F/1000 = 1.35 \text{ A}$$

The inductance of the coil may be obtained from equation (1.23)

$$L = \frac{N\phi}{I} = \frac{N B A}{I}$$

$$L = \frac{1000 \cdot 1.2 \cdot 10 \cdot 10^{-4}}{1.35} = 0.89 \text{ H}$$

The complete set of answers for the different air gaps is given in the table below:

g_{mm}	F AT	I amps	ϕ Wb	L henry
0.5	1350	1.35	$1.2 \cdot 10^{-3}$	0.89
1.0	2300	2.3	$1.2 \cdot 10^{-3}$	0.52
2.0	4200	4.2	$1.2 \cdot 10^{-3}$	0.29
5.0	9900	9.9	$1.2 \cdot 10^{-3}$	0.12

It can be seen from the table that as the air gap is increased, the current required to produce a given flux density is increased, whereas the inductance is decreased. Why then are air gaps introduced in iron cored inductors? They are introduced to provide a measure of linearity, i.e. to linearise the B-H curve so that the inductance remains constant over a range of operating currents.

1.4 Electrostatic System Equations

The ancient Greeks were aware that when amber was rubbed by a suitable material, it acquired the power to attract small objects. The term "electricity" is, in fact, derived from the Greek word for amber. In the sixteenth century, Dr. Gilbert, physician to the Queen of England, discovered other substances which behaved in a similar manner. A little later a French Colonel called Charles

Coulomb designed and built a delicate torsion balance on which he performed a series of elaborate experiments to determine, qualitatively, the force exerted between two objects each holding a static charge of electricity. His published result is known as COULOMB'S LAW, and states that the force between two very small objects separated in free space by a distance which is large compared to their size is proportional to the charge on each and inversely proportional to the square of the distance between them.

$$F = K\frac{q_1 q_2}{R^2} \qquad 1.26$$

where q_1 and q_2 are the magnitudes of the charges, R is the separation distance, and K the proportionality constant.

In the M.K.S. system of units, q is in coulombs, R in metres, F in Newtons and $K = \frac{1}{4\pi\varepsilon_o}$, where ε_o is the PERMITTIVITY of free space and is numerically equal to $\frac{1}{36\pi} \times 10^{-9}$. The direction of the force between the two charges is on the line joining the charges, noting that like charges repel and unlike charges attract.

The question asked by many students is "What is charge?" This question cannot be answered directly. All that can be said is the term charge is the name of a certain concept that accounts for the physical behaviour of electrostatic systems. The same difficulty is encountered with Gravity which is the name of the concept used to account for the fact that all objects fall to the ground.

Consider two metallic plates of cross sectional area A m^2, separated by a distance ℓ metres, with a p.d. of V volts applied to them. An electrostatic flux ψ is set up between the plates, and therefore substituting in equation 1.1 we have

$$\psi = \frac{\varepsilon_o A}{\ell} \cdot V \qquad 1.27$$

where ε_o is the permittivity of free space, i.e. the medium factor. The term $\frac{\varepsilon_o A}{\ell}$ is termed the CAPACITANCE of the system.

The capacitance of a parallel plate condenser is

$$C = \frac{\varepsilon_o A}{\ell} \quad \text{farads} \qquad 1.28$$

Figure 1.11

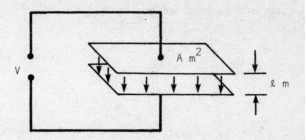

Equation 1.27 may be re-written as

$$\frac{\psi}{A} = \varepsilon_o \frac{V}{\ell}$$

or $\qquad D = \varepsilon_o E \qquad\qquad\qquad\qquad 1.29$

where D = electrostatic flux density coulombs/m^2

E = potential gradient V/m

Gauss's Law states that the flux emanating from a closed surface is equal to the charge enclosed by the surface.

$$\psi = Q$$

where Q is the total charge in coulombs. Therefore equation 1.27 becomes

$$Q = CV \qquad\qquad\qquad 1.30$$

1.4.1 Capacitors in Series and Parallel

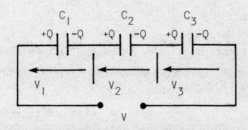

Figure 1.12

20 Basic Electrical Engineering and Instrumentation for Engineers

When a voltage is applied to capacitors connected in series the charge distribution is as shown in figure 1.12

Now $\quad V = V_1 + V_2 + V_3$

thus $\quad \dfrac{Q}{C} = \dfrac{Q}{C_1} + \dfrac{Q}{C_2} + \dfrac{Q}{C_3}$

and hence $\quad \dfrac{1}{C} = \dfrac{1}{C_1} + \dfrac{1}{C_2} + \dfrac{1}{C_3}$ \hfill 1.31

where C is the resultant capacitance of the series combination.

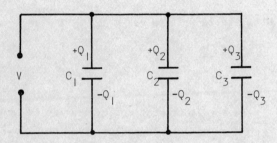

Figure 1.13

The total charge of the capacitors connected in parallel is

$$Q = Q_1 + Q_2 + Q_3$$

Thus $\quad CV = C_1V + C_2V + C_3V$

and hence $\quad C = C_1 + C_2 + C_3$ \hfill 1.32

where C is the resultant capacitance of the parallel combination.

1.4.2 Relative permittivity or Dielectric constant

The majority of capacitors used in practice are made by interleaving alternate strips of metal foil and insulating material (paper or mica) which is then rolled to form a cylinder.

Considering the simple case of the parallel plate capacitor of figure 1.11, if the space between the plates is filled with a material whose relative permittivity or dielectric constant is ε_r, then the capacity is redefined as

$$C = \varepsilon_o \varepsilon_r \frac{A}{\ell} \quad \text{farads} \qquad 1.33$$

The relative permittivity or dielectric constant of a material is defined as the ratio of the capacitance of a capacitor having that material interspaced between the plates, to the capacity of the same capacitor with a vacuum between the plates. The relative permittivities of waxed paper and mica are 5 and 6.5 respectively. It is possible to obtain relative permittivities of the order of 1000 by using high permittivity ceramics.

1.5 Energy stored in inductors and capacitors

Whenever the current through an inductor or capacitor is changing with respect to time there is a change in the stored energy. Consider that the current through an inductor changed from 0 to I in time t secs.

$$\text{Energy stored} = W = \int_o^t v\, i\, dt$$

$$W = \int_o^t i\, L\, \frac{di}{dt} \cdot dt$$

$$W = \int_o^I i\, L\, di$$

$$W = \tfrac{1}{2} L I^2 \qquad 1.34$$

Consider the charge on a capacitor to change from 0 to Q in time t secs.

$$\text{Energy stored} = W = \int_o^t v\, i\, dt = \int_o^Q v\, dq$$

$$W = \int_o^Q \frac{q}{C}\, dq$$

$$W = \tfrac{1}{2} \frac{Q^2}{C} \quad \text{or} \quad \tfrac{1}{2} C V^2 \qquad 1.35$$

Example

Calculate the capacitance of a pair of parallel metal plates, of area $120\, \pi\, \text{cm}^2$, separated by a distance of 2 mm, when the space between them is (a) filled with air $\varepsilon_r = 1$, (b) filled with a mica sheet $\varepsilon_r = 6$, of area 120π cm^2 and thickness 1 mm, (c) filled with a mica sheet $\varepsilon_r = 6$ area 60π cm^2, thickness 2 mm.

The capacitance of a parallel plate condenser is

$$C = \varepsilon_o \varepsilon_r \frac{A}{\ell}$$

(a) $C = \dfrac{10^{-9}}{36 \pi} \cdot 1 \cdot \dfrac{120 \pi \cdot 10^{-4}}{2 \cdot 10^{-3}} = 166.6 \cdot 10^{-12}$ farads

(b)

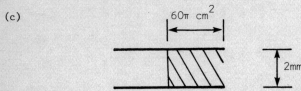

This capacitor is equivalent to two capacitors of area 120π cm^2 and thickness 1 mm, one filled with air, the other filled with mica, connected in series.

$C_{air} = \dfrac{10^{-9}}{36 \pi} \cdot 1 \cdot \dfrac{120 \pi \cdot 10^{-4}}{1 \cdot 10^{-3}} = 333.2 \cdot 10^{-12}$ farads

$C_{mica} = \dfrac{10^{-9}}{36 \pi} \cdot 6 \cdot \dfrac{120 \pi \cdot 10^{-4}}{1 \cdot 10^{-3}} = 1999.2 \cdot 10^{-12}$ farads

$\dfrac{1}{C} = \dfrac{10^{12}}{333.2} + \dfrac{10^{12}}{1999.2}$

$C = 285.71 \cdot 10^{-12}$ farads

(c)

This capacitor is equivalent to two capacitors of area 60 cm^2 and thickness 2mm, one filled with air, the other filled with mica, connected in parallel.

$$C_{air} = \frac{10^{-9}}{36\pi} \cdot 1 \cdot \frac{60\pi\, 10^{-4}}{2 \cdot 10^{-3}} = 83.3\, 10^{-12} \text{ farads}$$

$$C_{mica} = \frac{10^{-9}}{36\pi} \cdot 6 \cdot \frac{60\pi\, 10^{-4}}{2 \cdot 10^{-3}} = 500\, 10^{-12} \text{ farads}$$

$$C = 83.3 \cdot 10^{-12} + 500 \cdot 10^{-12}$$

$$C = 583.3\, 10^{-12} \text{ farads}$$

It should be noted that the unit of capacitance, i.e. the farad, is very large, and it is usual to break this down into microfarads and picofarads where

$$1 \text{ microfarad } (\mu F) = 10^{-6} \text{ farad}$$

$$1 \text{ picofarad } (pF) = 10^{-12} \text{ farad}$$

Examples

1. A straight conductor 10 cm long is placed between the faces of a permanent magnet. The flux density of the field produced by the permanent magnet is 1 tesla. If the mass of the conductor is 10 grms, what current must be passed through the conductor so that it just remains stationary? Sketch the field and current directions necessary for this condition to obtain.

(0.981)

2. A closed square 20 turn coil of side 20 cm and having a resistance of 10 Ω is placed with its plane at right angles to a magnetic field of flux density 0.1 tesla. Calculate the quantity of electricity passing through it when the coil is turned through 180° about an axis in its plane.

(16 millicoulombs)

3. A coil of 200 turns is wound on a soft iron core of cross sectional area 10 cm^2. A current of 1 ampere in the coil produces a flux of 2 mW in the core. Calculate the self inductance of the coil. An identical coil is placed on the core adjacent to the first such that 90% of the flux produced by the first coil links the second coil. Calculate the mutual inductance between the two coils. If the two coils are connected in series such that the flux they produce is additive, what is the effective inductance of the two coils?

(0.4 H, 0.36 H, 1.52 H)

4. A capacitor is formed from 10 sheets of metal foil each 2 cm by 10 cm. The metal sheets are separated by waxed paper 0.1 mm thick and with a dielectric constant of 2.03. Show how the metal foil can be connected to give a total capacitance of 40 picofarad or 3240 picofarad.

5. Three parallel conductors a, b, and c lie in the same plane, with b in the centre. The distances from b to a and c are 6 cm and 8 cm respectively. The currents in a and b are 1000 A and 700 A respect-

ively, flowing in opposite directions. The resultant force on b is 2 N per metre run acting towards c. Determine the magnitude and direction of the current in C.

(190.5 A)

6. The permittivity of the dielectric material between the plates of a parallel plate capacitor varies uniformly from ε_1 to ε_2. Show that the capacitance is

$$C = \frac{A(\varepsilon_2 - \varepsilon_1)}{d \ln \frac{\varepsilon_2}{\varepsilon_1}}$$

where d is the distance between the plates and
      A is the area of the plates.

CHAPTER TWO

Electric Circuits

2.1 Power and Energy

One major use of electric circuits is the transmission of power and energy from one place to another. Power is defined as the rate of doing work, or the rate of change of energy. Now the voltage across any two points in an electric circuit is defined as the work done in moving unit positive charge from the point of lower to the point of higher potential, thus the energy required to transfer a charge of q coulombs across a potential difference of v volts is

$$w = qv$$

or in incremental form

$$dw = dq\, v$$

therefore the power is given by

$$p = \frac{dw}{dt} = \frac{dq}{dt} v = iv \qquad 2.1$$

Using Ohm's law, the power supplied to a resistor is therefore

$$p = iv = i(iR) = i^2 R \qquad 2.2$$

Using Faraday's law, the power supplied to an inductor is

$$p = iv = iL\frac{di}{dt}$$

and the energy supplied is

$$w = \int p\,dt = \int iL\,di = \tfrac{1}{2} L i^2 \qquad 2.3$$

Using equation 1.35 we see that the voltage across a capacitor is

$$v = \frac{q}{C}$$

or in incremental form

$$dv = \frac{1}{C} i\,dt$$

therefore the power supplied to a capacitor is

$$p = iv = vC\frac{dv}{dt}$$

and the energy supplied is

$$w = \int p\,dt = \int Cv\frac{dv}{dt} = \tfrac{1}{2} C v^2 \qquad 2.4$$

In the case of the resistance there is an energy conversion process in that the power supplied to the resistance is converted into heat. However, with an inductor or capacitor there is no conversion but the energy supplied is stored and is available to be returned if the circuit conditions are changed.

2.2. Solution of Circuits

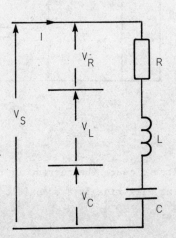

Figure 2.1

Consider the circuit shown in figure 2.1.

$$v_s = v_R + v_L + v_C \qquad 2.5$$

From equations 1.7 and 1.26 the voltages across the resistance and the inductance are

$$v_R = iR \text{ and } v_L = L\frac{di}{dt}$$

and from the previous section we showed that

$$dv_C = \frac{1}{C} i\,dt$$

therefore
$$v_c = \frac{1}{C} \int i \, dt$$

The complete voltage equation for the RLC circuit is therefore

$$v_s = iR + L\frac{di}{dt} + \frac{1}{C}\int i \, dt \qquad 2.6$$

The method used to determine the current from equation 2.6 depends to a great extent on the form of the supply voltage. If the supply voltage is time variant, equation 2.6 is solved as a differential equation. If the supply voltage varies sinusoidally with time, certain specialised techniques are developed, which are discussed in Chapter 3. For constant supply voltages (steady-state d.c. voltages) inductance and capacitance is meaningless and therefore steady-state d.c. circuits are purely resistive.

2.3 Solution of simple RL and RC circuits under transient conditions

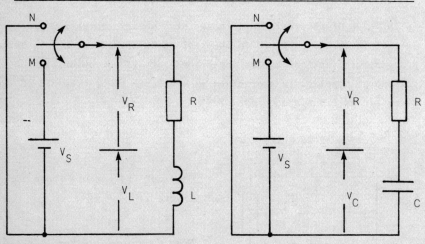

Figure 2.2

Consider the switch in position m.

RL circuit

At the instant just preceding the switch closure the current is zero. After the switch is closed, the current will rise in a transient manner to the steady-state value of v_s/R.

$$v_s = v_R + v_L$$

$$v_s = iR + L\frac{di}{dt} \qquad 2.7$$

RC circuit

At the instant just preceding the switch closure the voltage across the capacitor is zero. After the switch is closed the voltage v_c will rise in a transient manner to the steady-state value of v_s

$$v_s = v_R + v_c$$

also $q = C v_c$, and for a small increment in time dt

$$i\, dt = C\, dv_c$$

therefore

$$v_s = RC\frac{dv_c}{dt} + v_c \qquad 2.8$$

Equations 2.7 and 2.8 are of exactly the same form and may be solved by the technique of separating the variables and integrating.

From equation 2.7

$$I - i = \frac{L}{R}\frac{di}{dt}$$

where I is the final steady-state current and is equal to v_s/R

$$\frac{R}{L} dt = \frac{di}{I-i}$$

Integrating both sides

$$\frac{Rt}{L} = -\ln(I-i) + K$$

at $t = 0$, $i = 0$ so that $K = \ln I$

thus

$$\frac{Rt}{L} = \ln \frac{I}{(I-i)}$$

$$\frac{I-i}{I} = e^{\frac{-Rt}{L}}$$

$$i = I(1 - e^{\frac{-Rt}{L}}) \qquad 2.9$$

Now
$$v_L = L \frac{di}{dt}$$

therefore
$$v_L = RI e^{\frac{-Rt}{L}}$$

or
$$v_L = v_s e^{\frac{-Rt}{L}} \qquad 2.10$$

From equation 2.8
$$v_s - v_c = RC \frac{dv_c}{dt}$$

$$\frac{dt}{RC} = \frac{dv_c}{v_s - v_c}$$

$$\frac{t}{RC} = -\ln(v_s - v_c) + K$$

at $t = 0$, $v_c = 0$ so that
$$K = \ln v_s$$

$$\frac{t}{RC} = \ln \frac{v_s}{(v_s - v_c)}$$

$$\frac{v_s - v_c}{v_s} = e^{\frac{-t}{RC}}$$

$$v_c = v_s (1 - e^{\frac{-t}{RC}}) \qquad 2.11$$

Now
$$i = C \frac{dv_c}{dt}$$

therefore
$$i = \frac{v_s}{R} e^{\frac{-t}{RC}}$$

or
$$i = I e^{\frac{-t}{RC}} \qquad 2.12$$

Figure 2.3

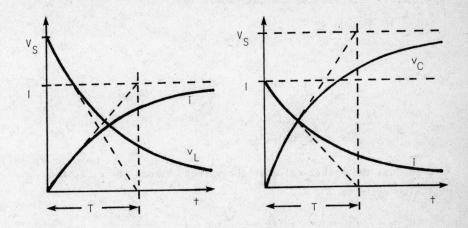

If the initial rate of rise or decay of current or voltage were maintained the steady-state value would be reached in T secs. This interval in time is termed the TIME CONSTANT of the circuit.
In the case of the RL circuit the initial current is zero and the initial growth of current is

$$\frac{di}{dt} = \frac{I}{T} \text{ amperes/sec}$$

therefore

$$v_s = \frac{L\,I}{T}$$

hence

$$T = \frac{L}{R} \text{ seconds}$$

In T seconds the actual value of the current will have risen to

$$i = I(1 - e^{\frac{-RT}{L}})$$

$$i = I(1 - e^{-1})$$

$$\underline{i = 0.632\,I}$$

In the case of the RC circuit the initial voltage across the capacitor is zero and the initial growth is

$$\frac{dv_c}{dt} = \frac{v_s}{T} \text{ volts/sec}$$

therefore

$$v_s = RC \frac{v_s}{T}$$

hence

$$T = RC \text{ seconds}$$

In T seconds the actual value of the voltage across the capacitor will have risen to

$$v_c = v_s (1 - e^{\frac{-T}{RC}})$$

$$v_c = v_s (1 - e^{-1})$$

$$\underline{v_c = 0.632 \, v_s}$$

If the switch position is instantaneously changed from m to n, then for the RL circuit

$$i + L \frac{di}{dt} = 0$$

$$-\frac{di}{i} = \frac{R}{L} dt$$

$$-\ln i = \frac{Rt}{L} + K$$

at $t = 0$, $i = I$ thus

$$K = -\ln I$$

$$\ln \frac{I}{i} = \frac{Rt}{L}$$

$$\frac{I}{i} = e^{\frac{Rt}{L}}$$

$$i = I e^{\frac{-Rt}{L}} \qquad 2.13$$

Also

$$v_L = L \frac{di}{dt}$$

$$v_L = -v_s e^{\frac{-Rt}{L}} \qquad 2.14$$

and for the RC circuit

$$v_c + RC \frac{dv_c}{dt} = 0$$

$$\frac{-dv_c}{v_c} = \frac{dt}{RC}$$

$$-\ln v_c = \frac{t}{RC} + K$$

at $t = 0$, $\quad v_c = v_s$

thus

$$K = -\ln v_s$$

$$\ln \frac{v_s}{v_c} = \frac{t}{RC}$$

$$\frac{v_s}{v_c} = e^{\frac{t}{RC}}$$

$$v_c = v_s e^{\frac{-t}{RC}} \qquad 2.15$$

$$i = C \frac{dv_c}{dt}$$

$$i = -I e^{\frac{-t}{RC}} \qquad 2.16$$

The transient solution of circuits involving inductance and capacitance requires the introduction of a second order differential equation which is outside the scope of this text.

An interesting feature of an RC circuit is its ability to (a) differentiate, and (b) integrate an input signal.

Figure 2.4

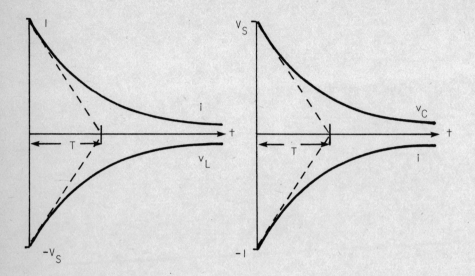

2.3.1 RC differentiator

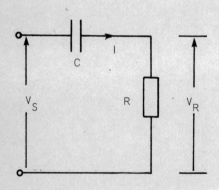

Figure 2.5

If the voltage drop across the capacitor v_c is very much greater than the voltage drop across the resistor v_R, then the source voltage v_s is approximately equal to v_c, therefore

$$v_s \approx \frac{1}{C} \int i \, dt$$

or

$$i \approx C \frac{dv_s}{dt}$$

If the output is taken acrosss the resistor then

$$v_R = iR \approx RC \frac{dv_s}{dt}$$

i.e. the output voltage is the differential of the input voltage, times the time constant of the circuit.

2.3.2 RC Integrator

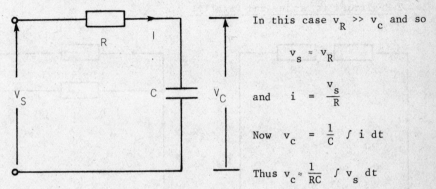

In this case $v_R \gg v_C$ and so

$$v_s \approx v_R$$

and $i = \dfrac{v_s}{R}$

Now $v_C = \dfrac{1}{C} \int i\, dt$

Thus $v_C \approx \dfrac{1}{RC} \int v_s\, dt$

Figure 2.6

i.e. the output voltage v_C is the integral of the input voltage v_s divided by the time constant of the circuit.

2.4 Solution of steady-state d.c. circuits

Some basic circuit theorems will now be developed which simplify the solution of electrical networks. They are introduced in this section for the sake of clarity but they are not solely confined to the steady-state d.c. case. They can be used for the solution of steady-state a.c. problems provided the phasor relationships are taken into account, see Chapter 3.

2.4.1 Kirchoff's Laws

First law - The algebraic sum of the currents meeting at a junction (or node) is zero.

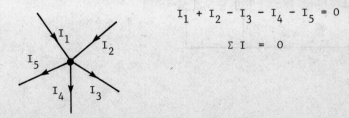

$$I_1 + I_2 - I_3 - I_4 - I_5 = 0$$

$$\Sigma I = 0$$

Second law - The algebraic sum of the voltage drops and e.m.f's around any closed circuit is zero.

$$\Sigma E - I R = 0$$

2.4.2 Resistors in Series and Parallel

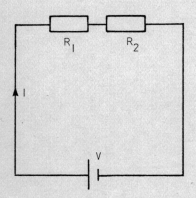
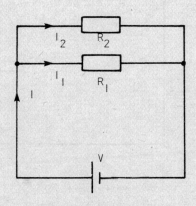

$V = I R_1 + I R_2$

If the total circuit resistance is R_T

$$\frac{V}{I} = R_T = (R_1 + R_2)$$

$$R_T = R_1 + R_2$$

$I = I_1 + I_2$

If the total circuit resistance is R_T

$$\frac{V}{R_T} = \frac{V}{R_1} + \frac{V}{R_2}$$

$$\frac{1}{R_T} = \frac{1}{R_1} + \frac{1}{R_2}$$

Example 1
Find the current drawn from the battery

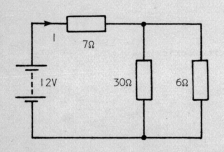

The 30 Ω and 6 Ω resistors may be replaced by an effective resistance R_e where

$$\frac{1}{R_e} = \frac{1}{30} + \frac{1}{6}$$

$$R_e = 5$$

The total circuit resistance is therefore 12 Ω and the current is
$\frac{12 \text{ V}}{12 \text{ Ω}}$ = 1 A.

Example 2
Find the current in each branch of the network

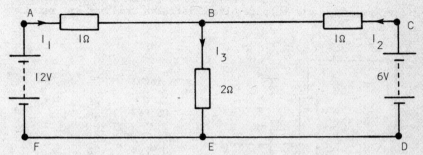

Let the current in the branches be I_1 I_2 and I_3 as shown.
Applying Kirchoff's first law at junction B

$$I_3 = I_1 + I_2$$

Applying Kirchoff's second law around the closed circuit ABEF

$$12 = 1.I_1 + 2.I_3$$

substitute for I_3

$$12 = 3I_1 + 2I_2 \qquad 2.17$$

Applying Kirchoff's second law around the closed circuit CBED

$$6 = 1.I_2 + 2.I_3$$

substitute for I_3

$$6 = 2I_1 + 3I_2 \qquad 2.18$$

Solving equations 2.17 and 2.18 simultaneously gives

$$I_1 = 4.8 \text{ A}, \ I_2 = -1.2 \text{ A}, \ I_3 = 3.6 \text{ A}$$

This means that the assumed direction of I_2 was wrong and the current I_2 is actually flowing into the positive terminal of the 6 V battery.

38 Basic Electrical Engineering and Instrumentation for Engineers

2.4.3 Potential Dividers

Potential dividers are used extensively as a method of obtaining a reduced voltage, especially in low power circuits. The potential divider is a resistance which may be tapped at any intermediate point, figure 2.7. The input voltage is connected across the ends of the resistor and the output voltage is taken from one end and the tapping point.

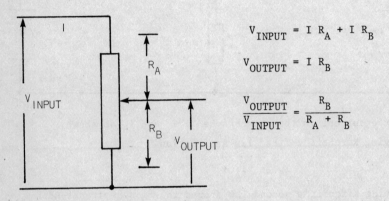

$$V_{INPUT} = I R_A + I R_B$$

$$V_{OUTPUT} = I R_B$$

$$\frac{V_{OUTPUT}}{V_{INPUT}} = \frac{R_B}{R_A + R_B}$$

Figure 2.7

Example 3

A student wishing to install a 9 V cassette player in his car decided to run the player from the 12 V car battery, via a potential divider. The potential divider he constructed to give an output voltage of 9.2 V is shown in figure 2.8.

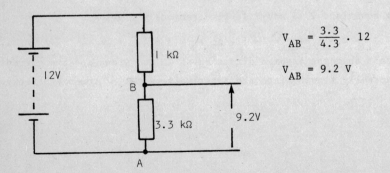

$$V_{AB} = \frac{3.3}{4.3} \cdot 12$$

$$V_{AB} = 9.2 \text{ V}$$

Figure 2.8

However on connecting the cassette to the potential divider the measured output voltage was only 7.32 volts.   Why?

The reason for the drop in voltage was that the cassette, which had an effective resistance of about 3 kΩ, loaded the potential divider, as shown in figure 2.9

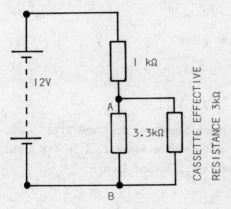

The resistance between A and B is

$$R_{AB} = \frac{3 \cdot 3.3}{3 + 3.3} = 1.572 \text{ k}\Omega$$

Thus

$$V_{AB} = \frac{1.572}{2.572} \cdot 12$$

$$V_{AB} = 7.32 \text{ V}$$

Figure 2.9

2.4.4 Mesh or Loop Analysis

The method of solution by mesh currents is best introduced by an example.   Let two mesh or loop currents i_a and i_b be drawn in the network of figure 2.10.

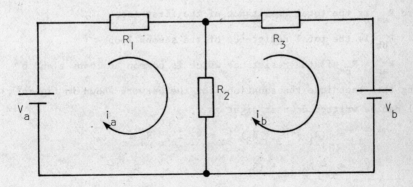

Figure 2.10

Using Kirchoff's second law around the first loop in the direction of the current i_a we have

$$V_a = R_1 i_a + R_2 i_a - R_2 i_b \qquad 2.19$$

and for the second loop in the direction of the current i_b

$$-V_b = R_2 i_b + R_3 i_b - R_2 i_a \qquad 2.20$$

Rewriting equations 2.19 and 2.20 gives

$$V_a = (R_1 + R_2) i_a - R_2 i_b \qquad 2.21$$

$$-V_b = -R_2 i_a + (R_2 + R_3) i_b \qquad 2.22$$

i_a and i_b can be obtained by solving equations 2.21 and 2.22 simultaneously. Using the parameters of example 2, $i_a = 4.8$ A and $i_b = 1.2$ A, and the branch currents I_1, I_2 and I_3 are

$$I_1 = i_a = 4.8 \text{ A}, \quad I_2 = -i_b = -1.2 \text{ A},$$

$$I_3 = i_a - i_b = 3.6 \text{ A}$$

Using the double subscript notation for the resistances, equations 2.21 and 2.22 may be written in the general form as

$$v_a = R_{aa} i_a - R_{ab} i_b \qquad 2.23$$

$$v_b = -R_{ab} i_a + R_{bb} i_b \qquad 2.24$$

where R_{aa} is the total resistance of the first loop

R_{bb} is the total resistance of the second loop

$R_{ab} = R_{ba}$ is the resistance which is common to loops a and b

Using this technique the equations for the network shown in figure 2.11 can be written down at sight as

$$V_a = (R_1 + R_2 + R_3) i_a - R_2 i_b - R_3 i_c$$

$$V_b = -R_2 i_a + (R_2 + R_4 + R_5) i_b - R_5 i_c$$

$$-V_c = -R_3 i_a - R_5 i_b + (R_3 + R_5 + R_6 + R_7) i_c$$

Electric Circuits   41

Figure 2.11

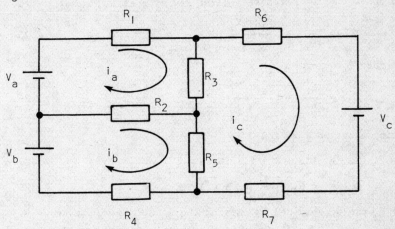

2.4.5 Nodal Analysis

Nodal analysis is the dual of mesh analysis, i.e. whereas mesh analysis uses Kirchoff's second law for obtaining the loop voltage equations, nodal analysis utilises Kirchoff's first law to obtain the node current equations.   Consider the network shown in figure 2.12.

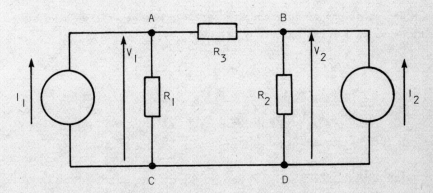

Figure 2.12

Nodes C and D are at the same potential and can be considered a single node. Using Kirchoff's first law at node A

$$I_1 = \frac{V_1}{R_1} + \frac{V_1 - V_2}{R_3} \qquad 2.25$$

and at node B

$$I_2 = \frac{V_2}{R_2} + \frac{V_2 - V_1}{R_3} \qquad 2.26$$

rewriting equation 2.25 and 2.26

$$I_1 = (G_1 + G_3) V_1 - G_3 V_2 \qquad 2.27$$
$$I_2 = -G_3 V_1 + (G_2 + G_3) V_2 \qquad 2.28$$

V_1 and V_2 can be obtained by solving the equations simultaneously. Using the double subscript notation equations 2.27 and 2.28 could be written as

$$I_1 = G_{aa} V_1 - G_{ab} V_2 \qquad 2.29$$
$$I_2 = -G_{ba} V_1 + G_{bb} V_2 \qquad 2.30$$

where G_{aa} is the total conductance connected to node A

G_{bb} is the total conductance connected to node B

$G_{ab} = G_{ba}$ = the conductance connecting nodes A and B

The nodal equations of more complex networks may be written down at sight as shown for the network of figure 2.13.

$$I_A = (G_1 + G_4 + G_6)V_A - G_4 V_B - G_6 V_C$$
$$0 = -G_4 V_A + (G_2 + G_4 + G_5)V_B - G_5 V_C$$
$$I_C = -G_6 V_A - G_5 V_B + (G_3 + G_5 + G_6)V_C$$

2.4.6 Thevenin's Theorem

Thevenin's theorem states that any network consisting of resistors and voltage sources can be replaced at any pair of terminals by a voltage source in series with a resistance.

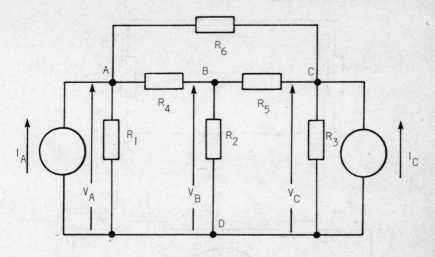

Figure 2.13

Figure 2.14

        a) Original Network   b) Thevenin Equivalent

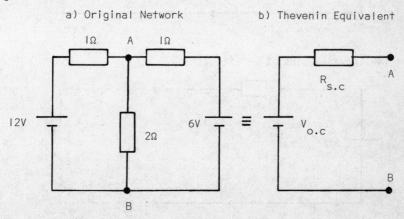

Across the terminals AB the network shown above in figure 2.14(a) may be replaced by an equivalent network figure 2.14(b), where $R_{s.c.}$ is termed the short circuit resistance and is obtained by considering all the voltage sources short circuited and calculating the resulting resistance of the circuit "looking in" at the terminals AB, and $V_{o.c.}$ is the open circuit voltage at AB with the 2 Ω resistor removed.

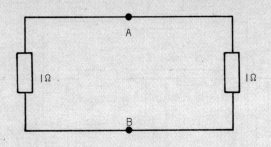

$R_{s.c}$ is therefore 0.5Ω

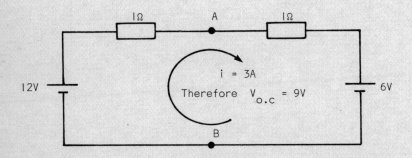

Thus the Thevenin equivalent circuit is

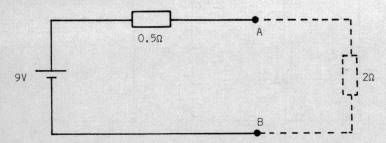

and the current in the 2 Ω resistor connected across AB is

$$I = \frac{9}{2.5} = 3.6 \text{ A}$$

which checks with the current found by Kirchoff's laws in example 2, and by mesh current analysis in section 2.4.4.

2.4.7 Norton's Theorem

Norton's theorem states that any network consisting of resistors and voltage sources can be replaced at any pair of terminals by a current source in parallel with a resistance, as shown in figure 2.15.

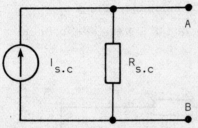

Where $R_{s.c.}$ is the short circuit resistance as before and $I_{s.c.}$ is the short circuit current and is equal to $V_{o.c.}/R_{s.c.}$

Figure 2.15

For the network of figure 2.14(a)

$$R_{s.c.} = 0.5; \quad I_{s.c.} = \frac{9}{0.5} = 18 \text{ A}$$

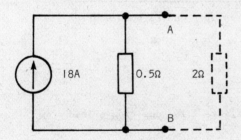

Therefore the current through the 2 Ω resistor connected across terminals AB is

$$I = \frac{0.5}{2+0.5} \cdot 18 = 3.6 \text{ A}$$

2.4.8 Superposition Theorem

The principle of superposition states that the response of a linear network containing several voltage or current sources is the sum of the responses of each source taken in turn, with the other voltage sources short circuited, and current sources open circuited.
Using the network of figure 2.14(a) as an example, the current in the 2 Ω resistor due to the 12 V battery, with the 6 V battery short circuited, is

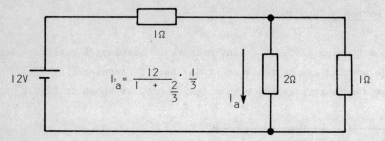

The current in the 2 Ω resistor due to the 6 V battery, with the 12 V battery short circuited is

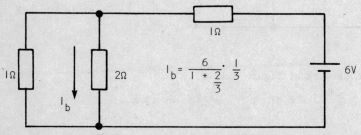

Total current through the 2 Ω resistor due to both sources is 18/5 A = 3.6 A, as before.

2.4.9 Maximum Power Transfer

Consider the battery shown in figure 2.16 supplying a load resistance R which is variable

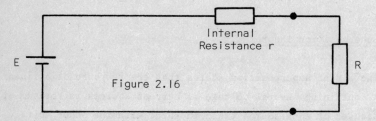

Figure 2.16

The power supplied to the load is $I^2 R$, where

$$I = \frac{E}{R+r}$$

therefore
$$P = \frac{E^2 R}{(R+r)^2} \qquad 2.31$$

Differentiating with respect to R

$$\frac{dP}{dR} = -\frac{2E^2 R}{(R+r)^3} + \frac{E^2}{(R+r)^2} \qquad 2.32$$

Equate $\frac{dP}{dR}$ to zero to find the maximum or minimum value of R.

$$\frac{2E^2 R}{(R+r)^3} = \frac{E^2}{(R+r)^2}$$

thus $\qquad R = r$

$$\frac{d^2 P}{dR^2} = \frac{6E^2 R}{(R+r)^4} - \frac{4E^2}{(R+r)^3} \qquad 2.33$$

Substituting for $R = r$ in equation 2.33 makes $\frac{d^2 P}{dR^2}$ negative, showing that $R = r$ is the condition of maximum power transfer. The above may be extended to any piece of electrical equipment in that the maximum power transfer occurs when the loading resistance (or impedance in the case of alternating current systems) is equal to the internal resistance (or impedance).

48 Basic Electrical Engineering and Instrumentation for Engineers

Examples

1. A plate condenser is charged to 40 µC at 100 V. It is then connected to a similar condenser of four times the plate area. Find the charge on each condenser and the loss in energy.

(8 µC, 32 µC, 1.6 mJ)

2. Explain the meaning of the time constant in an R-C or R-L circuit. A capacitor may be represented by a capacitance C in parallel with a resistance R. The capacitor is charged to a voltage of 1000 V. When the charging source is removed the voltage drops to 500 V in 20 secs. A pure capacitance of 60 µF is then connected in parallel with the first, and the combination again charged to 1000 V. When the source is removed, the voltage drops to 500 V in 60 secs. What is the value of C and R?

(30 µF, 0.965 MΩ)

3. For the circuit shown below, calculate the current in the 3 Ω resistor by
(a) Mesh analysis,
(b) Superposition theorem,
(c) Thevenin equivalent circuit theorem.

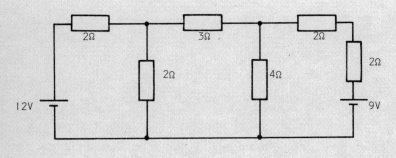

(0.25 A)

4. Using Norton's equivalent circuit, convert the circuit shown in question 3 into one involving current sources and conductances, and find the current in the 3 Ω resistor using nodal analysis.

CHAPTER THREE

Steady-state Alternating Currents

3.1 <u>Differential equation for RLC circuit</u>

A large proportion of the applications in electrical engineering involve voltages and currents which are of a sinusoidal nature. This is not surprising as the electricity authorities generate and transmit sinusoidal voltages. It is important, therefore, to consider as a special case the sinusoidal response of electric circuits.

In chapter 2 the voltage differential equation was established for a series RLC circuit, equation 2.6. For completeness the circuit diagram and differential equation are redrawn in Fig. 3.1.

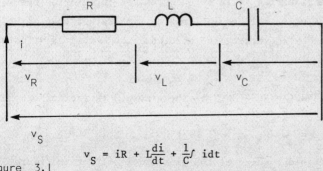

$$v_S = iR + L\frac{di}{dt} + \frac{1}{C}\int i\,dt \qquad 3.1$$

Figure 3.1

3.2 Definition of Quantities

The waveform of a sinusoidal voltage is shown in Figure 3.2.

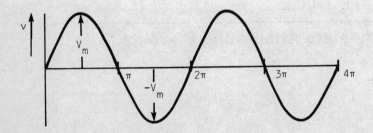

Figure 3.2

The AMPLITUDE, PEAK or MAXIMUM value is the maximum vertical component of the waveform, which occurs at $90°$, $270°$, $450°$, etc., and is designated plus or minus V_m.

The INSTANTANEOUS value of the waveform is the vertical component at any point and can be written as

$$v = V_m \sin \theta \qquad 3.2$$

It is seen that the waveform repeats itself after every $360°$ (2π radians). The period from 0 to $360°$ is termed one CYCLE. The FREQUENCY (f) is the number of cycles occurring in one second, and the PERIODIC TIME (T) is the time in seconds for one cycle.

In electrical engineering it is more convenient to describe the instantaneous value of voltage in terms of time rather than angle, and this is done by converting the abscissa into a time axis, using

$$\omega = \frac{2\pi}{T} = 2\pi f$$

where ω is the ANGULAR FREQUENCY of the sinusoid in radians/second, and,

$$\theta = \omega t$$

Thus the instantaneous value of the sinusoidal voltage in figure 3.2

may be expressed as

$$v = V_m \sin \omega t \qquad 3.3$$

Consider now two sinusoids as shown in figure 3.3

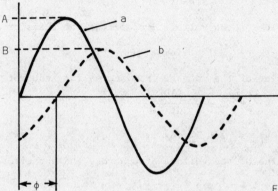

Figure 3.3

Sinusoid b is displaced from sinusoid a by angle ϕ.
This displacement is termed the PHASE DIFFERENCE, and the angle ϕ is termed the PHASE ANGLE.
If the instantaneous value of sinusoid "a" is

$$a = A \sin \omega t$$

then that of "b" is

$$b = B \sin (\omega t - \phi)$$

i.e. because the peak of "b" occurs later (to the right) of that of "a", it is said to LAG "a", and the phase angle is given a negative sign.

<u>NOTE</u>: Sinusoid "a" was used as the REFERENCE i.e. sinusoid "b" was defined with respect to "a". In electrical engineering it is usual to make the source voltage or current the reference sinusoid.
Suppose sinusoid "b" had been defined as the reference, what would the instantaneous value of "a" have been?
Reference $\qquad b = B \sin \omega t$
therefore because "a" leads "b" by angle ϕ

$$a = A \sin (\omega t + \phi)$$

3.3 Sinusoidal Response of R.L.C. Circuit

In the RLC circuit shown in figure 3.1 let the instantaneous current be

$$i = I_m \sin \omega t$$

The instantaneous value of the voltage drop across the resistor is

$$v_R = R I_m \sin \omega t \qquad 3.4$$

The instantaneous value of the voltage drop across the inductor is

$$v_L = L \frac{d(I_m \sin \omega t)}{dt}$$

$$v_L = \omega L I_m \cos \omega t \qquad 3.5$$

The instantaneous value of the voltage drop across the capacitor is

$$v_C = \frac{1}{C} \int I_m \sin \omega t \, dt$$

$$= -\frac{1}{\omega C} I_m \cos \omega t \qquad 3.6$$

It can be seen, from equation 3.4, that the voltage drop across a resistor is in phase with the current flowing through it, and from equation 3.5, that the voltage drop across the inductor LEADS the current flowing through it by $90°$ { $\cos \omega t = \sin (\omega t + 90)$ } and from equation 3.6, that the voltage drop across the capacitor LAGS the current flowing through it by $90°$ { $-\cos \omega t = \sin (\omega t - 90)$ }.
The voltage drops given in equations 3.4, 3.5 and 3.6 are all of the same form,
i.e. voltage drop = a constant term x current.
The constant ωL in equation 3.5 is termed the INDUCTIVE REACTANCE and is designated X_L.
The constant $\frac{1}{\omega C}$ in equation 3.6 is termed the CAPACITIVE REACTANCE and is designated X_C.

3.4 Phasor Diagrams

The series RLC circuit shown in figure 3.1 is one of the simplest electrical networks one can meet in practice. It is clear, therefore that as the network becomes more involved, by the combination of

series and parallel connections of resistors, inductors and
capacitors, the use of instantaneous values to determine the sinusoidal response is inconvenient. A simplified pictorial approach is to
represent a sinusoid by its magnitude and phase (with respect to some
reference datum) on a diagram. Addition and subtraction can then be
done diagramatically.

A sinusoid represented in this way is called a PHASOR and the diagram
on which it is drawn is termed a PHASOR DIAGRAM.

Consider the RLC circuit of figure 3.1. The current is common to
all the circuit elements, therefore the current is used as the
reference. (With circuits connected in parallel the source voltage
is often used as the reference.) The voltage drops across the passive elements v_R v_L and v_C are represented in magnitude and phase,
with respect to the current. The source voltage v_s is the
"vectorial" addition of the phasors $\bar{V}_R$ $\bar{V}_L$ and $\bar{V}_C$.

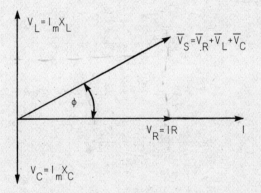

Figure 3.4

From figure 3.4 it is seen that the source voltage leads the current
by an angle of ϕ degrees, implying that the circuit is effectively
inductive. This is because the volt drop across the inductance was
arbitrarily chosen to be greater than the volt drop across the
capacitor.

Phasor values will be designated throughout the text by a capital

54 Basic Electrical Engineering and Instrumentation for Engineers

with a bar.

3.5 Root Mean Square Value

In the phasor diagram of figure 3.4 the magnitude of the sinusoidal current and voltages was represented by their maximum or peak value. It is more usual, however, to use the ROOT MEAN SQUARE (r.m.s.) value of a sinusoid in the analysis of a.c. circuits.

The r.m.s. or effective value of an alternating current is that current which when passed through a resistance will produce the same heating effect as a direct current passed through the same resistance. The heating effect of a current passed through a resistor is

$$i^2 R$$

The AVERAGE heating effect of a sinusoidal current, $i = I_m \sin \theta$, passed through a resistor R is

$$\frac{1}{\pi} \int_0^{\pi} i^2 R \, d\theta = \frac{1}{\pi} \int_0^{\pi} I_m^2 \sin^2 \theta \, R \, d\theta$$

$$= \frac{1}{\pi} \int_0^{\pi} \frac{I_m^2 R}{2} (1 - \cos 2\theta) \, d\theta$$

$$= \frac{I_m^2 R}{2\pi} \left[\theta - \frac{\sin 2\theta}{2} \right]_0^{\pi}$$

$$= \frac{I_m^2 R}{2}$$

Let I be the direct current which, when passed through the same resistor, will produce the same heating effect, thus

$$I^2 R = \frac{I_m^2 R}{2}$$

$$\text{or} \quad I = \frac{I_m}{\sqrt{2}}$$

Thus the r.m.s. value of a sinusoidal current (or voltage) is 0.707 times the maximum value.

3.6 The Use of Complex Numbers or "j" Notation

(See Appendix 3 regarding the manipulation of complex numbers.)

The operator j is a unit operator which, when multiplying a phasor, shifts it by $90°$ in an anticlockwise direction.

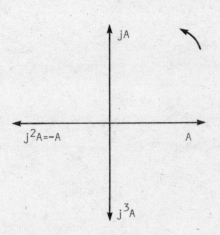

Figure 3.5

Let A be a phasor along the horizontal axis, figure 3.5. Mutliplying A by j gives a phasor jA along the vertical axis (A has been shifted by $90°$ anticlockwise). Multiplying jA by j gives a phasor j^2A along the horizontal axis in the opposite direction to A.

Therefore
$$j^2A = -A$$
$$\text{or } j^2 = -1$$

Multiplying j^2A by j gives a phasor along the vertical axis in the opposite direction to jA.

Therefore
$$j^3A = -jA$$
$$\text{or } j^3 = -j$$

NOTE: With this notation the horizontal axis is usually termed the REAL axis, and the vertical axis the IMAGINARY axis.

Consider now the application of this notation to the RLC circuit of figure 3.1, and the phasor diagram of figure 3.4. Using r.m.s. values and again taking the current as a reference, then, remembering that v_L leads I by $90°$ and v_C lags I by $90°$,

$$v_R = IR \qquad v_L = jIX_L \quad \text{and} \quad v_C = -jIX_C$$

$$V_s = IR + jIX_L - jIX_C$$
$$V_s = I\{R + j(X_L - X_C)\} \qquad 3.7$$

The term inside the curly bracket in equation 3.7 is termed the IMPEDANCE of this circuit, and is designated Z. Equation 3.7 can be written as

$$\bar{V}_s = \bar{I}\,\bar{Z} \qquad 3.8$$

which is Ohm's Law for a.c. circuits.

The phase angle between the source voltage V_s and the current I can be obtained from equation 3.7, i.e.

$$\phi = \tan^{-1} \frac{(X_L - X_C)}{R} \qquad 3.9$$

Consider the parallel RLC circuit shown in figure 3.6.

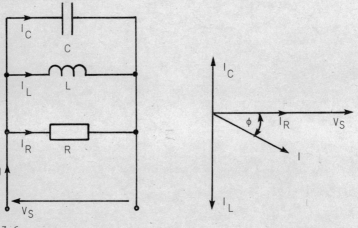

Figure 3.6

The magnitudes of the currents through the individual elements are obtained by using OHM's LAW,

$$I_R = \frac{V_s}{R} \qquad I_L = \frac{V_s}{X_L} \qquad I_C = \frac{V_s}{X_C}$$

Using the supply voltage as the reference and remembering that the current through an inductor lags the voltage across it by 90°, whilst the current through a capacitor leads the voltage across it by 90°, (see phasor diagram), then using Kirchoff's first law the total current I is given by the vector sum of the branch currents,

$$\bar{I} = \bar{I}_R + \bar{I}_L + \bar{I}_C$$

in complex form

$$I = \frac{V_s}{R} - j\frac{V_s}{X_L} + j\frac{V_s}{X_C}$$

$$I = V_s \left\{ \frac{1}{R} + \frac{1}{jX_L} - \frac{1}{jX_C} \right\} \qquad 3.10$$

$$I = V_s \left\{ \frac{-X_L X_C + j X_C R - j X_L R}{- R X_L X_C} \right\} \qquad 3.11$$

The phase angle of the current with respect to the voltage V_s is

$$\phi = \tan^{-1} \frac{R(X_L - X_C)}{X_L X_C} \qquad 3.12$$

Worked Example

In the circuit shown in figure 3.7 determine the magnitude and phase of the current drawn from the supply. The supply voltage is 20 V at 1000 Hz.

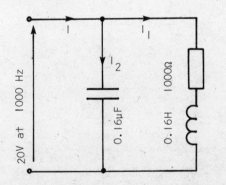

$X_L = 2\pi f\, L = 2\pi . 1000 . 0.16$

$X_L = 1000\ \Omega$

$X_C = \dfrac{10^6}{2\pi f\, C}$ where C is in μF

$X_C = \dfrac{10^6}{2\pi . 1000 . 0.16}$

$X_C = 1000\ \Omega$

Figure 3.7

$$I_2 = \frac{20}{-j1000} = j20\,10^{-3}\ A$$

$$I_1 = \frac{20}{1000 + j1000} = \frac{20(1000 - j1000)}{2.10^6}$$

$$I_1 = (10 - j10) . 10^{-3}\ A$$

By Kirchoff's first law

$$\overline{I} = \overline{I}_1 + \overline{I}_2$$
$$I = (10 - j10 + j20) \cdot 10^{-3} \text{ A}$$
$$I = (10 + j10) \cdot 10^{-3} \text{ A}$$
$$|I| = 14.14 \text{ mA}$$
$$\phi = \tan^{-1} \frac{10}{10} = \tan^{-1} 1$$
$$\phi = 45°$$

3.7 Power in a.c. Circuits

Let the voltage across and the current flowing through a particular a.c. circuit be,

$$v = V_m \sin(\omega t + \alpha)$$
$$i = I_m \sin(\omega t + \beta)$$

The instantaneous power is,

$$p = vi = V_m I_m \{\sin(\omega t + \alpha) \cdot \sin(\omega t + \beta)\}$$
$$= \frac{V_m I_m}{2} \{\cos(\alpha - \beta) - \cos(2\omega t + \alpha + \beta)\} \quad 3.13$$

The average power in the circuit is

$$P_{AV} = \frac{V_m I_m}{2} \cos(\alpha - \beta) \quad\quad 3.14$$

Because the term $\cos(2\omega t + \alpha + \beta)$ averaged over one cycle is zero whereas the term $\cos(\alpha - \beta)$ is a constant (α and β are constant angles).

Let $\phi = (\alpha - \beta)$ = the phase angle between the current and voltage, then

$$P_{AV} = \frac{V_m}{\sqrt{2}} \frac{I_m}{\sqrt{2}} \cos \phi$$

or, in terms of r.m.s. values

$$P_{AV} = V I \cos \phi \quad\quad 3.15$$

The term $\cos \phi$ is termed the POWER FACTOR.
The power in an a.c. circuit is equal to the r.m.s. voltage times the r.m.s. current times the power factor.

3.7.1 Calculation of Power using "j" Notation

The phasors of the instantaneous voltage and current may be drawn as shown in figure 3.8.

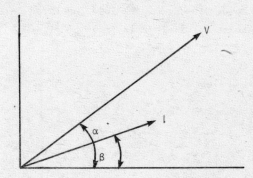

Figure 3.8

In complex form the voltage and current are,

$$V = V \cos \alpha + j V \sin \alpha = a + jb$$
$$I = I \cos \beta + j I \sin \beta = c + jd$$

It has already been shown that the power is,

$$P = VI \cos(\alpha - \beta) = VI \{\cos \alpha \cos \beta + \sin \alpha \sin \beta\}$$

thus

$$P = ac + bd$$

This result can be obtained by taking the REAL part of the product of the complex expression for voltage and the CONJUGATE of the complex expression for current, i.e.

$$P = \text{Re}\{VI^*\} \qquad 3.16$$

$$P = \text{Re}\{(a + jb)(c - jd)\} = R\{(ac + bd) + j(bc - ad)\}$$

thus

$$P = ac + bd$$

The same result could have been obtained by taking the real part of the conjugate of voltage multiplied by the current, i.e.

$$P = \text{Re}\{V^* I\} \qquad 3.17$$

3.7.2 Power Triangle

When dealing with electrical transformers and generators it will be found that their rating is given in kVA (kilo volt amperes), which is a measure of their current carrying capacity at a given voltage level. The power that can be drawn from these devices depends on their kVA rating and on the power factor of the load connected to them. The power triangle shows this simple relationship, figure 3.9.

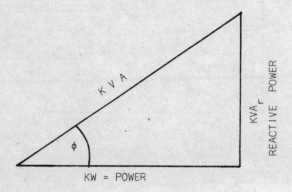

Figure 3.9

$$\begin{aligned} \text{kVA} &= V.I.10^{-3} \\ \text{kW} &= V I \cos\phi .10^{-3} = \text{kVA}.\cos\phi \\ \text{kVAr} &= V I \sin\phi .10^{-3} = \text{kVA}.\sin\phi \qquad 3.18 \end{aligned}$$

3.8 Frequency Response of Circuits

So far in this chapter the frequency of the source voltage has been assumed constant. When analysing the operation of equipment which is designed to operate over a range of frequencies, it is necessary to determine the performance at various frequencies.

The FREQUENCY RESPONSE of a circuit is obtained by plotting (a) the ratio of the output voltage, current or power to the input voltage, current or power against frequency, and (b) the phase of the ratio against frequency.

The frequency range for many circuits is very large. For instance, the response of an audio amplifier ranges from 30 Hz to 15000 Hz. On a linear frequency scale the bass range of 30 to 200 Hz will be confined to about 1% of the frequency spectrum. To overcome this cramping of information frequency response curves are normally plotted on logarithmic graph paper so that the space allocated to each decade is the same.

Normally the ratio of output voltage, current or power to the input voltage, current or power is also plotted in logarithmic form, the special unit given to the ratio being the DECIBEL (dB).

$$\text{voltage ratio in dB} = 20 \log_{10} \left| \frac{V_{out}}{V_{in}} \right| \qquad 3.19$$

$$\text{power ratio in dB} = 10 \log_{10} \left| \frac{P_{out}}{P_{in}} \right| \qquad 3.20$$

Consider now the frequency response of the RLC network of figure 3.10.

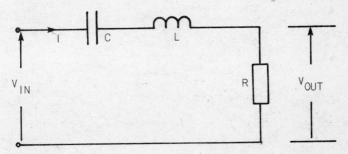

Figure 3.10

$$V_{out} = IR$$
$$V_{in} = I \left\{ R + j \left(\omega L - \frac{1}{\omega C} \right) \right\}$$

$$\frac{V_{out}}{V_{in}} = \frac{R}{R + j \left(\omega L - \frac{1}{\omega C} \right)} \qquad 3.21$$

$$\frac{V_{out}}{V_{in}} = \frac{R}{\left[R^2 + \left(\omega L - \frac{1}{\omega C} \right)^2 \right]^{\frac{1}{2}}} \qquad 3.22$$

phase or angle of

$$\frac{V_{out}}{V_{in}} = -\tan^{-1} \frac{\omega L - \frac{1}{\omega C}}{R} \qquad 3.23$$

Typical frequency response plots of an RLC circuit, obtained from equations 3.22 and 3.23 are shown in figure 3.11.

It is clear from equation 3.22 that the maximum value of V_{out}/V_{in} is unity, and this occurs at the frequency ω_o which makes $\omega L - \frac{1}{\omega C} = 0$. This is termed the RESONANT condition. The ratio is only unity if the inductor is pure, i.e. does not possess any resistance. This is never the case in practice and thus the ratio V_{out}/V_{in} even at resonance is slightly less than unity.

The BANDWIDTH is defined as

$$W = \omega_1 - \omega_2 \qquad 3.24$$

where ω_1 and ω_2 are the frequencies at which the voltage ratio has dropped by 3 dB.

The 3 dB point is chosen arbitrarily, because it represents a halving in the power, as shown below.

Let voltages V_1 and V_2 be two voltages across a resistor R such that V_2 produces half the power that V_1 produces.

$$\frac{\frac{|V_1|^2}{R}}{\frac{|V_2|^2}{R}} = 2$$

or $\qquad V_1 = \sqrt{2}\, V_2$

Then the voltage ratio $|V_2|/|V_1|$ in dB from equation 3.19 is

$$20 \log_{10} 0.707 = -3 \text{ dB}$$

The Q FACTOR or quality of a resonant circuit is the ratio of its centre frequency to its bandwidth.

$$Q = \frac{\omega_o}{W} \qquad 3.25$$

3.9 Three-Phase Circuits

Historically, the first electricity was supplied by d.c. generators and used mainly for lighting. As the demand increased more powerful generators were required operating at higher voltages. It was

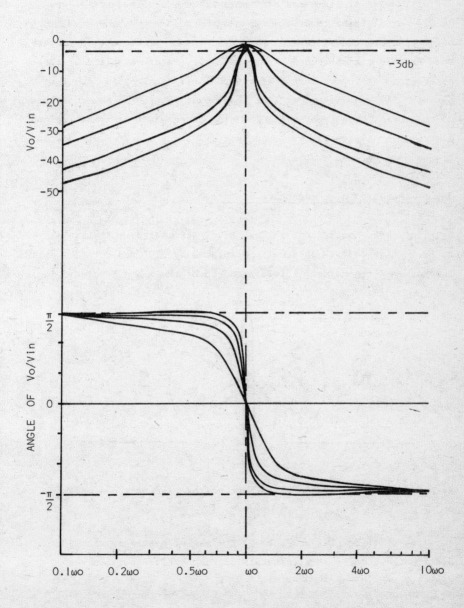

Figure 3.11

found that the commutator, an integral part of any d.c. machine, severely limited the maximum voltage that could be generated, and attention turned to a.c. generators or alternators. These initially were single-phase or single circuit devices, again used mainly for lighting loads. Over the years, however, the development of alternators tended towards three-phase machines because these were more efficient and because three-phase a.c. motors supplied by these alternators performed much better than their single-phase counterparts. All a.c. motors greater than about five horse-power are now of the three phase type. Gradually the three-phase transmission and distribution system was built up, and it is now accepted world-wide as the most optimum a.c. system.

3.9.1 Generation of Three-Phase Voltages

Consider three coils, Red, Yellow, Blue, designated $R_I R_O$, $Y_I Y_O$, $B_I B_O$ in figure 3.12, physically displaced by $120°$ and rotating anti-clockwise in the magnetic field produced by the N and S poles.

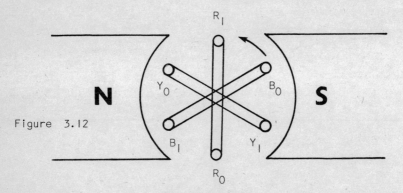

Figure 3.12

By Faraday's Law, equation 1.15, the voltage induced in a coil is,

$$e = N \frac{d\phi}{dt}$$

Assuming that the flux produced by the magnet system is sinusoidally distributed (practical alternators approach this condition very closely) and is equal to

$$\phi = \phi_m \sin \omega t$$

then the voltages induced in the coils are

$$e_R = \omega N \phi_m \cos \omega t$$
$$e_Y = \omega N \phi_m \cos (\omega t - 120)$$
$$e_B = \omega N \phi_m \cos (\omega t - 240) \qquad 3.26$$

as shown in figure 3.13.

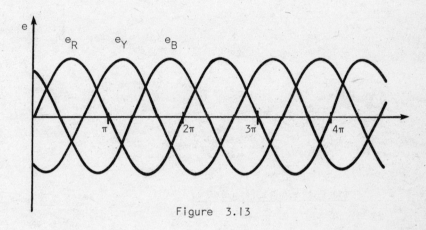

Figure 3.13

It would appear that to transmit the voltages from the three coils would require six conductors, a costly and physically undesirable condition. It is possible, however, to reduce the number of conductors required to four, or, in some cases, three, by connecting the coils in a symmetric fashion as shown below.

3.9.2 Star and Delta Connections

Basically the three coils shown in figure 3.12 can be connected symmetrically in only two ways, STAR (or WYE) connected as shown in figure 3.14(a), or DELTA (or MESH) connected as shown in figure 3.14(b). The star connection may, of course, be made by connecting R_O Y_O and B_O together, and the delta connection by connecting R_O to Y_I, Y_O to B_I, and B_O to R_I.

66 Basic Electrical Engineering and Instrumentation for Engineers

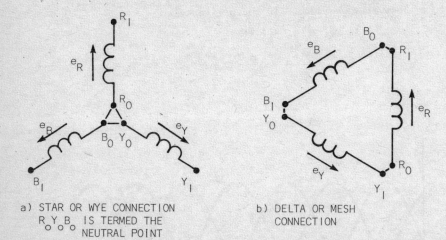

a) STAR OR WYE CONNECTION
   $R_O Y_O B_O$ IS TERMED THE
   NEUTRAL POINT

b) DELTA OR MESH
   CONNECTION

Figure 3.14

3.9.3 Voltage and Current Relationships

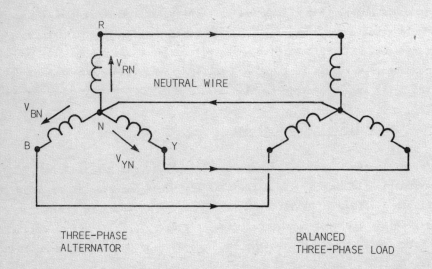

THREE-PHASE
ALTERNATOR

BALANCED
THREE-PHASE LOAD

Figure 3.15

Figure 3.15 shows a three phase star connected alternator supplying currents I_R I_Y I_B to a balanced (all three phases are equal) resistive-inductive load.

Figure 3.16 shows the phasor diagram for the circuit shown in figure 3.15. The phase voltages, or voltages with respect to the neutral point, are V_{RN}, V_{YN}, and V_{BN}. For the balanced case they are equal in magnitude and displaced by 120°. The currents supplied to the load, I_R I_Y and I_B lag the respective phase voltages by some phase angle ϕ. The neutral current I_N is, by Kirchoff's first law, equal to the phasor addition of the phase currents, i.e.

$$\overline{I}_N = \overline{I}_R + \overline{I}_Y + \overline{I}_B$$

which, for a balanced system, is always zero. The voltages between the terminals RY and B are termed the LINE voltages. The voltage of the RED LINE with respect to the YELLOW LINE is written V_{RY} and is given by,

$$\overline{V}_{RY} = \overline{V}_{RN} - \overline{V}_{YN}$$

or $\quad \overline{V}_{RY} = \overline{V}_{RN} + \overline{V}_{NY}$ \hfill 3.27

This phasor addition is shown on the phasor diagram of figure 3.16 and it can be seen that if the phase voltages are equal, then the line voltages are given by

$$V_{LINE} = 2 \cdot V_{PHASE} \cos 30°$$

or $\quad \underline{V_L = \sqrt{3} \cdot V_P}$ \hfill 3.28

It can be seen from figure 3.15 that the currents in the lines are equal to the currents in the phases, for the star connection, i.e.

$$\underline{I_L = I_P} \quad\quad 3.29$$

Now consider that the alternator of figure 3.15 is reconnected in delta, as shown in figure 3.17.

The voltage of line 1 with respect to line 2 is equal to the red phase voltage, the voltage of line 2 with respect to line 3 is equal to the yellow phase voltage, etc. Therefore, for a delta connection,

$$V_L = V_P \qquad 3.30$$

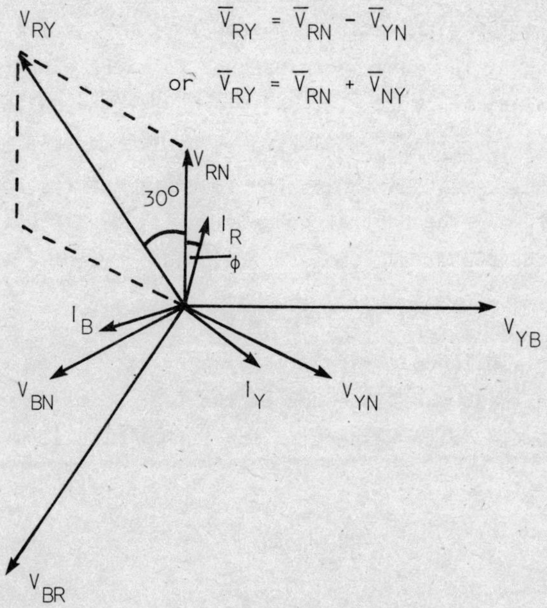

$$\overline{V}_{RY} = \overline{V}_{RN} - \overline{V}_{YN}$$

$$\text{or} \quad \overline{V}_{RY} = \overline{V}_{RN} + \overline{V}_{NY}$$

Figure 3.16

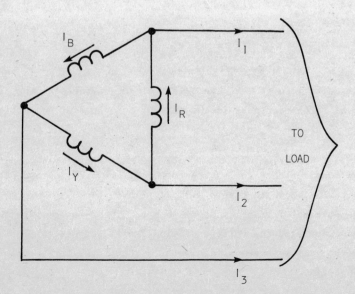

Figure 3.17

The line current I_1 is obtained by Kirchoff's first law, and is given by,

$$\overline{I}_1 = \overline{I}_R - \overline{I}_B$$
$$\overline{I}_1 = \overline{I}_R + (-\overline{I}_B)$$

This phasor addition is shown on the phasor diagram of figure 3.18, and it is again seen that if the phase currents are equal, then the line currents are given by,

$$I_{LINE} = 2 \cdot I_{PHASE} \cos 30°$$
$$\underline{\underline{I_L = \sqrt{3} \, I_P}} \qquad 3.31$$

To summarise,

STAR CONNECTION
$$V_L = \sqrt{3} \, V_P$$
$$I_L = I_P$$

DELTA CONNECTION
$$V_L = V_P$$
$$I_L = \sqrt{3} \, I_P$$

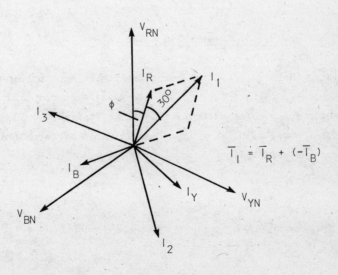

Figure 3.18

3.9.4 Power in Three-Phase Circuits

If V_P is the r.m.s. phase voltage, I_P is the r.m.s. phase current, and ϕ is the phase angle between V_P and I_P, then the power per phase is given by equation 3.15, as

$$P/_{PHASE} = V_P I_P \cos \phi$$

Thus the total power for a three phase circuit is

$$P = 3 V_P I_P \cos \phi \qquad 3.32$$

Now for a star connection

$$P = 3 \cdot \frac{V_L}{\sqrt{3}} \cdot I_L \cdot \cos \phi$$

$$P = \sqrt{3} V_L I_L \cos \phi$$

and for a delta connection

$$P = 3 \cdot V_L \cdot \frac{I_L}{\sqrt{3}} \cdot \cos \phi$$

$$P = \sqrt{3} V_L I_L \cos \phi$$

Therefore, in terms of the LINE values the power in either a star or delta connected system is,

$$\underline{P = \sqrt{3} V_L I_L \cos \phi \qquad 3.33}$$

3.9.5 Measurement of Three-Phase Power

Providing there is a neutral point available the simplest way to measure the power supplied to a three-phase balanced load is to connect a dynamometer wattmeter* as shown in figure 3.19.
The total power supplied to the load is given by three times the wattmeter reading.

---

*For details of dynamometer wattmeter, see chapter 13.

Figure 3.19

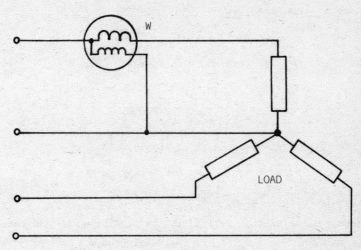

However, for a three-wire system with a balanced or unbalanced load the usual technique for measuring the total power supplied is by the two-wattmeter method as shown in figure 3.20.

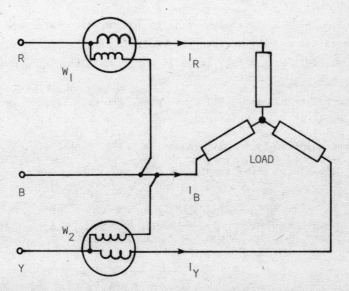

Figure 3.20

In instantaneous form the total power supplied to a three-phase load (balanced or unbalanced) is

$$v_R i_R + v_Y i_Y + v_B i_B$$

For the star connected load shown in figure 3.20

$$i_R + i_Y + i_B = 0$$

$$\therefore i_B = -i_R - i_Y$$

Hence the total instantaneous power supplied is

$$v_R i_R + v_Y i_Y + v_B (-i_R - i_Y)$$
$$= (v_R - v_B) i_R + (v_Y - v_B) i_Y$$

The average value of $(v_R - v_B) i_R$ is the reading on W_1 and the average value of $(v_Y - v_B) i_Y$ is the reading on W_2.

Therefore the sum of the two wattmeter readings is the total average power supplied to the load.

Worked Examples

1. A 500 kVA, 6600 V alternator is delta connected. Calculate the magnitude of the current in each phase of the alternator when it delivers its full output to a balanced three-phase load at a power factor of 0.8 lagging.

NOTE: The rating of almost all three-phase electrical plant is given in terms of the LINE values.

Total power supplied by the alternator = $\sqrt{3} \, V_L \, I_L \cos \phi$

also Total power supplied = total kVA . cos ϕ
= 500 . 0.8
= __400 kW__

Thus 400,000 = $\sqrt{3}$. 6600 . I_L . 0.8

I_L = __43.8 A__

For a delta connection

$$\text{the phase current} = I_L / \sqrt{3}$$

Thus $\quad I_P = \dfrac{43.8}{1.73} = \underline{25.3 \text{ A}}$

2. A 4 wire cable supplies a star connected load consisting of Red Phase $5 - j\,10\,\Omega$, Yellow Phase $3 - j\,6\,\Omega$, Blue Phase $2 + j\,4\,\Omega$. If the cable voltage is balanced and equal to 415 V, calculate the magnitude and phase of the line and neutral currents, and draw the phasor diagrams. Calculate also the total power supplied to the load.

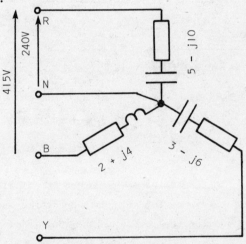

The line-neutral voltage or the phase voltage V_P is

$$V_P = \dfrac{415}{\sqrt{3}} = 240 \text{ V}$$

Figure 3.21

Take the phases in order and calculate the phase (and line currents for a star system) by Ohm's law using the Red phase voltage as the reference phasor.

RED PHASE

$$I_R = \dfrac{240}{5 - j\,10} = \dfrac{240\,(5 + j\,10)}{25 + 100}$$

$$I_R = 9.6 + j\,19.2 \text{ A} = 21.4\,\underline{/63°\,24'} \text{ A}$$

YELLOW PHASE

The yellow phase voltage with respect to the red phase voltage is

$$V_{YN} = 240\,\underline{/-120} = 240\,\{\cos(-120) + j\sin(-120)\}$$

74 Basic Electrical Engineering and Instrumentation for Engineers

$$V_{YN} = -120 - j\ 207.8\ V$$

$$I_Y = \frac{-120 - j\ 207.8}{3 - j\ 6} = \frac{-(120 + j\ 207.8)(3 + j\ 6)}{45}$$

$$\underline{I_Y = 19.7 - j\ 29.85\ A = 35.7\ /\!-57°\ 36'\ A}$$

BLUE PHASE

The blue phase voltage with respect to the red phase voltage is

$$V_{BN} = 240\ /\!-240 = 240\ \{\cos(-240) + j\sin(-240)\}$$

$$V_{BN} = -120 + j\ 207.8\ V$$

$$I_B = \frac{-120 + j\ 207.8}{2 + j\ 4} = \frac{(-120 + j\ 207.8)(2 - j\ 4)}{20}$$

$$\underline{I_B = 29.56 + j\ 44.78\ A = 53.5\ /57°\ 36'\ A}$$

By Kirchoff's first law the neutral current is given by

$$\overline{I}_N = \overline{I}_R + \overline{I}_Y + \overline{I}_B$$

$$I_N = 9.6 + j\ 19.2 + 19.7 - j\ 29.85 + 29.56 + j44.78$$

$$\underline{I_N = 58.86 + j\ 34.15\ A = 67.9\ /30°\ 6'\ A}$$

Because the currents are not balanced the power cannot be calculated from equation 3.33. The total power for an unbalanced three-phase system must be obtained by summing the powers in the individual phases.

$$\text{Power in Red phase} = V_P\ I_R\ \cos\phi_R$$

The angle between the current and voltage in the Red phase is $63°\ 24'$.

Therefore

$$P_R = 240\ .\ 21.4\ .\ \cos 63°\ 24'$$

$$\underline{P_R = 2300\ W}$$

$$\text{Power in Yellow phase} = V_P\ I_Y\ \cos\phi_Y$$

The angle between the current and voltage in the Yellow phase is $120 - 57°\ 36' = 63°\ 24'$.

$$P_Y = 240\ .\ 35.7\ \cos 63°\ 24'$$

$$\underline{P_Y = 3840\ W}$$

Power in the Blue phase = $V_P I_B \cos \phi_B$

The angle between the current and voltage in the Blue phase is $120 - 57° 36' = 63° 24'$.

$$P_B = 240 \cdot 53.5 \cdot \cos 63° 24'$$

$$\underline{P_B = 5750 \text{ W}}$$

$$\underline{\text{Total power} = 11,890 \text{ W}}$$

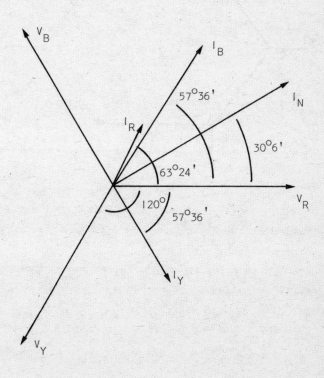

Figure 3.22

76 Basic Electrical Engineering and Instrumentation for Engineers

Example

A small engineering works has its power supplied from a ring main system and the current taken by various items of plant is as shown below. Calculate the current in each branch and the voltage at point P.

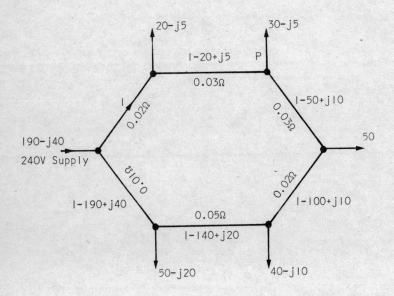

Let the current in the first branch be I and use Kirchoff's first law at the subsequent junctions.

Use Kirchoff's second law around the closed loop assuming the resistance of the cables to be as shown.

$$0 = I(0.02) + (I - 20 + j5)(0.03) + (I - 50 + j10)(0.03) +$$
$$(I - 100 + j10)(0.02) + (I - 140 - j20)(0.05) +$$
$$(I - 190 + j40)(0.01)$$

$$0 = 0.61\,I - 13 + j2.05$$

$$\underline{I = 81.25 - j12.81 \text{ A}}$$

The currents in each branch are therefore,

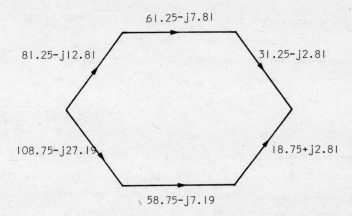

The voltage at point P is

    240 − (81.25 − j12.81)(0.02) + (61.25 − j7.81)(0.03)

= 240 − 3.47 + j0.49

= <u>236.53 − j0.49 V</u>

Examples

1. A coil takes a current of 4 A when 24 V d.c. is applied and for the same power on a 50 Hz supply the applied voltage is 40 V. Explain the reason for the difference in applied voltage and calculate (a) the reactance of the coil, (b) the inductance of the coil, (c) the phase angle, (d) the a.c. power supplied.

$$(8 \ \Omega, \ 0.0255 \ H, \ 53° \ 7', \ 96 \ W)$$

2. In order to use three 110 V, 60 W lamps on a 230 V, 50 Hz supply, they are connected in parallel and a condenser put in series with the group. Find (a) the capacitance required to give the correct voltage across the lamps, (b) the power factor of the circuit.

$$(25.8 \ \mu F, \ 0.478 \ \text{leading})$$

3. A coil of resistance 12 Ω and inductance 0.12 H is connected in parallel with a capacitor of 60 μF, to a 100 V variable frequency supply. Calculate the frequency at which the circuit becomes purely resistive, and calculate the value of the effective resistance. Sketch a phasor diagram.

$$(57.2 \ Hz, \ 167 \ \Omega)$$

4. A factory has a load of 150 kVA at a power factor of 0.8 lagging. Calculate the value of capacitance which, when connected in parallel with this load, will improve the power factor to 0.9 lagging. The supply voltage is 500 V at 50 Hz.

$$(405 \ \mu F)$$

5. Three coils each of 48 mH inductance and 20 Ω resistance are connected in star to a 400 V, three-phase, 50 Hz supply. Calculate (a) Line Current, (b) Power Factor, (c) Total Power Supplied.

$$(9.24 \ A, \ 0.8 \ \text{lagging}, \ 5.12 \ kW)$$

6. A three-phase 240 V supply is connected with ABC phase rotation to a delta connected load consisting of $Z_{AB} = 10\ \Omega$, $Z_{BC} = 8.66 + j5\ \Omega$ $Z_{CA} = 15\underline{/-30}\ \Omega$. Calculate the line currents and draw a phasor diagram showing voltages and currents ($V_{BC} = 240\underline{/0}$).

$$(I_A = 38.7\underline{/108°19'},\ I_B = 46.5\underline{/-45°},$$
$$I_C = 21.2\underline{/190°54'})$$

7. A balanced three-phase load consists of a three-phase induction motor operating at a power factor of 0.866 lagging and delivering 59.68 kW at an efficiency of 0.89, together with a three-phase synchronous motor taking an input of 20 kVA at a power factor of 0.3 leading. Find the reading on each of two wattmeters connected to measure the total power.

(31 kW, 42 kW)

CHAPTER FOUR

Devices

4.1 Semiconductors

Materials may be roughly classified by their electrical resistivity. The metals are, in general, good conductors, a typical resistivity value being of the order of 10^{-8} ohm metres, while at the other extreme a good insulator would have a resistivity value of the order of 10^{10} or 10^{12} ohm metres. Semiconductors are materials whose resistivity falls between these extreme limits. There are many materials which may be classified as semiconductors, including elements and intermetallic compounds. However, from the electrical component point of view, the elements germanium and silicon have found the widest use. Of these two, the earlier transistors, and other devices, were mainly of germanium, but at the present time silicon devices have almost completely replaced those of germanium. Each individual atom of germanium, or silicon, comprises a positively charged nucleus, and a number of negatively charged orbiting electrons. In the normal state, the negative charge of the electrons is equal in magnitude to the positive charge on the nucleus; the atom is, therefore, electrically neutral. Both germanium and silicon are Group IV elements, having four electrons in their outer orbit. The atomic structure of a pure germanium or silicon crystal can be represented schematically as shown in Figure 4.1. The four outer valency electrons of each atom form covalent bonds with one electron of each neighbouring atom, so forming an even crystal structure. At an absolute temperature of $0°K$ all the atomic electrons are bound in this manner into the crystal lattice. At room temperature, however, some electrons acquire sufficient energy to be able to break free from their correct positions, and are free to move throughout the

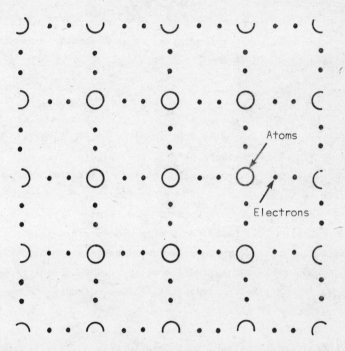

Figure 4.1 Two dimensional schematic representation of the crystal lattice.

lattice. The free electrons can, therefore, be moved by the application of an electric field, and the material has some conducting properties. Although the resistivity of the material is infinite at $0°K$, its value decreases with an increase in temperature. This contrasts with the effect of temperature on metals, the resistivity normally increasing with increasing temperature.
When an electron, of negative charge e coulombs, is freed from its

attachment to the nucleus, the nucleus is left with a net positive charge of +e coulombs. A vacancy now exists in the lattice which may be filled by any other available electron. It is useful to consider the available space in the lattice as a hole, and to associate the net positive charge of the nucleus in the lattice with this hole. A quantity of energy is thus able to produce a hole-electron pair, producing a free negatively charged electron, and effectively, a positively charged hole. The amount of energy required to so generate a hole-electron pair is a function of the type of material. For a given material, the number of electrons per unit volume available for conduction at an absolute temperature of T^oK is given by

$$n \simeq \text{constant} \times e^{-E_g/kT} \qquad 4.1$$

where k = Boltmann's coefficient (1.38×10^{-23} watt second/^oK). The value E_g represents the amount of energy required to excite an electron from its covalent state (the valence band) to its free state (the conduction band) and is about 0.7 electron volt (eV) for germanium and 1.1 eV for silicon. At any temperature, therefore, there are far more free electron-hole pairs in a germanium crystal than in a silicon crystal. This fact accounts for the preference for silicon devices rather than germanium devices since, at room temperature, the currents which flow due to the thermally produced electron-hole pairs, and which constitute undesirable leakage currents are much smaller in silicon devices.

A pure semiconductor in this state has equal numbers of free electrons and holes, and is said to be in the intrinsic state. Conduction in the semiconductor is due to both the free electrons and the holes. The concept of a moveable hole may be better understood by considering a simple analogy. If a patron at a cinema wishes to sit in an empty seat at the opposite end of a row of occupied seats, he can either walk along the row past all the full seats, (this being analogous to a moving electron) or he can ask all the seat occupants to move along one seat so effectively moving the empty seat along the row to him (this being analogous to a moving hole). The hole in the crystal lattice thus can be considered to move when in actual fact its effective movement is due to the original hole being re-occupied by another electron, so producing a further hole in a different place. In effect, free electrons and holes are continually being produced, and are continually recombining, the number of electrons and holes in the crystal at any instant being a balance between the two effects. The actual life time of a free electron or hole before recombination takes place is dependent upon the concentration of the electrons and the holes, and would normally be a very small fraction of a second. During this lifetime the electron or hole would be able to travel a distance which is termed its 'mean free path'. If a potential difference is applied across such a semiconductor crystal, there will be a drift of the free electrons towards the positive terminal, and of the holes towards the negative terminal. Under the action of the electric field E volts/metre, established by the applied potential difference, the electrons acquire an average drift velocity which is related to the applied electric field by the equation

$$\text{average drift velocity} = v_e = -\mu_e E \qquad 4.2$$

The quantity μ_e is known as the electron mobility. The minus sign indicates that the direction of the drift velocity is in the opposite direction to that of the electric field, i.e. is towards the positive electrode.

Similarly, the average drift velocity for the holes is given by

$$v_h = \mu_h E \qquad 4.3$$

where μ_h is the hole mobility, and no minus sign is required because the hole velocity is opposite in direction to that of the electrons. Note that the values for μ_e and μ_h are not the same, and that they are temperature dependent quantities.

Note also that as the unit of the drift velocity is metres per second, the unit of mobility is

$$\mu = \frac{v}{E} \text{ metres}^2 \text{ per volt second}$$

The current density produced by the electron drift is

$$J_e = nev_e \text{ amperes/square metre} \qquad 4.4$$

where n is the number of free electrons per cubic metre.

Similarly for the holes, current density is

$$J_h = pev_h \text{ amperes/square metre} \qquad 4.5$$

where p is the number of free holes per cubic metre.

As explained, the holes and electrons drift in opposite directions under the influence of the electric field. Their contributions to the total current density, however, add directly.

The total current density is, therefore,

$$J = J_e + J_h = e(nv_e + pv_h)$$
$$= e(n\mu_e E + p\mu_h E) \qquad 4.6$$

The resistivity ρ of the crystal is then

$$\rho = \frac{E}{J} = \frac{1}{e(n\mu_e + p\mu_h)} \text{ ohm metres} \qquad 4.7$$

In an intrinsic material, the numbers of electrons and holes are, of course, equal,

$$\text{i.e. } n = p = n_i \qquad 4.8$$

Thus we may write,

$$\text{resistivity} = \frac{1}{en_i(\mu_e + \mu_h)} \qquad 4.9$$

4.2 Doped semiconductors

The production of very pure intrinsic silicon forms the first part of the process used in the preparation of silicon transistors and other devices. The conduction property of the silicon is then markedly altered by the controlled addition of minute quantities of impurity. The impurity added is either from the GROUP V elements, such as antimony, phosphorus or arsenic, if n type semiconductor is required, or from the GROUP III elements, perhaps boron or aluminium, if p type semiconductor is required. The GROUP V elements have five orbital electrons in their outer shell. When a trace of such an element is added to intrinsic silicon, and a crystal forms, the impurity atoms take the place of silicon atoms in the lattice, with four of their five outer electrons forming the required covalent bonds with the neighbouring silicon atoms. At $0^\circ K$ the extra electron is loosely held to its parent atom, but requires only a small additional energy to release it into the conduction band. Compared with the energy gap of silicon of about 1.1 eV, the energy required for the extra impurity electron is of the order of 0.01 eV. At room temperature, therefore, the resistivity of the doped silicon is reduced substantially from that of the intrinsic silicon, and can be controlled carefully by control of the amount of added impurity. Doped semiconductor is given the name 'extrinsic semiconductor'. An impurity atom added as described to provide an extra source of free conducting electrons is called a donor atom, the resulting doped semiconductor being n type.

It is also possible to produce doped or extrinsic semiconductor with extra conducting holes. This p type semiconductor requires the addition of a GROUP III impurity element, which has only three electrons in its outer shell. When the crystal lattice is formed, a hole is produced by the shortage of valency electrons at the impurity atom centres. As with the donor impurity, only small additional energy is required to allow conduction via the extra holes. An impurity added in this way to produce p type semiconductor is called an 'acceptor' impurity.

Whereas the intrinsic semiconductor contains equal numbers of electrons and holes, the addition of impurities upsets this balance. In the n type semiconductor, for example, the presence of the additional free electrons increases the probability that a hole will recombine with a free electron. The number of holes is thus much smaller in the doped n type semiconductor than in the intrinsic semiconductor. A similar effect occurs in a doped p type semiconductor where the numbers of free electrons is greatly reduced. In doped semiconductors, therefore, conduction is due mainly to one type of charge carrier, i.e. electrons in n type material, and holes in p type material. This carrier is referred to in each case as the 'majority' carrier, while, of course, the other carrier is the 'minority' carrier. Thus, in p type material, holes form the majority carriers, while electrons form the minority carriers.

We have already considered the mechanism of current flow due to the drift of charge carriers under the influence of an applied electric field. This is referred to as current flow due to carrier drift.

Conduction can also occur due to carrier diffusion. This effect occurs where the concentration of a charge carrier is not the same throughout the material. A drift of carriers occurs away from areas of high concentration towards areas of low concentration. The magnitude of the current density produced by diffusion is given by

$$J_e = eD_e \frac{dn}{dx} \quad \text{for electrons} \qquad 4.10$$

and

$$J_h = -eD_h \frac{dp}{dx} \quad \text{for holes} \qquad 4.11$$

where D_e and D_h are the diffusion constants, and $\frac{dn}{dx}$ and $\frac{dp}{dx}$ represent the magnitude of the change of carrier concentration with distance.

4.3 The pn junction diode

A pn junction diode is formed when the doping in a crystal is such that the semiconductor changes from p type to n type in a very small distance, typically of the order of 10^{-6} or 10^{-7} metres. Figure 4.2(a) shows two isolated sections of semiconductor, one p type and one n type. Omitting, for the sake of clarity, the basic semiconductor lattice, the p type material is represented as fixed negatively charged impurity atom nuclei, together with free positively charged holes, these being the majority carriers in the p type material. There are also, of course, a relatively small number of free electrons, the minority carriers. Remember, however, that the piece of semiconductor is electrically neutral, that is it contains equal amounts of positive and negative charge. Similarly, the n type material is represented in the diagram by fixed positively charged impurity atom centres, together with free negatively charged electrons, the majority carrier. Again, a small number of free holes, the minority carriers, are also present.

A pn junction is not formed by merely bringing together the two sections of semiconductor. In order to realise the rectifying properties of a junction diode, it is necessary that the crystal structure is continuous across the transition from p type material to n type material. The junction is, therefore, produced by the addition of the necessary impurities to a single crystal of the basic semiconductor.

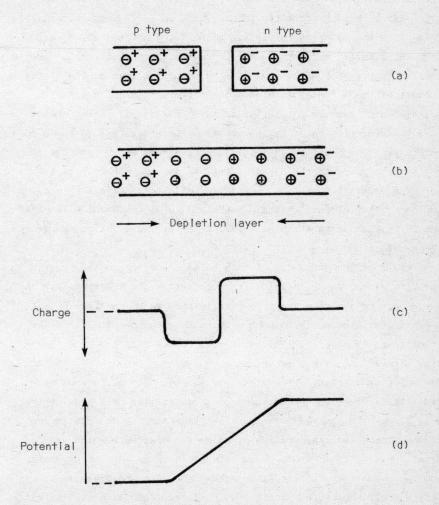

Figure 4.2 The p-n junction.

Let us consider the conditions at the transition between the n and p type materials in a correctly formed junction. In the p material, holes form the majority carriers, while in the n material, holes form the minority carriers. The concentration of holes is, therefore, much greater on the p side of the junction than on the n side. Because of the difference in hole concentration, a diffusion of holes occurs from the p material to the n material. This results in a transferral of positive charge from the p material to the n material. Consider now also the free electrons which form the majority carriers in the n type material, and the minority carriers in the p material. The concentration of electrons is much greater in the n material than in the p material; a diffusion of electrons occurs from the n to the p material, transferring negative charge to the p material.

The result of the diffusion of both holes and electrons is, therefore, to produce a potential difference across the junction with the n side of the junction positive with respect to the p side. This potential difference is in such a sense as to oppose the diffusion of carriers which originally caused it; it is effectively a potential barrier across the junction.

The majority carriers which cross the junction become, of course, minority carriers in the semiconductor material of the opposite type. Once across the junction, the carriers take part in the continuous recombination and regeneration of holes and electrons which is normal in a semiconductor. The existence of the barrier potential and the resulting electric field across the junction produces a narrow region about the junction which is depleted of charge carriers. The charge distribution and the potential distribution are illustrated in Figure 4.2(c) and 4.2(d).

Although the electric field in the depletion layer is in the sense to oppose the continued crossing of the junction by majority carriers, it is, of course, in the correct sense to accelerate minority carriers across the junction. Any minority carriers entering the depletion layer and coming under the influence of the electric field are, therefore, rapidly transferred across the junction. In an isolated pn junction, without external circuit connections, the net current, i.e. the mean rate of movement of electrical charge must, of course, be zero. A state of balance is, therefore, produced in the junction whereby the potential barrier is of the exact magnitude to allow sufficient majority carriers to cross the junction to equal and cancel the charge carried across the junction by the minority carriers.

The electric field in the depletion layer is produced as we have seen by the removal of free charge carriers across the junction, leaving the charged donor or acceptor impurity atom centres fixed in the crystal lattice. The width of the depletion layer is, therefore, dependent upon the actual concentration of the impurity atoms on each side of the junction. Thus, a relatively high concentration of impurity atoms will result in a narrow depletion layer. Again, if the impurity concentrations on each side of the junction are equal, the depletion layer will be equally spaced in both the p and the n types of material. However, if the impurity concentration is lower in, for example, the p material, than in the n material, the depletion layer will extend further into the p material than into the n material.

Outside the depletion region, however, the net charge is zero, i.e. the material is electrically neutral. The electric field in the material away from the depletion layer is, therefore, also zero.

4.3.1 The biassed pn junction diode

If a voltage is applied across the junction via connections made to the p and n sides, the conditions in the junction are altered. If the negative connection is made to the p type material and the positive connection is made to the n type material, the applied voltage is in such a direction as to increase the potential barrier across the depletion layer. A small external applied voltage is found to be sufficient to increase the barrier potential to a level which effectively prevents majority carriers from crossing the junction. Charge flow across the junction is then due solely to minority carriers which continue to be accelerated across the junction. Under these conditions the junction is said to be reverse biassed.

All minority carriers which enter the depletion layer region are accelerated across the junction and the current in the reverse biassed junction is due only to the minority carriers. The current in the junction, and the external circuit, is, therefore, very small, and over a large range is independent of the magnitude of the applied reverse voltage, and, because it is fixed by the number of thermally generated minority charge carriers, is temperature dependent.

Forward bias is applied to a pn junction by connecting the external source in such a direction to make the p material positive with respect to the n material. The external voltage now has the effect of reducing the magnitude of the barrier potential, so allowing a great increase in the number of majority carriers crossing the junction. The forward current thus increases very rapidly for small applied forward bias voltages.

The junction current voltage relationship is found to be of the form

$$I = I_s (e^{\frac{eV}{kT}} - 1) \qquad 4.12$$

where I is the junction current
  V is the applied voltage
  k is Boltzmann's constant
  T is the absolute temperature
  I_s is the reverse saturation current

due, as previously explained, to the thermally generated minority carriers.

Figure 4.3 shows a current/voltage characteristic typical of a silicon diode. The figure shows clearly the basic rectifying characteristic of the diode, i.e. the very high impedance presented by the diode to an applied voltage of the reverse polarity and the very low impedance presented by the diode to an applied voltage of the forward polarity. The diode is, in effect, a switch which is controlled by the polarity of the applied voltage, the switch being closed, or on, for forward voltages, and open, or off, for reverse voltages.

Note that, in the figure, the voltage and current scales are different for the forward and reverse bias conditions. The reason for this is that the reverse current I_s, which may typically be of the order of a few nA, would be indistinguishable from zero if the

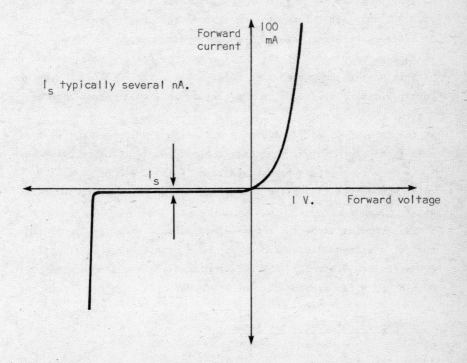

Figure 4.3 Typical silicon diode characteristic.

reverse current scale was not magnified in relation to the forward current scale.

At room temperature (approximately $300^{\circ}K$) the value of the term $\frac{e}{kT}$ is approximately 40. With an applied reverse voltage of a little over 0.1 volt, therefore, the reverse current has risen to within 1% of the saturation value I_s. The reverse current is then virtually independent of the applied reverse voltage until breakdown of the junction commences at a voltage V_{BO} which, depending upon the construction of the diode, may be as low as 1 volt, or as high as several thousand volts.

In the forward direction again, a bias of a little over 0.1 volt will result in a forward current of about 100 I_s. However, because I_s is so very small, a forward voltage of the order of 0.7 - 1.0 volt is required to achieve the normal forward current range of the diode.

For a germanium diode the reverse saturation current is higher than that of a silicon diode, and the forward bias voltage is slightly lower (0.3 to 0.6 volt). For both diodes the reverse current will double for a temperature rise of the order of 7 to $8^{\circ}C$.

4.3.2 Depletion layer capacitance

The junction depletion layer consists of a region of semiconductor without free charge carriers, sandwiched between two sections of conducting material, the p and n type semiconductors. A capacitance therefore, exists across the junction of a size determined by the width of the junction, the area of the material, and the effective permittivity of the material. This capacitance has a significant effect upon the characteristic of the junction when it is subject to a.c. signals. In particular, as we have seen that the width of the depletion layer is dependent upon the applied voltage, increasing with increasing voltage, the effective value of the junction capacitance may be changed electrically. Use is made of this effect in many automatic tuning systems in which a pn junction is employed solely as a voltage-adjustable capacitor.

4.3.3 Point-contact diodes

The junction capacitance associated with a pn junction diode is often too large to allow the required speed of operation necessary for very high frequency signals. For very high frequency operation, for example, in microwave circuits, or in the radio frequency circuits of radio and television receivers, point contact diodes are employed rather than junction diodes.
A point contact diode consists of a small piece of semiconductor with an electrical connection which forms one lead of the diode. The other connection is via a tungsten wire whose point is held against the semiconductor. Point contact diodes are fabricated using either silicon or germanium semiconductors.

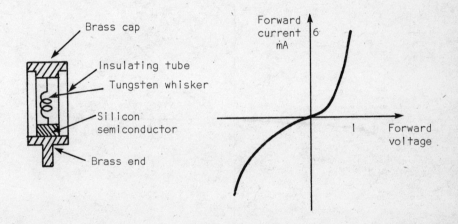

Figure 4.4 Silicon point contact diode.

Figure 4.4 shows the construction of a silicon point contact diode. The importance of silicon point contact diodes lies in their ability to operate at very high frequencies, as high as several thousand megahertz. At these frequencies, in the microwave range, conduction normally occurs in waveguides rather than in normal wires, and the diode construction is designed to fit directly into the waveguide. A typical characteristic is also shown in the figure. Notice in particular the much greater reverse current obtained in the point contact diode in comparison with that of the junction diode. The frequency response of a germanium diode is not as high as that of the silicon diode, but may be up to about 100 MHz. Its main use is as a circuit element in radio and television receivers, and the diode is, therefore, normally encapsulated in a glass container with wire ends.

A germanium point contact diode characteristic is illustrated in Figure 4.5.

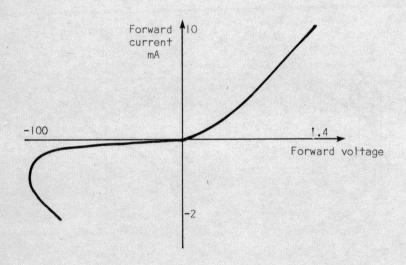

Figure 4.5 Germanium point contact diode characteristic.

Comparison of the characteristics of the silicon and germanium diodes shows that the forward voltage drop, and the allowable reverse voltage, are both greater for the germanium diode.

4.4 The diode in a circuit

The current/voltage characteristic of the junction diode was given in the previous section in the form

$$I = I_s (e^{\frac{eV}{kT}} - 1) \qquad 4.12$$

This equation relates the current flowing in the diode to the voltage across the diode. Such a relationship is a function only of the device itself, and is fixed, for any device, by its construction. In use in a circuit, the current flowing will depend also upon the other devices in the circuit.

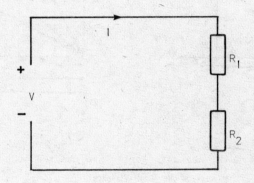

Figure 4.6

Consider the circuit of Figure 4.6. This shows two resistors R_1 and R_2 in series across a direct voltage supply. Analysis of this circuit is relatively simple because of the simple nature of the characteristic of each resistor. The current/voltage characteristic for a pure resistor is a straight line as shown in Figure 4.7(a). Over its working range, the slope of the characteristic is constant; the resistor current is proportional to the voltage across it and Ohm's Law applies.

Thus, the circuit current is given by

$$I = \frac{V}{R_1 + R_2}$$

Figure 4.7(b) shows, on the same axes, the individual characteristics of both the resistors R_1 and R_2. When connected in series, the current I causes individual volt drops across the resistors R_1 and R_2 of V_1 and V_2 respectively; the resistors operate at points

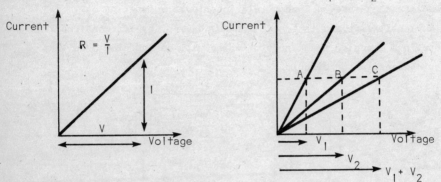

Figure 4.7

on their individual characteristics represented by points A and B in the figure. The sum of V_1 and V_2 must at all times equal the total voltage applied to the circuit; in other words point C represents the operating point of the circuit, and the locus of point C for different applied voltages will be the composite characteristic for the circuit.

The circuit represented in this example is simple, and it is not necessary, normally, to use a composite characteristic in this way to obtain a value for the circuit current. Resistors are particularly simple to deal with since they are linear devices, i.e. resistance value remains the same, irrespective of the magnitude of the current. There is obviously a limit to this statement in practice, since at large currents, overheating of the resistor will alter its resistance value. Within their designed range, however, resistors may be considered to be linear devices. Because of this fact, the applied voltage is always shared between the resistors in the same ratio. One further point should be noted regarding the characteristics. These have been plotted for only one quadrant, i.e. for positive voltages and currents only. This is quite common practice because of the symmetry between the positive and negative halves of the complete characteristic. In other words, the positive and negative halves are identical; there is, therefore, no 'right way round' for a resistor or similar linear device.

When dealing with a circuit containing a non-linear device, such as a diode, these ideas must be revised. There may, even with a non-linear device, be symmetry between the positive and negative halves, but in general this is not so, and the characteristic must include all four quadrants.

Consider again a typical diode characteristic. If we define the resistance of the diode in the same way that was used for the linear resistor, i.e.

$$\text{resistance} = \frac{\text{voltage across the device}}{\text{current flowing through the device}}$$

we immediately meet the difficulty that the value of resistance obtained varies with the value of the current flowing. As an example, consider the diode in its reverse biassed condition, where the application of a relatively large voltage results in a very small reverse current. The effective diode resistance is obviously very large, and for silicon diodes, may be of the order of 100 MΩ. Conversely, in the forward direction, the application of about 0.7 volts will result in about the normal forward current of the diode and its resistance would be of the order of 0.1 to several ohms, depending upon the type of diode. It is obvious that it is not of practical value to attempt to define a diode in terms of its resistance value, and a full current/voltage characteristic is required. If a resistance value is not known for the diode, how then can we determine the circuit current flowing in the diode resistor network of Figure 4.8. The circuit symbol commonly used to represent diodes is shown in this figure.

The terms anode and cathode, which are still used to name the two connections to a diode, derive from the two electrodes of the thermionic diode, which has, to a great extent, been superseded by the semiconductor diode. The arrowhead in the diode symbol indicates the direction of forward current flow; thus, when in forward conduction, the diode anode is positive with respect to its cathode.

98 Basic Electrical Engineering and Instrumentation for Engineers

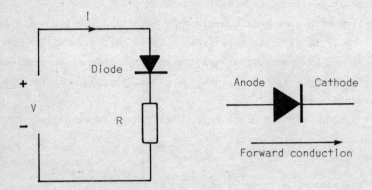

Figure 4.8

A simple analysis of the series diode-resistor network is not possible. It is rather like the chicken and the egg situation, in that, in order to calculate the circuit current, the diode voltage must be known, but in order to obtain the diode voltage from its characteristic, the circuit current must be known. A simple method of determination of the current is given, however, by use of a composite characteristic. Figure 4.9 shows the individual characteristics of the diode, and the resistor, together with the composite characteristic formed as described in Figure 4.7(b). The resistor characteristic is, of course, a straight line through the origin, bearing in mind that any change of scales between the positive and negative axes would cause a change in the slope of the resistor characteristic as it passes through the origin. Because of the very high resistance of the diode when reverse-biassed, the composite characteristic is virtually identical to the diode characteristic for applied reverse voltages.

An alternative method of determining the d.c. current and voltage conditions in a diode resistor network is illustrated in Figure 4.10. This represents again the circuit of Figure 4.8, and shows the diode characteristic, together with the resistor characteristic plotted as

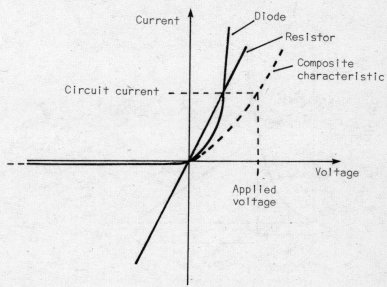

Figure 4.9

a 'load line'. The load line is most conveniently drawn, by connecting with a straight line, two points;
(a) the point on the voltage axis which represents the voltage of the supply, V,
(b) the point on the current axis which represents the current which would flow in the resistor if it was connected alone across the supply, i.e. $I = \frac{V}{R}$.

The gradient of the load line is then equal to the reciprocal of the resistor value.

The point of intersection of the load line, and the diode characteristic, is known as the 'operating point' of the network.

In a particular circuit, the choice of the operating point is often governed by the maximum allowed power dissipation in the device.

At the operating point, the power dissipation in the diode is given by

$$\text{Power} = \text{diode voltage} \times \text{diode current}$$
$$= V_D \times I \text{ watts}$$

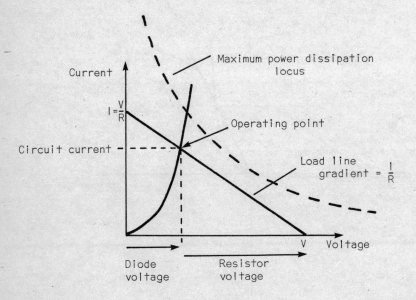

Figure 4.10

The maximum power rating for a particular diode is specified by the manufacturer, as is the maximum allowable reverse voltage, and the maximum allowable forward current.

Thus, maximum power dissipation is

$$W = V_D \times I \qquad 4.13$$

For any given diode voltage, therefore, there is a maximum value of allowable diode current. The locus of the extreme limit of the operating point can thus be drawn on the characteristic as a rectangular hyperbola. This is also shown in Figure 4.10. The operating point must lie to the left of the maximum power dissipation curve.

4.4.1 The use of an equivalent circuit

An alternative approach to the problem of non-linear devices in circuits is to try to produce an equivalent circuit which may

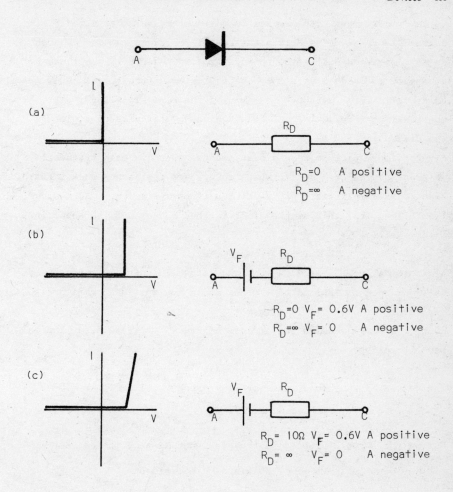

Figure 4.11

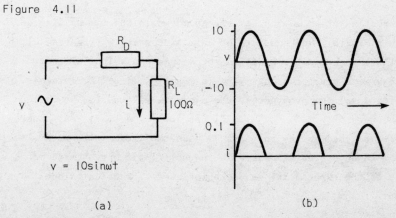

Figure 4.12

be used to represent the device in a linear manner over a restricted range of conditions. Within the restricted range the use of such an approximation will allow reasonably simple analysis. It may be necessary to produce more than one equivalent circuit, each one being valid for a different range of input conditions.

Consider, as an example, possible equivalent circuits to represent the diode.

A very simple picture of a diode, and one which is used frequently in practice, is as a device which is of zero resistance when forward biassed, and infinite resistance when reverse biassed. An equivalent circuit for this simple diode is thus a resistor, as shown in Figure 4.11(a).

Example

A sinusoidal voltage source of peak voltage 10 volts and zero internal impedance is applied to a diode and a 100 ohm load resistor in series.

Using the simple equivalent circuit of Figure 4.11(a), determine the waveform of the current in the load resistor.

The circuit is shown in Figure 4.12(a).

The current in the load resistor is given by

$$i = \frac{v}{100 + R_D}$$

During the forward biassed half-cycle, when $R_D = 0$, the current is

$$i = \frac{1}{100} \cdot 10 \sin \omega t = 0.1 \sin \omega t$$

During the reverse biassed half cycle, when $R_D = \infty$ the current is zero.

The current waveform is, therefore, as shown in the timing diagram of Figure 4.12(b), and consists of half sine waves. Figures 4.11(b) and 4.11(c) show modifications to the simple equivalent circuit which can give greater accuracy. In Figure 4.11(b), for example, the forward voltage drop across the diode is allowed for by including a battery of e.m.f. V_F. In Figure 4.11(c) a small resistance value is assigned to the resistor R_D when the diode is forward-biassed, in an attempt to make the characteristic approximate more closely to the exponential characteristic of a practical diode.

The actual value would have to be chosen to suit the range of current values for which it is intended that the equivalent circuit should be valid.

By the use of more and more complex equivalent circuits, greater accuracy can be achieved; however, the aim is to keep the equivalent circuit as simple as possible, consistent with allowing a sufficient accuracy.

The diode equivalent circuits so far discussed have been intended for use over the whole diode current and voltage range; They are, in fact, 'large signal' equivalent circuits.

It is also interesting to consider the case of an applied voltage which is varied by a small amount about its steady value.

Figure 4.13(a) shows the characteristic for a diode with an applied steady forward bias voltage V_B, which results in a diode current of I_B. Only the forward conduction quadrant of the characteristic has been shown.

The point A on the characteristic is the effective operating point for this value of applied voltage. The effective resistance of the diode at this operating point, and remember this applies only at this operating point, is given by

$$R_{DC} = \frac{V_B}{I_B} \text{ ohms} \qquad 4.14$$

and is represented by the reciprocal of the slope of the chord OA of the characteristic.

Let us consider now the addition of a small alternating voltage to the steady diode voltage V_B such that the voltage applied to the diode can be expressed as

$$v = V_B + V_S \sin \omega t$$

The conditions in the diode are now represented by the characteristic shown in Figure 4.13(b). The current produced in the diode is effectively the same current I_B as before, but with a varying current added to it. Although the varying voltage added to the steady voltage V_B was sinusoidal, the varying component of the diode current cannot be sinusoidal unless the diode characteristic between the points X and Y on the characteristic is exactly straight, i.e. unless a linear relationship exists between the alternating compon-

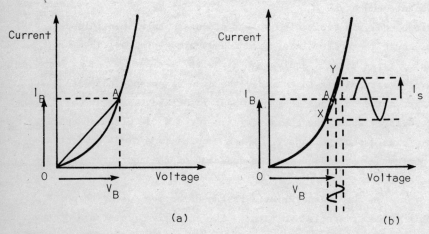

Figure 4.13

ent of the applied voltage, and the alternating component of the resulting current.

For all practical diodes, however, such a linear relationship does not exist, and the curvature of the characteristic results in a distorted current waveform, that is in the production of current components at frequencies which are harmonically related to the frequency $\frac{\omega}{2\pi}$ of the alternating voltage. The effect of such curvature is reduced, however, if the magnitude of the alternating component of the voltage is reduced, and almost linear operation can be approached if the magnitude is reduced to a very low level. Such operation is called 'small signal' operation.

Under small signal conditions, for an applied voltage of

$$v = V_B + V_S \sin \omega t \qquad (V_S \ll V_B)$$

the diode current will be very nearly

$$i = I_B + I_S \sin \omega t \qquad (I_S \ll I_B)$$

The direct components of current and voltage will be related to each

other, as before, by
$$R_{DC} = \frac{V_B}{I_B} \text{ ohms}$$
where R_{DC} represents the effective d.c. resistance of the diode at the operating point A. The alternating components of the current and voltage will be related to each other by
$$R_{AC} = \frac{V_S}{I_S} \text{ ohms} \qquad 4.15$$
where R_{AC} is the effective a.c. resistance of the diode at the operating point A and is effectively the reciprocal of the gradient of the diode characteristic at the operating point.
Alternative names used for the a.c. resistance are the 'incremental' resistance, or 'dynamic' resistance.

The use of a direct 'bias' voltage in this manner, together with a small applied alternating voltage is very common; the function of the direct bias voltage is to effectively select the position of the operating point A upon the diode characteristic, and hence effectively to control the value of the a.c. resistance represented by the device to the alternating signal voltage.

It is also convenient to use a 'small signal' equivalent circuit which will represent the diode characteristic in its effect upon the signal voltage. Referring again to Figure 4.13(b) the a.c. resistance at the operating point A was shown to be equivalent to the reciprocal of the gradient of the characteristic at point A.

The current/voltage characteristic is
$$I = I_S (e^{\frac{eV}{kT}} - 1)$$
whence
$$\frac{dI}{dV} = \frac{e}{kT} I_S e^{\frac{eV}{kT}} \simeq \frac{e}{kT} I$$
The a.c. resistance is thus very nearly
$$\frac{dV}{dI} = \frac{kT}{e} \cdot \frac{1}{I}$$
At room temperature
$$\frac{dV}{dI} = \frac{kT}{e} \frac{1}{I} \simeq \frac{1}{40I} \text{ ohms}$$

i.e.

$$\frac{dV}{dI} = \frac{25}{I} \text{ ohms} \qquad 4.16$$

if the current I is expressed in milliamperes.

The value of the a.c. resistance is thus fixed by the position of the operating point upon the characteristic, and this, in turn, is determined, as previously explained, by the value of the biasing direct voltage.

A 'small signal' equivalent circuit would, in its simplest form, merely consist of a resistor of value $\frac{25}{I}$ ohms (I in milliamperes). At other than low frequencies, however, the small signal equivalent circuit should also include the effects of the depletion layer capacitance, and also should make allowance for the ohmic resistance of the semiconducting material, and of the diode internal connections. The resulting equivalent circuit is shown in Figure 4.14.

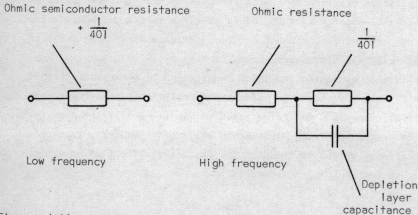

Figure 4.14

4.5 The Zener diode

From the diode characteristic of Figure 4.3 it can be seen that when the reverse voltage is increased beyond a certain value, V_{BO}, the reverse current increases rapidly. The diode is said to break down. The critical voltage V_{BO} is termed the 'breakdown' or Zener voltage. The sudden and large increase in reverse current is caused by either or both of two different effects, depending upon the impurity doping levels in the semiconductor.

(a) The large value of reverse voltage applied to the diode results in an electric field in the depletion layer sufficiently high to break the covelent bonds, so producing free electron-hole pairs. This effect is known as the Zener effect.

(b) Mobile charge carriers are accelerated by the electric field until they collide with atoms in the crystal lattice, moving, on average, a distance known as the 'mean free path'. If, between collisions, they are accelerated so that their kinetic energy is sufficient to ionise atoms with which they collide, then free carriers will be produced, which in turn will be accelerated, and can cause ionising collisions. This results in an 'avalanche breakdown' effect.

The mechanism of breakdown in a given diode depends upon the doping levels in the semiconductor, and which determine the actual value of the breakdown voltage V_{BO}. In fact, both mechanisms may occur at the same time.

The breakdown characteristic of correctly designed diodes is utilised in a range of diodes termed Zener diodes, which are used to provide stable reference voltages. They are manufactured to have breakdown voltages in the range from about 3.0 volts to 20.0 volts. The lower voltage reference diodes rely mainly upon the Zener effect, while the higher voltage diodes operate mainly due to the 'avalanche effect'. The devices are used in practice to provide a voltage drop which is, to a large extent, independent of the current flowing through the diode. The symbol used to represent a Zener diode is shown in Figure 4.15.

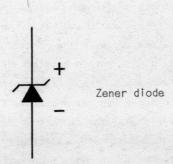

Figure 4.15

108   Basic Electrical Engineering and Instrumentation for Engineers

Because the diode is used in a reverse biased condition, the voltage polarity is as indicated in the figure.

4.6 The junction transistor

The development of the transistor, from its beginnings in 1948, to the cheap, mass-produced device of today, has revolutionised the field of electronics, and affected most aspects of life.   The junction transistor is a bipolar device, in that current flow in the transistor is due to movement of both holes and electrons.
The transistor consists, essentially, of two pn diodes formed with one common section, so giving a three element device, as illustrated in Figure 4.16.

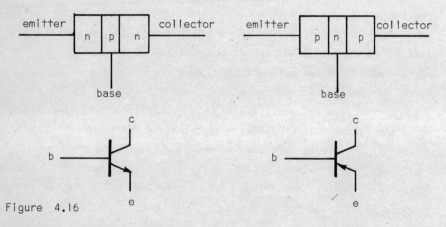

Figure 4.16

The common section, called the 'base', may be of either p or n type semiconductor, resulting in two basic types of transistor known as the npn and the pnp types.   The central base section, which is extremely thin, separates the emitter and collector sections of the transistor.   Connecting leads are provided to all three sections. The figure also shows symbols which are commonly used to represent the two types of transistor in circuit diagrams.   In describing the operation of the bipolar transistor, reference will be made to the npn transistor.   The mode of operation of the pnp transistor is exactly the same as that of the npn transistor, with the exception that the polarities of all applied voltages and currents, and that of the charge carriers, are reversed.

In normal use, as a linear amplifying device, the transistor is used with the emitter to base junction forward biased, and with the collector to base junction reverse biased. In the case of the npn transistor, this means that the emitter is negative with respect to the base, while the collector is positive with respect to the base. Because the collector-base junction is reverse biased, and assuming, for the moment, that the emitter connection is open circuit, then the only current flowing in the collector connection will be the small reverse current of the collector-base junction. This current is termed the collector-base leakage current, and is given the symbol I_{CBO}.

When the emitter-base diode is forward biased, the current flowing in the junction will be due to electrons crossing from the emitter to the base, and also to holes crossing from the base to the emitter. By deliberately using lightly doped p semiconductor for the base material, and relatively heavily doped n semiconductor for the emitter material, the junction current may be designed to be mainly due to electron flow from the emitter to the base region. (Similarly, in considering the pnp transistor, the emitter-base junction current is designed to be mainly due to holes crossing from the emitter to the base.) The majority carriers of the emitter region are said to be 'injected' into the base region by the emitter-base bias.

The emitter current may thus be written

$$I_E = I_{electron} + I_{hole}$$

where $I_{electron}$ and I_{hole} are respectively the components due to electron flow and hole flow. Remember that the passage of negatively charged electrons from the emitter to the base is represented by a conventional current from base to emitter.

Once the electrons cross into the p type base region they become minority carriers, and if they remained in the base for more than a very short time, they would disappear, due to recombination with the majority carrier holes. However, the increased concentration of electrons in the base region near the emitter-base junction causes a diffusion of the electrons across the base, and, because the base region is very thin, most reach the collector side of the base region without recombination. The very small number of electrons which do recombine with holes in the base region gives rise to a small component of base current, because the positive charge removed in the recombination must be replaced in order to maintain steady conditions.

On reaching the collector-base junction, the electrons come under the influence of the electric field across the junction which, because the junction is reverse biased, is in such a direction to accelerate the carriers into the collector region. The carriers reaching the collector then form the collector current.

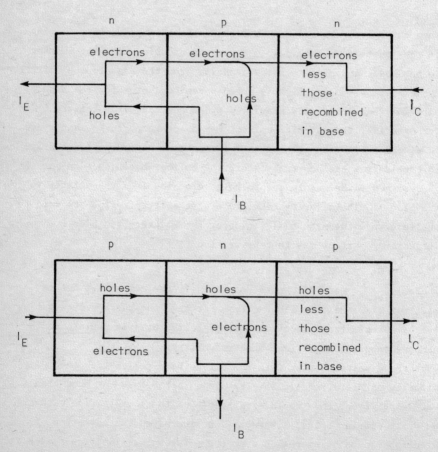

Figure 4.17

Figure 4.17 shows diagrammatically the transistor action in both npn and pnp transistors, omitting the effects of the collector base leakage current I_{CBO}.

Summarising, the current due to the electrons injected from the emitter into the base of the npn transistor, forms a fraction θ_1 of the total emitter current. A further fraction θ_2 of the electrons injected into the base actually reaches the collector base junction and crosses into the collector region.

The steady collector current I_C is thus a fraction of the steady emitter current I_E. The fraction is known as the current gain between collector and emitter, and is given the symbol h_{FB}, or, very commonly, α, where

$$h_{FB} = \alpha = \theta_1 \theta_2$$

In a well designed transistor, the value of h_{FB} will be very close to unity, typical values lying in the range 0.95 to 0.995.

The transistor is, of course, a three terminal device, so that the difference between the emitter current and the collector current must form the base current.

The basic equations which characterise the transistor may thus be written

$$I_E = I_C + I_B \qquad 4.17$$

$$I_C = h_{FB} I_E \qquad 4.18$$

The equation 4.18 describes the basic transistor action, while equation 4.17 relates the actual currents in the three transistor leads.

4.6.1 The common base characteristics

Figure 4.18 shows an npn transistor connected so that its common base characteristics may be determined. The emitter current is determined by the value of the resistor R_1, together with the value of the supply voltage. Notice that it would be impractical to vary the emitter current by means of a variable voltage supply alone. This is so because, as the base emitter junction is forward biased, an increase in emitter supply voltage of from about 0.6 to 0.7 volt

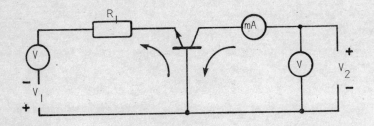

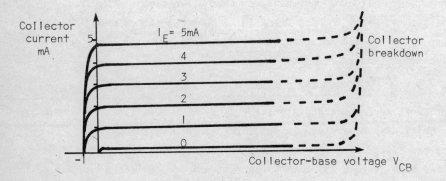

Figure 4.18

would increase the emitter current from a very low value almost through its full range. The use of a series resistor and a high supply voltage, effectively makes the emitter current almost independent of the base emitter diode characteristic.

Thus
$$I_E = \frac{V_1 - V_{BE}}{R_1}$$

and if V_{BE} is small in comparison with V_1 then
$$I_E \simeq \frac{V_1}{R_1} \qquad 4.19$$

The common base collector current/voltage characteristics are also shown in Figure 4.18.

The following points should be noted from the characteristics

(a) Over a wide voltage range, the collector current is very nearly independent of the collector voltage, i.e. the collector circuit may be said to be of very high impedance. The collector circuit is almost a constant current source.

(b) The value of the collector current is determined by the value of the emitter current.

(c) The collector base voltage must actually be reversed (collector negative with respect to the base) in order to reduce the collector current to zero.

(d) With zero emitter current, the collector current is very small, and is, in effect, the reverse current of the collector base junction I_{CBO}.

(e) If the collector base voltage is increased beyond a certain level collector junction breakdown begins, and the collector current increases rapidly.

The mechanism of collector breakdown is similar to that described in Section 4.5, and may be due to either or both avalanche and Zener effects. A further effect which occurs is that as the collector base voltage is increased, the width of the depletion layer also increases, so decreasing the width of the already very narrow base region. As the base width decreases, the time spent by the carriers in crossing the base also decreases, giving less chance of the minority carriers recombining. As a result, the current gain of the transistor increases. At relatively high collector-base voltages, the depletion layer width extends well into the base region, and may approach the emitter base junction. At excessive voltages, 'punch through' occurs, whereby carriers injected from the emitter move directly across into the collector; the resulting current may cause permanent damage to the transistor.

The characteristics shown in Figure 4.18 are termed the common base characteristics, because the base connection is common to both the emitter current circuit and to the collector current circuit.

4.6.2 The common emitter characteristics

Figure 4.19(a) shows an npn transistor connected with its emitter lead common to both the base current circuit and to the collector current circuit. Using this circuit the collector characteristics for the common emitter connection may be obtained. These characteristics are shown in Figure 4.19(b). Comparison with the common base collector characteristics of Figure 4.17 reveals several differences.

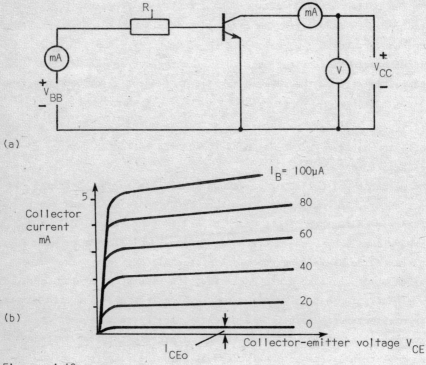

Figure 4.19

(a) The collector-emitter voltage V_{CE} must be positive to produce positive collector current. At low values of V_{CE} the collector current I_C is itself low, but after the voltage V_{CE} has been increased to exceed a 'knee' voltage, the characteristics assume an almost constant current state, i.e. the characteristics become almost horizontal.

(b) The characteristics are not quite as near to horizontal as are those of the common base connection. The effective collector impedance, i.e. the reciprocal of the gradient of the characteristic is, therefore, not as high in the common emitter connection as in the common base connection.

(c) With the base current I_B at zero, the collector current is not zero, but has a value I_{CEO} which is larger than the equivalent leakage current in common base, I_{CBO}.

(d) The current gain in the common emitter connection, with the symbol h_{FE} or sometimes β is defined as

$$h_{FE} = \frac{I_C}{I_B} = \frac{\text{collector current}}{\text{base current}}$$

From the figure it can be seen that the characteristics are not as evenly spaced, nor as parallel as in the case of the common base connection. This shows that the current gain, h_{FE}, varies more with a variation of the collector voltage, than was the case with the common base connection.

Let us now make an estimate of the value of the common emitter gain. The two basic equations which characterise the transistor are

$$I_E = I_C + I_B \qquad 4.17$$

$$I_C = h_{FB} I_E \qquad 4.18$$

Eliminating I_E from these two equations, we have

$$I_C = h_{FB}(I_C + I_B)$$

whence

$$h_{FE} = \frac{I_C}{I_B} = \frac{h_{FB}}{1-h_{FB}} \qquad 4.20$$

For a transistor whose steady state (d.c.) current gain in common base is 0.97, the current gain in common emitter is

$$h_{FE} = \frac{0.97}{1-0.97} = 32.3$$

Notice also that a small change in the value of h_{FB} from 0.97 to 0.98 gives a relatively large change in the value of h_{FE} from 32.3 to 49.0, i.e. the current gain h_{FE} is much more sensitive to second order effects, for example, changes in collector voltage, collector current, or perhaps changes in temperature, than is the common base current gain, h_{FB}.

4.7 The transistor in a circuit

Figure 4.20(a) shows an npn transistor connected in the common emitter mode, and with a resistor R_C as its collector load. As with the diode resistor network of section 4.4, various methods may be used to determine the d.c. conditions in the circuit. Figure 4.20(b) shows the common emitter collector characteristics with a load line for the resistor R_C superimposed. In the diode circuit, the operating point is given by the intersection of the load line with the diode characteristic. In the transistor circuit, we are presented with a choice of operating points, the point of intersection of the load line with the collector characteristic may be varied by varying the transistor base current. In the figure, the load line for the resistor is drawn using the method described in section 4.4 by connecting the point V_{CC} on the voltage axis to the point $I_C = \frac{V_{CC}}{RC}$ on the current axis. The operating point A is then selected by setting the base current I_B to be 40 µA.

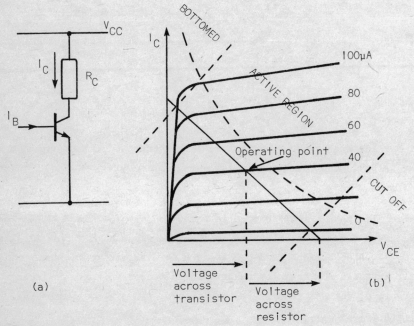

Figure 4.20

The maximum allowable power dissipation curve given by the locus of
$$V_{CE} \, I_C = W \text{ watts}$$
is also drawn on the characteristics; the operating point must lie to the left of this curve.

By varying the value of the base current I_B, the value of the collector current can be changed by a larger amount, the ratio of the two currents being, of course, the circuit current gain.

Notice, however, that the extent of the change in the collector current is limited. Thus, as the base current I_B is reduced, eventually to zero, the operating point moves down the load line to reach the point B. At this point the transistor is said to be cut-off; the only current flowing in the collector circuit is the collector emitter leakage current I_{CEO} which is a relatively small current of, at the most, several microamperes. There is, therefore, very little voltage drop across the load resistor R_C, and the collector emitter voltage V_{CE} is very nearly equal to the collector supply voltage V_{CC}.

Alternatively, if the base current I_B is increased, the operating point A moves upwards along the load line to reach the point C at which the collector current has increased to such an extent that the voltage drop in the load resistor R_C is almost equal to the supply voltage V_{CC}. The collector emitter voltage cannot be reduced any further, and the transistor is said to be 'bottomed' or 'saturated'. In a typical transistor, the bottoming voltage may be as low as 0.1 to 0.2 volt. Notice that as the base emitter voltage in this heavily conducting condition will be of the order of 0.6 to 0.7 volt, then the transistor collector has become more negative than the base; i.e. the collector base junction is now so saturated with carriers that it has become forward biased.

The 'cut-off' and 'bottomed' region of the characteristics are of great interest when the transistor is being used as an on-off switch, because the two conditions represent a very high impedance, and a very low impedance state of the transistor, analogous with a switch open and a switch closed. However, when in use as a linear amplifier, the operating point is positioned in the centre of the active region of the characteristics.

As in dealing with diode circuits, it is convenient to derive an equivalent circuit to represent the transistor in circuits, which will allow relatively easy calculation of the steady state currents in the circuit.

Figure 4.21 shows equivalent circuits which may be used to represent both npn and pnp transistors. As can be seen, the form of the circuits differs only in the polarity of the voltages, and in the dir-

118 Basic Electrical Engineering and Instrumentation for Engineers

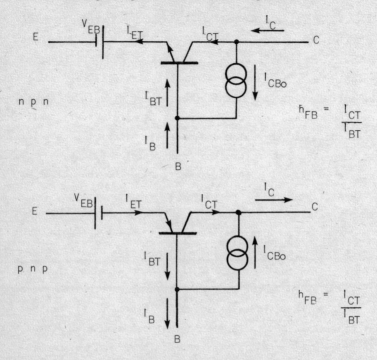

Figure 4.21 D.c. equivalent circuits for transistors.

ection of the currents. A trap often fallen into is to attempt to visualise the operation of such an equivalent circuit in terms of electrons or holes as current carriers. An understanding of the different types of charge carrier is necessary when dealing with the internal operation of the transistor. However, an equivalent circuit is, like any other circuit, best dealt with in terms of conventional current, flowing from the positive terminal of each voltage source to the negative terminal, irrespective of whether the actual current consists of holes flowing from positive to negative, or electrons flowing from negative to positive.

Each equivalent circuit consists of a transistor, which may be considered to be ideal in that it is characterised completely by the two equations

$$I_{E_T} = I_{C_T} + I_{B_T} \qquad 4.17$$

and

$$I_{C_T} = h_{FB} I_{E_T} \qquad 4.18$$

The emitter base diode voltage V_{EB} is shown as a battery in series with the emitter lead, while the collector base junction leakage current I_{CBO} is shown as a constant current generator across the collector base junction.

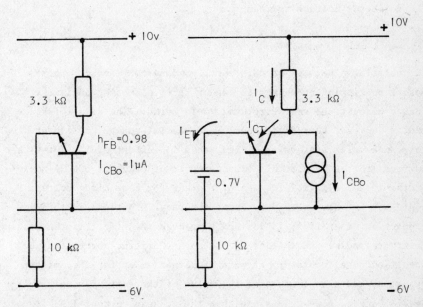

Figure 4.22

Example

In the circuit of Figure 4.22(a) determine the transistor collector potential. Take the transistor current gain, h_{FB} to be 0.98, the collector base leakage current I_{CBO} to be 1 µA, and the emitter base diode voltage, V_{EB}, to be 0.7 volt.

Figure 4.22(b) shows the same circuit with the transistor replaced by the steady state equivalent circuit.

Considering the emitter base circuit, the emitter current may be written

$$I_E = \frac{6 - 0.7}{10 \times 10^3} \times 10^3 = 0.53 \text{ mA}$$

The collector current is then
$$I_C = h_{FB} I_E + I_{CBO}$$
where the collector current is considered to be the current flowing in the transistor collector lead, i.e. in the 3.3 kΩ resistor.
$$I_C = 0.98 \times 0.53 + .001 \text{ mA}$$
$$= 0.519 + .001 = 0.520 \text{ mA}$$
The collector voltage is then
$$V_C = 10 - 3.3 \times 10^3 \times \frac{0.520}{10^3} = 8.284 \text{ volts}$$

In analysing a network of this sort, commonsense must be applied before specifying an answer to two or three places of decimals. In such a circuit the resistors used would normally be of a tolerance ±20%. Unless, therefore, there is a special reason why closer tolerance resistors are specified, then the effect of I_{CBO}, i.e. a current of 1 µA in a total collector current of about 500 µA would certainly be negligible. Again, the emitter base diode voltage V_{EB} would not be known to a very great accuracy. Notice, in this connection, that if it is desired to reduce the effect of the uncertainty regarding the voltage V_{EB}, then the emitter current may be specified more accurately by using a larger value for the emitter resistor, and also increasing the value of the emitter supply voltage. Do not, however, assume that the leakage current I_{CBO} is negligible in all circuits; each different system must be considered on its merits.

Example

In the circuit of Figure 4.23(a), determine the transistor collector current.

Take h_{FB} to be 0.97 and the collector base junction leakage current I_{CBO} to be 1 µA.

The circuit of Figure 4.23(a) shows an npn transistor connected with a collector-emitter voltage V_{CE} of 10 volts and with its base open circuit. Figure 4.23(b) shows the same circuit with the transistor replaced by its equivalent circuit.

With the transistor base open circuit, the only path available to the leakage current I_{CBO} is into the base connection of the ideal

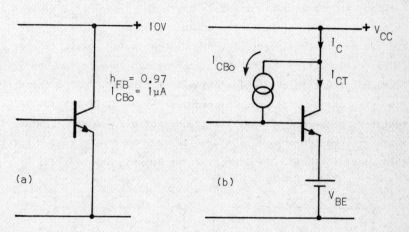

Figure 4.23

transistor, and across the base-emitter junction to the negative terminal of the supply voltage. This current across the base-emitter junction, therefore, causes, by transistor action, a current in the collector of the ideal transistor of

$$I_{CT} = h_{FE} I_{CBO}$$

Note that the symbol I_{CT} here is being used to denote the collector current of the ideal transistor used in the equivalent circuit. Of course, the current generator used to denote the source of the leak-current I_{CBO} is also internal to the complete transistor; the actual collector current in the transistor collector lead is thus

$$I_C = I_{CT} + I_{CBO}$$

or
$$I_C = h_{FE} I_{CBO} + I_{CBO}$$
$$= I_{CBO} (1 + h_{FE}) \qquad 4.21$$

The collector current is, therefore,

$$I_C = \frac{1}{10^6} (1 + h_{FE})$$

and as $\quad h_{FE} = \dfrac{0.97}{1-0.97} = 32.3$

then $\quad I_C = 33.3\ \mu A$

Looking again at the characteristic of Figure 4.19, the collector current with zero base current is given the symbol I_{CEO}. This is termed the collector-emitter leakage current. As the example shows, I_{CEO} is due to the effect of the collector base leakage current I_{CBO}, in effect amplified by the transistor action to give an even larger leakage current in the collector-emitter circuit. If a low resistance path can be provided, for example, by a relatively low value resistor, connected between the transistor base and the common line, which will reduce the value of the leakage current actually entering the transistor base, the effect of the leakage may be reduced.

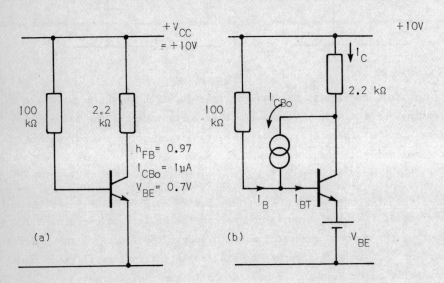

Figure 4.24

Example

In the circuit of Figure 4.24(a) calculate the collector voltage. Take h_{FB} to be 0.97, I_{CBO} to be 1 μA, and V_{BE} to be 0.7 volt. The base resistor current can be calculated from the equation

$$V_{CC} - V_{BE} = I_B \times R_B$$

Devices 123

or $$I_B = \frac{10.0 - 0.7}{10^5} = \frac{93}{10^6} \text{ A}$$

The total base current, including the leakage current I_{CBO} is then

$$I_{BT} = I_B + I_{CBO}$$
$$= \frac{93}{10^6} + \frac{1}{10^6} = \frac{94}{10^6} \text{ A}$$

The total collector current I_C is given by

$$I_C = I_{CT} + I_{CBO}$$
$$= h_{FE} I_{BT} + I_{CBO}$$

But
$$h_{FE} = \frac{0.97}{1-0.97} = 32.3$$

whence
$$I_C = 32.3 \times \frac{94}{10^6} + \frac{1}{10^6} = 3.037 \text{ mA}$$

Collector voltage
$$V_C = V_{CC} - I_C \times R_C$$
$$= 10.0 - 3.037 \times 2.2$$
$$= 3.32 \text{ volts}$$

4.8 The field effect transistor (FET)

The operation of the FET is based upon completely different principles to those of the bipolar transistor. It is often described as a unipolar device in that conduction through it takes place predominantly by means of one type of charge carrier, i.e. either by electrons or by holes depending upon its type. This may be contrasted with the method of operation of the bipolar transistor in which conduction takes place due to the movement of both holes and electrons.

4.8.1 The junction field effect transistor (JFET)

The JFET consists, basically, of a short thin bar of semiconductor which forms a channel between its two end connections, termed the 'source' and the 'drain'. The device is known as an 'n channel' device if n type semiconductor is used in its construction, or a 'p channel' device if p type semiconductor is used.

Considering an n channel device, when a direct voltage source is connected between the end connections, conduction occurs by electron flow from the negative of the voltage supply along the channel from the 'source' to the drain and back to the positive of the supply. Thinking in terms of 'conventional current flow', in the n channel JFET, current flow occurs from the drain to the source. The magnitude of the current flow is determined by the magnitude of the applied voltage, V_{DS}, and by the resistance of the channel. For any conductor, its resistance is given by

$$R = \frac{1}{\sigma} \times \frac{\text{length}}{\text{area}}$$

where σ is the conductivity of the material.

If the channel is formed of uniformly doped semiconductor, then the conductivity σ will be constant throughout its length, and the channel will be, in effect, a linear resistor.

The operation of the device is based upon the control of the current flow through the device by effectively controlling the thickness of the channel. Control of the thickness is achieved by forming at both sides of the channel a p type region (n type region for a p channel FET). The basic construction is illustrated in Figure 4.25. The two gate p regions, together with the n type channel, form pn junctions which, in normal use, are reverse biased by a d.c. potential V_{GS}. Because they are reverse biased, only a very small leakage current flows; the gate inputs are, therefore, extremely high impedance inputs. As in any other pn junction, the depletion layer formed at the junction is almost free of charge carriers, and has, therefore, an extremely high resistivity. By using a lightly doped n semiconductor for the channel, the gate depletion layer may be made to extend well into the channel width. As the depletion layer width is also a function of the voltage applied across the junction, increasing with increasing voltage, the gate voltage has a controlling effect upon the extent of the depletion layer penetration into the channel, and hence upon the effective remaining channel thickness. The result is a device in which a control voltage, the gate-source voltage V_{GS}, effectively controls the resistance of the path between

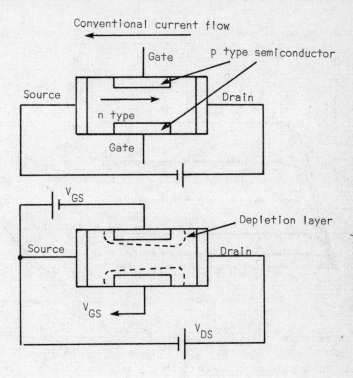

Figure 4.25

the source and the drain. Figure 4.25 shows the depletion layer at each of the gate-channel junctions. The voltage difference between the gate and the channel is greater at the drain end of the gate than at the source end of the gate, because of the voltage gradient along the channel, this resulting in a greater extension of the depletion layer into the channel at the drain end.

The characteristic curves for an n channel junction FET are shown in Figure 4.26. For a given value of gate-source voltage an increase of drain-source voltage, from zero, initially produces a relatively linear rise in drain current. Because of the potential gradient along the channel, the channel width is a minimum near the drain end of the gate junction. As the drain voltage is increased, the depletion layer at the gate extends further into the channel and starts to 'pinch off' the drain current. The drain current then turns the 'knee' of the characteristic and becomes virtually independent of

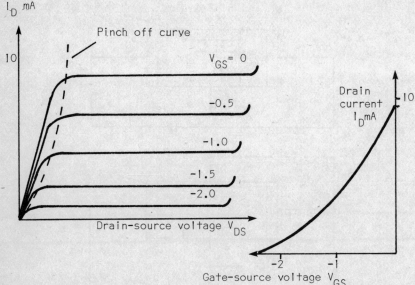

Figure 4.26

the drain voltage V_{DS}, until the voltage exceeds the value at which drain-gate junction breakdown commences.

Figure 4.26 also shows the characteristic plotted in terms of the relationship between the drain current I_D and the controlling gate-source voltage V_{GS}. This curve applies only for the constant current region between the 'pinch off' region and the breakdown region. Symbols used to represent junction field effect transistors are shown in Figure 4.27. As in the case of npn and pnp transistors, p and n channel FETs are indicated by the different directions of the arrowhead used in the gate symbol.

4.8.2 The metal oxide semiconductor transistor (MOST)

The metal oxide semiconductor transistor (MOST), which is also known as the insulated gate FET (IGFET), or the MOSFET, utilises a different mechanism to control the resistance of the source to drain channel. Figure 4.28 shows a schematic diagram of the construction of an n channel MOST. A p type semiconductor is used for the basic

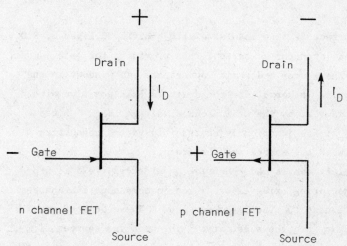

Figure 4.27

substrate, and has two strongly doped n type regions formed in it; these are the source and drain regions.

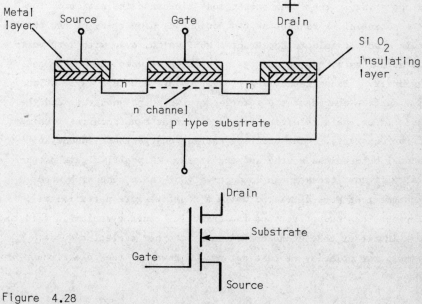

Figure 4.28

The complete surface is then insulated with a diffused layer of SiO_2, and a metallised layer formed on top, this providing the gate connection, and also the source and drain connections via windows in the insulating layer. In normal use the drain is held positive with respect to the source, by the drain-source voltage, V_{DS}. Because the drain to substrate junction is reverse biased, no conducting path exists between the source and the drain.

When the gate electrode is made positive an electric field is produced in the insulating oxide layer, which induces negative charges in the substrate near the insulated surface. These charges consist of electrons drawn into the substrate. The negative charges form an induced conducting channel between the source and the drain; the number of charges and hence the conductivity of the channel is controlled by the voltage applied to the gate electrode. Because the gate electrode is insulated from the substrate, the impedance of the gate circuit is very high indeed, the gate current being typically of the order of 10^{-12} A. A MOST operating in this manner is said to be operating in the 'enhancement mode', because the action of the gate potential is to enhance the conducting properties of the channel. It is also possible to construct a MOST which, even with zero voltage at the gate, has conducting properties between the source and the drain. These conducting properties are due to a layer of positive surface charge at the interface between the substrate and the oxide layer, this resulting from a different manufacturing treatment of the substrate. The positive surface charge thus induces an n channel between the source and the drain. A positive gate potential will now increase the conductivity of the n channel by the inducement of more carriers, while a negative gate potential will decrease the conductivity of the channel. Such a device is said to work in either the 'enhancement' mode, or the 'depletion' mode.

With either polarity of gate potential, however, the gate circuit impedance is very high; this contrasts with the operation of the gate circuit of the junction FET, which, being a pn junction, cannot be forward biased without becoming low impedance. The characteristics for a device of this type are shown in Figure 4.29.

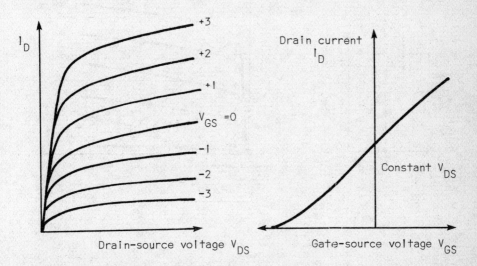

Figure 4.29

4.9 Switching devices

The use of the bipolar transistor in switching circuits was mentioned briefly in section 4.7. The desirable properties of an ideal switch may be summarised as follows:
(a) Infinite impedance when in the open circuit state.
(b) Zero impedance when in the closed state.
(c) Zero transition time between the two states.
(d) Zero power dissipation at the control input, i.e. at the input which initiates the change of state of the switch.

A transistor switch with a collector load resistor R_L is shown in Figure 4.30, together with the load line for resistor R_L drawn on the collector characteristics. The diagram shows the effect of increasing the transistor base current I_B from zero to a value of 800 µA. With zero base current, the transistor is in its OFF

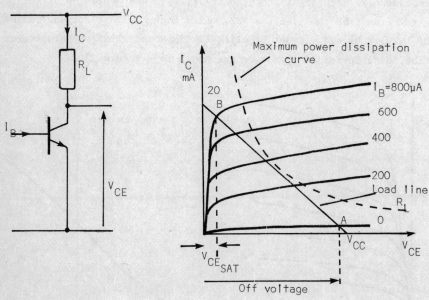

Figure 4.30

state, the collector voltage V_{CE} being a little less than the supply voltage V_{CC}, because of the voltage drop across the load resistor R_L, due to the collector-emitter leakage current. The operating point in the off state is point A in the diagram; because this point is below the maximum power dissipation curve, the transistor may be operated continuously in this state.

A base current I_B of 800 µA will cause the operating point to move to point B, giving a very low value of collector voltage, represented in the figure by the ON voltage, $V_{CE_{SAT}}$. The transistor is said to be 'saturated', or 'bottomed', and the collector voltage can be as low as 0.1 to 0.2 volt.

For a supply voltage of 20 volts the collector voltage could be expected to change, therefore, from about 0.1 volt to well over 19.9 volts for all reasonable values of collector load resistor, and for a silicon transistor.

Switching operations of this sort may be performed using either bipolar transistors or field effect transistors. Specially designed bipolar switching transistors are available which will allow switching times as low as several nanoseconds.

Switching operations are of many sorts, and while some applications may require very low switching times, other applications may be more concerned with the switching of very large current and voltage levels rather than with the achievement of very high switching speeds.

There are several devices available which are specially designed for switching applications.

4.9.1 The thyristor

This device, which is also known as the Silicon Controlled Rectifier, is in very wide use as a power switch. Units are available with current ratings of several hundred amperes and voltage ratings of several thousand volts. The thyristor is a unidirectional device with three connections, known as the anode, cathode and the control gate. Current flow is from the anode to the cathode only. Thus, with the cathode positive, with respect to the anode, the device has a very high impedance. With the anode positive with respect to the cathode, the device has two stable states:

(a) a high impedance state (order of megohms),

(b) a low impedance state (order of fraction of one ohm).

The operation of the thyristor can best be explained by considering it as consisting of two transistors, one an npn and one a pnp device, interconnected to form a four layer pnpn structure. This arrangement is illustrated in Figure 4.31. In the figure, the three pn junctions of the four layer device are labelled A, B and C.

Reverse polarity

When the cathode is positive with respect to the anode, junction B is forward biased, but junctions A and C are reverse biased. The current flow is very small, being the leakage current of junction B.

Forward polarity

When the anode is positive with respect to the cathode, junctions A and C are forward biased, but junction B is reverse biased. Almost all the applied voltage is across junction B. The current flow is very small, being the leakage current of junction B. Let the leakage current across junction B be I_{SAT}. Consider,

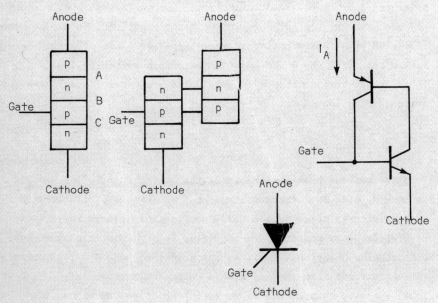

Figure 4.31

initially, only the hole component of the total current I. Let the hole component be $\gamma_1 I$. Thus, considering the pnp transistor section of the complete device, the hole current crossing junction A is $\gamma_1 I$. Of these holes, only a fraction β_1 reach junction B without recombining. The hole current reaching junction B is, therefore $\gamma_1 \beta_1 I$. At junction B, due to avalanche multiplication effect, the hole current crossing the junction is $M \gamma_1 \beta_1 I$, where M is an avalanche multiplication factor whose value is 1 or greater. In addition, there will be a small leakage current across junction B due to holes, of value I_{SATp}. The total hole current across junction B is $M\beta_1 \gamma_1 I + MI_{SATp}$. Similarly, the total electron current across junction B is $M\beta_2 \gamma_2 I + MI_{SATn}$.
Putting $I_{SATp} + I_{SATn} = I_{SAT}$

Total current across junction B is

$$I = M\beta_1\gamma_1 I + M\beta_2\gamma_2 I + MI_{SAT}$$

or

$$I = \frac{M I_{SAT}}{1 - M(\beta_1\gamma_1 + \beta_2\gamma_2)}$$

$\gamma_1\beta_1$ and $\gamma_2\beta_2$ are, in effect, the current gains of the two transistors making up the thyristor.
Thus

$$I = \frac{M I_{SAT}}{1 - M(h_{FE_1} + h_{FE_2})} \qquad 4.22$$

If conditions in the thyristor are such that

$$M(h_{FE_1} + h_{FE_2}) = 1$$

then the denominator is zero, and the device current is limited only by the impedance of the external circuit. The thyristor can be switched from its forward blocking (high impedance) state into its forward conducting (low impedance) state by any effect which can increase the value of the factor $M(h_{FE_1} + h_{FE_2})$, called the 'loop gain' G, to unity.

Once triggered into the conducting state, all the junctions in the device become forward biased, and the potential drop across the device is approximately equal to that of a single pn junction.
As the thyristor is constructed, the current gains h_{FE_1} and h_{FE_2} are low at low emitter currents, but increase fairly rapidly as the current is increased. The thyristor may, therefore, be 'turned on' by any mechanism which can cause a momentary increase in the effective 'emitter' currents.

1. Switch on by voltage breakdown.

If the forward voltage is increased, so increasing the value of the avalanche multiplication factor M, the leakage current carriers are sufficient to cause an avalanche multiplication of the current.
This effect occurs when the voltage is increased to the 'breakover voltage'.

2. Switch on by rate of change of voltage.

In section 4.3.2, the existence of a depletion layer capacitance across a pn junction was described. If a step voltage function is applied across the thyristor, when in its forward blocking condition, the charging current which flows in the capacitance of junction B effectively acts as a base current for the npn transistor. Due to the transistor action this allows the loop gain G to approach unity, switching on the device.

3. Switch on by transistor action.

The introduction of a short duration current into the gate connection (i.e. the base of the npn transistor), will switch the thyristor on.

4. Switch on by radiation.

Radiant energy within the correct spectral frequency can generate hole electron pairs. The consequent increase in the junction leakage current will cause switch on. This effect forms the basis of a range of optical pnpn switches, but is not applicable to normal thyristor devices.

Conduction in a thyristor will continue as long as the current in the device is maintained above the 'holding current'. It is necessary, therefore, in order to turn off the device, to remove the forward voltage from the device. In alternating current circuits, therefore, conduction normally ceases at the end of the forward conduction half-cycle of the supply voltage. For extremely rapid turn off of the device, it is advantageous to reverse the polarity of the voltage across the device.

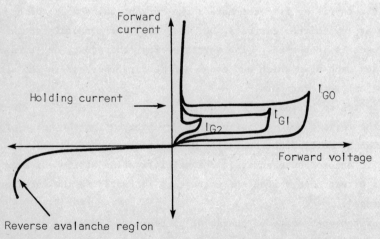

Figure 4.32

Figure 4.32 shows the current/voltage characteristic for a thyristor for different gate currents.

4.9.2 The triac

The triac (triode a.c. semiconductor switch) is similar in basic operation to a thyristor, but differs in that it can be switched into conduction in both directions. It is in essence equivalent to two thyristors mounted back to back, and finds application in systems requiring the control of conduction in both directions, for example, in full wave alternating current power systems.
The commonly used circuit symbol for a triac is shown in Figure 4.33.

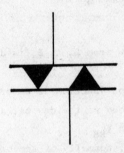

Figure 4.33

4.9.3 The unijunction transistor

The unijunction transistor is a three terminal device which may be considered as a piece of n type silicon semiconductor, with connections at each end, and with a third connection made via a pn junction. The two end connections are known as the base connections, B_1 and B_2, while the third connection is the emitter E. Between B_1 and B_2, the unijunction has the characteristics of a resistance, called the interbase resistance R_{BB}, and whose value is dependent upon the doping levels used in the silicon. R_{BB} is normally in the range 5 to 9 kΩ.
Under normal conditions, base B_2 is biased positive with respect to

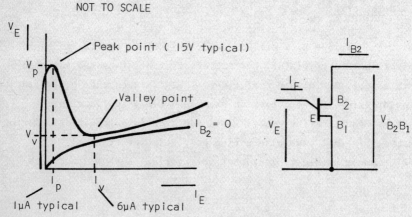

Figure 4.34 Static characteristic of unijunction transistor

B_1. The current flow from B_2 to B_1 is determined by the value of R_{BB}.

The potential at the pn junction is determined by the positioning of the junction on the body of the device, and is given by

$$V = \eta V_{BB} \qquad 4.23$$

where η is a fraction, normally of the order of 0.6 to 0.7. If the emitter potential is less positive than the potential V, the pn junction is reverse biased, and only a very small leakage current flows. If, however, the emitter potential is taken more positive than the voltage V, to a voltage V_P, where

$$V_P = \eta V_{BB} + V_D \qquad 4.24$$

then the emitter pn junction becomes forward biased, holes being injected into the body of the transistor. The voltage V_P is known as the 'peak point' voltage, and V_D is the emitter pn diode voltage, and is of the order of 0.5 volt. The injection of holes into the device cause the voltage between the emitter and base B_1 to fall rapidly to a low value, termed the valley point voltage V_V.

The unijunction is thus a device with two stable states which finds many applications in switching circuits. The static emitter voltage/emitter current characteristic is shown in Figure 4.34, together with the unijunction circuit symbol. The most important characteristic of the device is the peak point voltage V_p, this determining the critical switching voltage level in circuit applications. The fraction η is known as the 'intrinsic standoff ratio'.

Examples

1. In intrinsic germanium at room temperature the numbers of free electrons and holes is given by $n_i = 2.32 \times 10^{13}$ per cm^3. If the mobilities of electrons and holes are respectively $\mu_e = 3600$ $cm^2/$volt sec. and $\mu_h = 1800$ $cm^2/$volt sec., calculate the resistivity of the germanium.

(0.02 ohm cm)

2. The following is the characteristic of a non-linear resistor.

V (volts)	0	1	2	3	4	6	8	10	12	14	16	18	20
I (mA)	0	2.2	4.0	5.5	6.8	9.0	10.7	12.0	13.0	13.9	14.6	15.0	15.3

(a) This resistor is connected in series with a linear resistor of value 1.5 kΩ across a d.c. supply of 24 volts. Determine
  (i) the current in the non-linear resistor,
  (ii) the current in the linear resistor,
  (iii) the voltage across the non-linear resistor,
  (iv) the voltage across the linear resistor,
  (v) the power dissipated in the non-linear resistor.
(b) If the two components are connected in parallel to a 12 volt supply, what will be the total current flowing and the power taken from the supply?

(10.7 mA, 10.7 mA, 8 V, 16 V, 85.6 mW, 21 mA, 0.252 W)

3. The non-linear resistor of Question 1 is connected in series with a 1.5 kΩ linear resistor acrosss a d.c. supply. What voltage is required to give a current of 7 mA? If the supply voltage is increased to 24 volts, what extra resistor is required in series to limit the current to 7 mA? The non-linear resistor is connected in parallel with a linear resistor of 1 kΩ, across a 15 volt supply. What additional resistor is required in parallel to make the supply current 40 mA?

(14.7 V, 2.82 kΩ, 1.39 kΩ)

4. Show that, at 17°C, the a.c. resistance of a forward biassed diode is given by
$$r = \frac{25}{I} \text{ ohms}$$
where I is the diode current.

Assume that $I \gg I_{SAT}$ the diode reverse saturation current and take Boltzmann's constant to be 1.38×10^{-23} joules/°K and the electronic charge e to be 1.6×10^{-19} coulombs.

5. The I_C/V_{CE} characteristics of a transistor are

V_{CE} volts	1	4	8	10	
I_C	3.0	3.1	3.3	3.4	I_B = 20 µA
	6.8	7.5	8.4	8.9	40
	10.8	11.9	13.1	13.9	60
	14.5	15.9	17.8	18.8	80
	18.6	20.8	23.3	24.8	100
	22.9	25.2	28.7	30.2	120
	26.0	29.0	33.0	35.0	140
	29.1	32.9	37.8	40.1	160
	32.5	36.7	42.0	44.7	180
	36.0	40.5	46.5	49.5	200

Plot the characteristics and draw the curve representing a dissipation in the transistor of 100 mW.

What is the lowest value of collector load resistor which may be used with a collector supply voltage of 10 V if the dissipation in the transistor is not to exceed 100 mW for any value of base current? What base current is required with this value of collector load resistor to set the collector-emitter voltage to 5 V? Estimate the value of the d.c. current gain (h_{FE}) and the a.c. current gain (h_{fe}) at this operating point.

N.B. $h_{fe} = \frac{\delta I_C}{\delta I_B}$ with V_{CE} constant

(250 Ω, 95 µA, 210, 250)

6. A transistor with I_{CB_o} = 2 µA is connected in common emitter. When the base current is 40 µA the collector current is 3 MA. What will be the collector current when the base is open circuit? What doubtful assumption must be made in order to reach an answer?

(146 µA)

7. With reverse bias a p-n junction diode saturates at 2.5 µA at a temperature of $27°C$. Calculate the current for a forward voltage of 0.22 V.

(12.3 mA)

CHAPTER FIVE

Amplifiers

5.1 <u>Amplifier gain</u>

Throughout engineering practice, it is necessary to find the means whereby signals developed in a system can be increased in power to eventually be sufficiently powerful to perform their allotted tasks. Thus the mechanical force available from a man's arm can be used to raise from the ground a load weighing several tons.

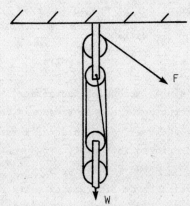

Figure 5.1

Consider a method by which a man may lift a very heavy weight. Figure 1 shows a 4 part pulley arrangement. Ignoring any frictional losses in the pulleys, the tension in the string in all parts is equal to the pull F newtons. Since there are, in all, four strings supporting the lower block and load, the system will be in equilibrium if the downward force due to the weight of the load W and of the lower block is equal to 4F newtons.

The pulley system is thus a force amplifier. Suppose now that, in order to raise the load, the pull F moves the string a distance of 1 metre. Since there are 4 strings supporting the lower block, the load will be raised by a distance $\frac{1}{4}$ metre. The work done by the pull is given by

$$\text{Work done} = F \times 1 \text{ joules} \qquad 5.1$$

The work done on the load is given by

$$\text{Work done} = 4F \times \tfrac{1}{4} = F1 \text{ joules} \qquad 5.2$$

In practice, of course, no pulley arrangement, or indeed any other form of machine, can be constructed perfectly, and there would be some energy loss in overcoming frictional forces at the pulley bearings. Even if, however, these losses are negligible, we can see that in using such a force amplifier nothing is gained that is not put into the system. There is a 'mechanical advantage' in amplifying the original available force of F newtons to a magnitude of 4F newtons, but all energy used must be supplied by the applied pull.

Since the input power is equal to the rate of supplying energy, and considering a time interval during the movement of δt seconds, then

$$\text{input power} = \frac{F1}{\delta t} \text{ watts} \qquad 5.3$$

$$\text{and} \quad \text{output power} = \frac{4F}{\delta t} \cdot \frac{1}{4} = \frac{F1}{\delta t} \text{ watts} \qquad 5.4$$

Thus even though we obtain a 'mechanical advantage' by using the force amplifier, the output power can never exceed the input power and would, in practice, be less due to internal losses.

An analogy may be drawn between the pulley arrangement and the electrical transformer discussed in Chapter 3.

Suppose that we have available an alternating voltage of r.m.s. value V volts, and that for a particular task a voltage of 4V volts is required. This could be arranged by using a transformer with a secondary winding consisting of 4 times the number of turns of the primary winding.

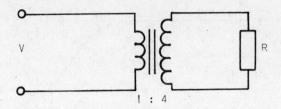

Figure 5.2

Figure 5.2 shows the transformer connected to a load resistor of R ohms, giving a current in the transformer secondary winding of

$$I_s = \frac{4V}{R} \text{ amperes} \qquad 5.5$$

The current in the transformer primary, because of the 1:4 turns ratio is equal to four times the secondary current and is, therefore,

$$I_p = \frac{16V}{R} \qquad 5.6$$

The power supplied to the load resistor is

$$P = V_s I_s = 4V \times \frac{4V}{R} = \frac{16V^2}{R} \text{ watts} \qquad 5.7$$

The power input to the transformer is

$$P = V_p I_p = V \times \frac{16V}{R} = \frac{16V^2}{R} \text{ watts} \qquad 5.8$$

The effect of the transformer is thus to amplify the voltage available, to the value required to supply the load. As in the pulley example, however, there is no gain of power, and in fact the input power to the transformer is exactly the same as that supplied to the load. Of course, we have again neglected the unavoidable small losses which would occur in the transformer due to winding resistance and losses in the iron core.

In the types of machine we have considered, the system is designed to give the operating conditions desired for a particular purpose, but no gain of power is possible. Consider, however, again, the problem of raising the heavy weight of W. If, instead of using the simple block and tackle, we obtain a motor driven crane, a different condition exists. All the energy used to raise the weight W will now be obtained from the electric mains supplying the crane motor.

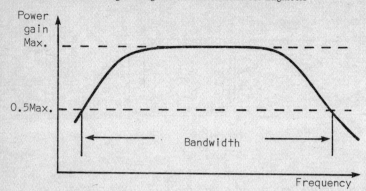
(a) Wideband amplifier

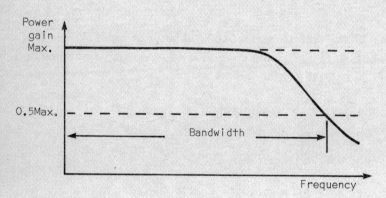
(b) D.c. coupled wideband amplifier

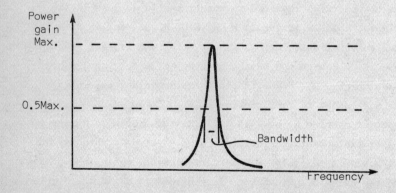
(c) Narrow band (tuned) amplifier
Figure 5.3

All that is required from the system 'input signal', i.e. from the crane operator, is a control signal to switch on the motor, to perhaps adjust the speed of winding, and to judge when to switch off. An energy balance between the system input, i.e. the crane operator and the system output, i.e. the work done in raising the load, is now different in meaning from that applied in the previous examples. The function of the system input is only to control the transfer of energy from a power supply to the system output. The actual work done by the operator in pushing the control switches will be very small, and so the power gain of the system when defined as

$$\text{Power gain} = \frac{\text{Useful power output applied to the load}}{\text{Power input to the control switches}} \qquad 5.9$$

is very high indeed.

The point to note is, of course, that as in the earlier examples, we have not succeeded in obtaining 'something for nothing'. All power supplied to the load is taken from the motor main supply and must be paid for.

A similar condition exists in the electronic amplifiers which are the subject of this chapter. They consist in general of electronic circuits which are supplied with energy from a d.c. source, which may be a battery or a rectified alternating current source (see chapter 6). A signal input to the circuit is used to control the transfer of energy from the power supply to the signal output. The normal requirement for an amplifier is that the output signal should be a higher power version of that applied to the input; thus the signal waveform at the output should be of exactly the same shape as that at the input. In particular, if the input signal waveform is sinusoidal, the output waveform should also be sinusoidal. Any deviation from this linear relationship is caused by the presence in the output signal of frequencies not present in the input signal, and which must have been generated internally in the amplifier. Such internally generated output signals are termed 'distortion' and the amplifier design should reduce such signals to a minimum.

The power gain of an amplifier is defined as

$$\text{Power gain} = A_p = \frac{\text{Output signal power}}{\text{Input signal power}} \quad 5.10$$

Consider, for example, an amplifier designed to drive a high fidelity loudspeaker system when driven by the pytput from a record player pick-up cartridge. The output from such a cartridge would be a signal containing frequencies in the range from about 20 Hz to perhaps 20 kHz. An input power level of about 1 microwatt (μW) would be available at a voltage of the order of 100 mV. A power output would be required of perhaps 30 watts, which, with a loudspeaker load resistance nominally of 8 ohms, means an amplifier output voltage of the order of

$$V = \sqrt{P_o \cdot R_L} = \sqrt{30 \times 8} \approx 15 \text{ volts r.m.s.}$$

The amplifier would, therefore, have an overall power gain of

$$A_p = \frac{30}{10^{-6}} = 30 \times 10^6$$

and a voltage gain of

$$A_v = \frac{15}{0.1} = 150$$

These values of power and voltage gain would be achieved, in practice, by a cascaded arrangement of amplifiers of lower gain. Amplifiers may be broadly classified by considering the frequency range over which they are designed to operate.

(a) Wideband amplifiers

These amplifiers are designed to provide power gain over a wide frequency range. In the example considered of an audio frequency amplifier, the requirement is for constant gain over the complete audio frequency spectrum, i.e. from approximately 20 Hz to 20 kHz. The operating bandwidth of an amplifier is normally specified as that frequency range over which the amplifier will give a power gain of not less than half of its maximum value. Figure 5.3(a) shows a power gain/frequency response curve typical for a wide band amplifier. Such a response is normally plotted on a logarithmic frequency scale because of the wide frequency range which must be considered.

A further classification depnds upon whether the amplifier bandwidth extends downward to include 0 Hz, i.e. whether the amplifier will respond to a d.c. input signal. In many applications a response to a d.c. signal is required; for example, a television receiver video amplifier requires a bandwidth of from 0 to about 5 MHz, while amplifiers designed for use with transducers in instrumentation systems often have a bandwidth of from 0 to 1.5 kHz. Figure 5.3(b) shows a response of this type.

(b) Narrow band (tuned) amplifiers

These amplifiers are designed to provide power gain over a very narrow band of frequencies about a nominal centre frequency, and are used to select a desired signal frequency from a range of signals. A typical frequency response is shown in figure 5.3(c). Amplifiers of this type form the tuning arrangements in radio and television receivers. Another application of narrow band amplifiers is as null detector amplifiers to increase the sensitivity of alternating current bridge measurements. The use of a narrow band amplifier in bridge measurements gives some protection against interfering signals of other frequencies when extremely sensitive measurements are being made.

5.2 Input and output impedance

An amplifier may be represented by the box symbol shown in figure 5.4. It has one input and one output, the input signal being applied between the input connection and a common connection, while the amplifier output is available between the output connection and the same common connection. Note that in this diagrammatic representation, no attempt is made to show the battery or power supply discussed in the previous section. We are concerned at this stage merely with producing a model which will enable us to consider the signal conditions in the system.

The impedance between the input connections, which is presented by the amplifier to the signal source providing the input signal, is called the amplifier 'input impedance' Z_I.

Figure 5.4

In general this impedance must be represented by resistance and either inductance or capacitance, but over a limited frequency range about the middle of the amplifier operating range may be represented as an input resistance R_i. Figure 5.5 shows an amplifier of input resistance R_i when driven by a signal source of voltage v_s and source resistance R_s.

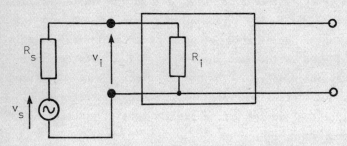

Figure 5.5

Only if the amplifier input resistance is infinite will the full source voltage v_s be applied to the amplifier input.
For any non-infinite value of input resistance the amplifier input voltage is given by

$$V_i = \frac{R_i}{R_i + R_s} \cdot v_s \qquad 5.11$$

and the source resistance should be made as low as possible when compared with the input resistance, in order to prevent a considerable reduction in overall voltage gain.

The amplifier output may be considered as an output voltage source, together with an output impedance Z_o. Again the output impedance may be taken to be resistive, and of value R_o over a limited mid-range of frequencies. Figure 5.6 shows the amplifier connected to a load resistor R_L. For maximum voltage gain, the output resistance R_o should be small when compared to the load resistor R_L.

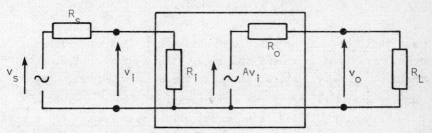

Figure 5.6

Example

An amplifier of open circuit voltage gain $A_v = 100$ (i.e. with $R_L = \infty$) has an input resistance of 2000 Ω and an output resistance of 20 Ω. The amplifier is driven by a signal source of e.m.f. 10.0 mV and source resistance 600 Ω. The load connected to the amplifier output is a 100 Ω resistor. Determine (a) the overall voltage gain, and (b) the power gain.

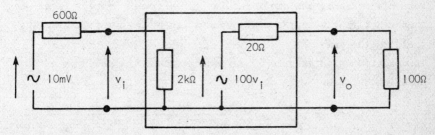

Figure 5.7

Figure 5.7 shows the circuit arrangement.
The amplifier input voltage is given by

$$v_i = \frac{2000}{2000 + 600} \times 10 \text{ mV} = 7.69 \text{ mV}$$

The open circuit output voltage is

$$A v_i = 100 \times 7.69 = 0.769 \text{ V}$$

The output voltage when driving a 100 Ω load is

$$v_o = \frac{R_L}{R_L + R_o} \cdot A\, v_i$$

$$= \frac{100}{120} \times 0.769 = 0.641 \text{ V}$$

Before determining the value of the overall voltage gain, some thought must be given to the meaning of this quantity. The open circuit e.m.f. of the signal source is known to be 10.0 mV. By using the amplifier described in the question we have achieved an output voltage of 0.641 V, i.e. an overall voltage gain of

$$\frac{0.641}{10.0} \times 1000 = 64.1$$

However, if the amplifier voltage gain was measured using voltmeters connected across the output (to determine v_o) and across the input (to determine v_i) then the gain would be found to be

$$\text{amplifier voltage gain} = \frac{v_o}{v_i} = \frac{0.641}{7.69} \times 1000 = 83.4$$

Obviously the exact meaning of 'voltage gain' must be considered carefully.

Note also that if the amplifier input resistance is very large when compared with the source resistance R_s then the two voltage gains would be the same.

The question also asks for a determination of the power gain. Defining the power gain as

$$\text{power gain } A_p = \frac{\text{power output to amplifier load}}{\text{power input to amplifier}}$$

then we obtain the following

$$\text{power output to load} = \frac{v_o^2}{R_L} = \frac{0.641^2}{100} \text{ watt} = 4.11 \text{ mW}$$

$$\text{power input to amplifier} = \frac{v_i^2}{R_i} = \frac{(7.69)^2}{(10^3)} \cdot \frac{1}{2000} \text{ watt}$$

$$= 0.0296 \text{ μW}$$

$$\text{power gain} = \frac{4.11}{10^3} \times \frac{10^6}{0.0296} = 138{,}800$$

It can be seen that a very large power gain is available between the output and input of such an amplifier.

A point which ought to be considered, however, is whether the amplifier is accepting from the source all the available signal power.

Let us consider a signal source of e.m.f. v_s and source resistance R_s connected directly to a load R. This is shown in Figure 5.8.

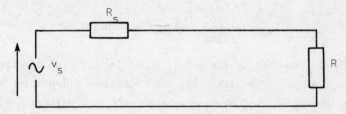

Figure 5.8

The current in the circuit is $i = \dfrac{v_s}{R_s + R}$ amperes

power in the load resistor = $i^2 R$

or $P = \dfrac{v_s^2}{(R_s + R)^2} \times R$ watts

For a given source, in which the source e.m.f. and the source resistance are fixed, the load resistor R may be varied to obtain maximum load power.

We can determine the maximum load power as follows:

$$\frac{dP}{dR} = v_s^2 \frac{(R_s+R)^2 - R \times 2(R_s+R)}{(R_s+R)^2}$$

Equating $\dfrac{dP}{dR}$ to zero, maximum load power will be obtained when

$$(R_s+R)^2 - 2R(R_s+R) = 0$$

i.e. when $R = R_s$ 5.12

This is a general result which is covered by the maximum power transfer theorem which states: 'For maximum power transfer from a source to a load, the load resistance should equal the source resistance.'

The maximum power obtainable from the course is therefore

$$P_{max} = \frac{v_s^2 R_s}{(R_s+R_s)^2} = \frac{v_s^2}{4R_s} \text{ watts} \qquad 5.13$$

Returning to the source in the example, the maximum power available is

$$P_{max} = \frac{v_s^2}{4R_s} = \frac{10^2}{10^3} \frac{1}{4.600} = 0.0417 \text{ μW}$$

As we determined previously, the actual power input to the amplifier is only 0.0296 μW. It is clear that the conditions which exist in an amplifier chosen to give maximum voltage gain, are different from those needed to give maximum power gain.

5.3 Determination of amplifier input and output resistance

The input resistance of an amplifier may be measured directly by applying a known input signal voltage and measuring the signal current passing into the amplifier input, as shown in Figure 5.9(a). The measurement must, however, be taken with conditions in the amplifier exactly as in normal use. Thus the input signal voltage will necessarily be quite small, and difficulty would, in general, be found in making the measurements, especially in the case of the input current, which would be very small indeed, unless the input resistance is very low. Measurements of signal voltages and currents are, perhaps, best undertaken using an oscilloscope, which, with its very high input impedance, is ideally suited to measurements on high impedance circuits. It must always be borne in mind that if a measuring instrument, such as a voltmeter, abstracts an appreciable current from the circuit to which it is connected, then the voltage and current values in the circuit will be altered and an incorrect reading obtained. In amplifier circuits, where resistors may have values of many thousands of ohms, only high impedance measuring circuits may be used.

A better method for determining the input resistance is illustrated in

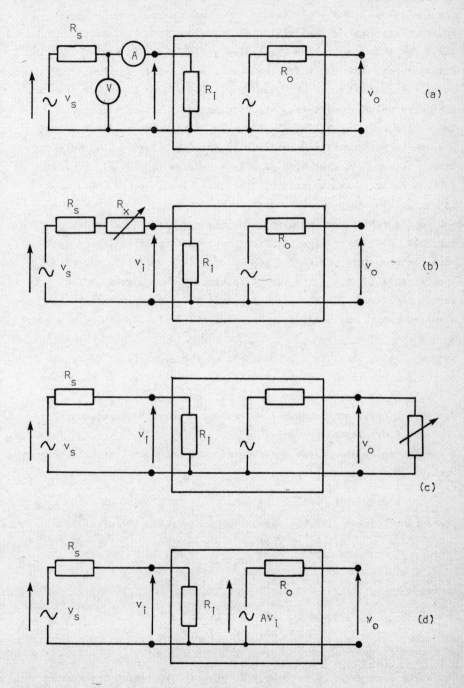

Figure 5.9

Figure 5.9(b). The amplifier is fed from a signal source whose source resistance is negligible in comparison with the amplifier input resistance R_i. The calibrated variable series resistor R_x is set at zero. The amplifier input voltage v_i will, in this condition equal the source e.m.f. v_s. The amplifier output voltage v_o can be monitored using a voltmeter, or, alternatively, an oscilloscope. The series resistor R_x is now increased, until the amplifier output voltage is reduced to one half of its initial value. The amplifier input voltage, v_i, must also have been reduced to one half of its initial value. Examination of the input circuit shows that, if R_s is negligible, this occurs when $R_x = R_i$.

The output resistance may also be determined using a calibrated variable resistor. Figure 5.9(c) shows the amplifier loaded by the variable resistor. With the amplifier driven normally, and with R_x open circuit ($R_x = \infty$), the output voltage v_o is measured. The series resistor R_x is then reduced until the output voltage v_o is also reduced to one half of its open circuit value, which occurs when $R_x = R_o$.

Notice that an absolute determination of the accurate value of the output voltage is not necessary; it is only necessary to know when the voltage has been reduced to one half of its value.

The output resistance could be determined by direct measurements at the amplifier output terminals.

Referring to Figure 5.9(d) the open-circuit output voltage, (i.e. with no load connected) is

$$V_{o_{oc}} = A v_i$$

The output current with the output terminals short-circuited is

$$i_{sc} = \frac{A v_i}{R_o}$$

The output resistance can then be calculated from

$$\text{output resistance} = \frac{v_{oc}}{i_{sc}} = \frac{A v_i}{\frac{A v_i}{R_o}} = R_o \qquad 5.14$$

Obviously care would be necessary in practice in taking the short-circuit current measurement. In many cases such a measurement would not be possible while still maintaining the normal operating conditions in the amplifier. However, if not always practically a feasible method, analytically this method is often very useful.

5.4 Logarithmic gain units - the decibel

Instead of expressing the gain of an amplifier as a simple ratio of the output power to the input power, advantages are found, in practice, by the use of a logarithmic unit.
Thus power gain is expressed as

$$\text{power gain} = \log_{10} \frac{P_o}{P_I} \text{ bels} \qquad 5.15$$

where P_o = output power

and P_I = input power

Since the bel is rather large for many applications, it is common practice to express the gain as

$$\text{power gain} = 10 \log_{10} \frac{P_o}{P_I} \text{ decibels (dB)} \qquad 5.16$$

The decibel unit may also be applied to voltage or current ratios. Consider two equal resistors, R_1 and R_2, each of value R ohms, with applied voltages V_1 and V_2 volts respectively. In resistor R_1

$$\text{power} = P_1 = \frac{V_1^2}{R} \text{ watts}$$

and in resistor R_2

$$\text{power} = P_2 = \frac{V_2^2}{R} \text{ watts}$$

The power ratio is thus given by

$$\text{power ratio} = 10 \log_{10} \frac{P_2}{P_1} \text{ dB}$$

$$= 10 \log_{10} \frac{V_2^2}{R} \cdot \frac{R}{V_1^2}$$

$$= 20 \log_{10} \frac{V_2}{V_1} \text{ dB} \qquad 5.17$$

Example

An amplifier, of input resistance 1 kΩ is driven by a source of e.m.f. 10 mV and negligible source resistance. The amplifier has an open circuit voltage gain of 200 and an output resistance of 1 kΩ. The amplifier load resistance is 1 kΩ.

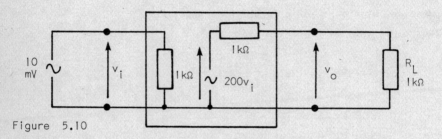

Figure 5.10

In Figure 5.10, the amplifier input voltage is

$$v_i = 10 \text{ mV}$$

The amplifier output voltage is

$$v_o = A v_i \cdot \frac{R_L}{R_o + R_L} = 200 \cdot \frac{10}{10^3} \cdot \frac{1 \times 10^3}{2 \times 10^3} \times 1.0 \text{ volt}$$

$$\text{voltage gain} = A_v = 20 \log_{10} \cdot \frac{1}{10^{-3}} = 60 \text{ dB}$$

$$\text{input power} = \frac{v_L^2}{R_i} = \frac{(10^{-3})^2}{10^3} = 10^{-9} \text{ watt}$$

$$\text{output power} = \frac{v_o^2}{R_L} = \frac{(1)^2}{10^3} = 10^{-3} \text{ watt}$$

$$\text{power gain} = A_p = 10 \log_{10} \frac{10^{-3}}{10^{-9}} = 60 \text{ dB}$$

Thus for equal resistance, the voltage ratio and the power ratio have the same numerical value when the decibel unit is used.
This logarithmic system is in widespread use, and is applied generally, even between load resistors of different values, giving

power and voltage (or current) gains which are not equal. Notice also that where a gain is less than unity, the gain expressed in decibels is a negative quantity.

Example

The attenuator network in Figure 5.11 is driven by a battery of e.m.f. 1.5 volt and negligible internal resistance. Determine the load voltage and the voltage and power gains in decibels.

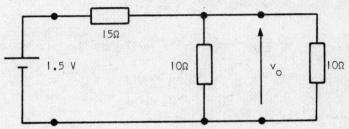

Figure 5.11

As the two 10 Ω resistors are effectively in parallel, the output voltage can be expressed as

$$v_o = \frac{5}{5 + 15} \times 1.5 = 0.375 \text{ volt}$$

Thus,

voltage gain $= \dfrac{0.375}{1.5} = 0.25$

or in decibels

$$\begin{aligned}
\text{voltage gain} &= 20 \log_{10} 0.25 \\
&= -20 \log_{10} \frac{1}{0.25} \\
&= -20 \times 0.602 \\
&= -12.04 \text{ dB}
\end{aligned}$$

input resistance $= 15 + 5 = 20 \ \Omega$

input power $= \dfrac{V_I^2}{R_i} = \dfrac{1.5^2}{20} = 0.1125 \text{ watt}$

$$\text{output power} = \frac{V_o^2}{R_L} = \frac{0.375^2}{10} = 0.0141 \text{ watt}$$

$$\text{power gain} = \frac{0.0141}{0.1125} = 0.1253$$

or, in decibels

$$\text{power gain} = 10 \log_{10} 0.1253$$

$$= -10 \log_{10} \frac{1}{0.1253} = -9.02 \text{ dB}$$

Notice also that the same method can be applied to the current gain.

$$\text{input current} = \frac{1.5}{20} = 0.075 \text{ ampere}$$

$$\text{output current} = \frac{0.375}{10} = 0.0375 \text{ ampere}$$

$$\text{current gain} = \frac{0.0375}{0.075} = 0.5$$

or, in decibels

$$\text{current gain} = 20 \log_{10} 0.5 = -6.02 \text{ dB}$$

The use of a logarithmic unit for measuring gains has several advantages.

For example, large ratios can be expressed, in decibel form, by a relatively small number. Thus a voltage ratio of 10^6 is, in decibel form

$$\text{voltage ratio} = 20 \log_{10} 10^6 = 120 \text{ dB}$$

Further, if two amplifiers are cascaded, the overall gain is given by the sum of the two gains expressed in decibels. Figure 5.12 shows two cascaded amplifiers, of gains A_1 and A_2 respectively.

Figure 5.12

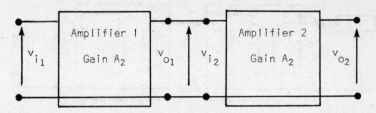

For an input voltage to amplifier 1 of v_{i_1} volts, the output voltage is

$$v_{o_1} = A_1 v_{i_1}$$

The overall gain is thus the product of the two individual gains

i.e.
$$\frac{v_{o_2}}{v_{i_1}} = A_1 A_2 \qquad 5.18$$

Expressing this in decibels,

$$\text{gain} = 20 \log_{10} A_1 A_2$$

$$= 20 \log_{10} A_1 + 20 \log_{10} A_2 \qquad 5.19$$

In decibel form, therefore, the overall gain is given by adding together the individual gains.

We have, however, made one assumption, that is, we have assumed that in connecting the two amplifiers in cascade, their individual gains are not altered. This is not always true in practice.

Example

Figure 5.13 shows two amplifiers in cascade, each with a voltage gain of 10, an output resistance of 1 kΩ, and an input resistance of 3 kΩ. The cascaded pair is driven from a signal source of source resistance 1 kΩ, and is loaded with a 3 kΩ load.

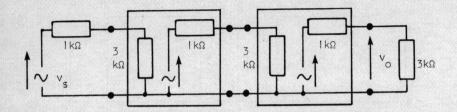

Figure 5.13

The open circuit gain of each individual amplifier is given as 10, i.e.

$$\text{gain} = 20 \log_{10} 10 = 20 \text{ dB}$$

At each coupling, however, between the source and the input of amplifier 1, between the two amplifiers, and also between amplifier 2 and its load, a voltage attenuation will occur, of ratio given by

$$\text{attenuation} = \frac{3 \times 10^3}{3 \times 10^3 + 1 \times 10^3} = 0.75$$

i.e.

$$\text{attentuation} = 20 \log_{10} 0.75 = -2.5 \text{ dB}$$

The overall gain is the sum of each individual gain, as follows:

(a) a gain of -2.5 dB due to the load effect of amplifier 1 on the source,
(b) a gain of 20 dB in amplifier 1,
(c) a gain of -2.5 dB due to the loading effect of amplifier 2 on amplifier 1,
(d) a gain of 20 dB in amplifier 2,
(e) a gain of -2.5 dB due to the loading effect of the load upon amplifier 2.

The overall gain is thus

$$-2.5 + 20 - 2.5 + 20 - 2.5 = 32.5 \text{ dB}$$

It is instructive to obtain this result directly using the gain ratios.

Thus we have

(a) a gain of 0.75 due to the loading effect of amplifier 1 on the

source,
(b) a gain of 10 in amplifier 1,
(c) a gain of 0.75 due to the loading effect of amplifier 2 on amplifier 1,
(d) a gain of 10 in amplifier 2,
(e) a gain of 0.75 due to the loading effect of the load upon amplifier 2.

The overall gain is thus

$$0.75 \times 10 \times 0.75 \times 10 \times 0.75 = 42.18$$

or, in decibel form

$$\text{gain} = 20 \log_{10} 42.18 = 32.5 \text{ dB}$$

5.5. Amplifier bandwidth

The bandwidths of various types of amplifier were considered briefly in section 5.1. The upper and lower cut off frequencies, i.e. the frequencies at which the gain has reduced by 3 dB, are fixed either by the characteristics of the active device used to make the amplifier, or, alternatively, by frequency dependent components in the amplifier circuit. Every active amplifying device, e.g. the transistor, has an upper limit to its operating frequency, often decided by the time of transit of charge through the device. However, many amplifiers have upper cut off frequencies far lower than the cut off frequency of the active device. In fact, unavoidable stray capacitance between the signal leads and earth has the effect of reducing the upper cut-off frequency. At the lower end of the frequency spectrum the devices will respond easily to frequencies down to 0 Hz. However, any slow change in device characteristics, due to ageing, or to temperature effects, may cause a corresponding change in the d.c. potential at the amplifier output terminal, which would be indistinguishable from a d.c. (0 Hz) output signal.

To remove these drift effects, a low frequency cut off characteristic is sometimes included, usually by the inclusion of capacitors in series with one or more input connections.

5.5.1 The effect of stray capacitance

Let us consider a signal source of e.m.f. v_s and source resistance R_s feeding a load resistor R_L, as shown in Figure 5.14. Across the load resistor is a capacitor C to represent the circuit stray capacitance.

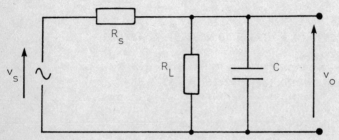

Figure 5.14

By applying Thevenin's theorem, the circuit may be redrawn as in Figure 5.15. Here the source resistance and the load resistance have been included together as

$$R = \frac{R_s R_L}{R_s + R_L} \qquad 5.20$$

while the generator now has the value

$$v = \frac{R_L}{R_s + R_L} \cdot v_s \qquad 5.21$$

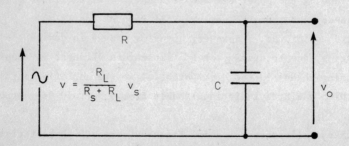

Figure 5.15

Amplifiers 163

The circuit is thus equivalent to a single resistor-capacitor network. If the capacitor C is negligibly small, then the circuit response is aperiodic, i.e. the circuit output is the same at all frequencies.

If, however, the capacitor is too large to neglect, then

$$\frac{v_o}{v} = \frac{\frac{1}{j\omega c}}{R + \frac{1}{j\omega c}} = \frac{1}{1 + j\omega CR}$$

By defining an angular frequency ω_o as

$$\omega_o = \frac{1}{CR}$$

then

$$\frac{v_o}{v} = \frac{1}{1 + j\frac{\omega}{\omega_o}} \qquad 5.22$$

To study the magnitude and phase of the output voltage, the expression should be rationalised and converted into the polar form

$$\frac{v_o}{v} = \frac{1}{1 + (\frac{\omega}{\omega_o})^2} \cdot 1 - j\frac{\omega}{\omega_o}$$

$$= \frac{\sqrt{1 + (\frac{\omega}{\omega_o})^2}}{1 + (\frac{\omega}{\omega_o})^2} \quad \underline{/\tan^{-1} - \frac{\omega}{\omega_o}}$$

$$= \frac{1}{\sqrt{1 + (\frac{\omega}{\omega_o})^2}} \quad \underline{/\tan^{-1} - \frac{\omega}{\omega_o}} \qquad 5.23$$

The form of the response is conveniently examined as follows:

(a) when $\omega \ll \omega_o$

If ω is small in comparison with ω_o, then the term $\frac{\omega}{\omega_o}^2$ may be neglected in comparison with 1.

The expression therefore becomes

$$\frac{v_o}{v} = \frac{1}{\sqrt{1}} \underline{/\tan^{-1} - \frac{\omega}{\omega_o}} = 1 \underline{/\tan^{-1} - \frac{\omega}{\omega_o}}$$

While this approximation holds, therefore, the output voltage is constant in magnitude, and in decibels

$$20 \log_{10} \frac{v_o}{v} = 20 \log_{10} 1 = 0 \text{ dB}$$

The output voltage lags the input voltage by a very small angle

$$\phi = \tan^{-1} \frac{\omega}{\omega_o}$$

(b) when $\omega = \omega_o$

At this particular frequency, the function becomes

$$\frac{v_o}{v} = \frac{1}{\sqrt{1+1}} \underline{/\tan^{-1} - 1} = \frac{1}{\sqrt{2}} \underline{-45°}$$

The magnitude of the output voltage has fallen to

$$v_o = \frac{1}{\sqrt{2}} v$$

i.e. the gain has reduced by

$$20 \log_{10} \sqrt{2} = 3 \text{ dB}$$

The output voltage now lags the input voltage by $45°$. This special frequency when $\omega = \omega_o = \frac{1}{CR}$ is therefore the upper cut off frequency.

(c) when $\omega \gg \omega_o$

At frequencies, when ω is large in comparison with the cut off frequency ω_o, then $(\frac{\omega}{\omega_o})^2$ is large in comparison with unity.

The expression becomes

$$\frac{v_o}{v} = \frac{1}{\sqrt{(\frac{\omega}{\omega_o})^2}} \underline{/\tan^{-1} - \frac{\omega}{\omega_o}}$$

$$= \frac{\omega_o}{\omega} \angle \tan^{-1} -\frac{\omega}{\omega_o}$$

The gain is now inversely proportional to the frequency; by doubling the frequency, therefore, the gain is halved, i.e. reduced by

$$20 \log_{10} 2 = 6 \text{ dB}$$

If $\omega \gg \omega_o$ therefore, the gain reduces by 6 dB per octave (20 dB per decade).

The output voltage now lags the input voltage by an angle greater than $45°$, and which would increase to $90°$ at infinite frequency.

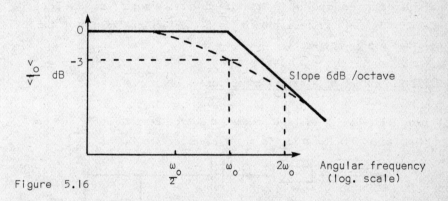

Figure 5.16

The magnitude response can be plotted in an approximate form as shown in Figure 5.16. If plotted as gain in decibels against frequency to a logarithmic scale, the response can be approximated to two straight lines corresponding to the two approximations,

$$\omega \ll \omega_o \quad \text{and} \quad \omega \gg \omega_o$$

For $\omega \ll \omega_o$, the response approximates to a straight line

$$\frac{V_o}{V} = 0 \text{ dB}$$

For $\omega \gg \omega_o$, the response approximates to a straight line, reducing

166 Basic Electrical Engineering and Instrumentation for Engineers

from 0 dB at $\omega = \omega_o$ at a constant slope of 6 dB per octave.

The true response is, of course, assymptotic to both of these lines, and must pass through the -3 dB point at $\omega = \omega_o$. The true response can be sketched in with reasonable accuracy. This form of response plot, with gain in decibels plotted against frequency to a logarithmic scale, is known as a Bode plot. Note that with a logarithmic frequency scale, equal lengths along the frequency axis do not correspond to equal frequency steps, but rather to equal ratios. Arising directly out of this form of plot, the cut off frequency ω_o is also known as the 'corner frequency'.

We have now considered the effect of one shunt capacitor across a load driven by a signal source. Stray capacitance will, of course, occur at all points throughout the amplifier circuit, and must be considered at each point. Its effect will be more serious, i.e. the cut off frequency will be lower, the higher are the values of the load and source resistors.

5.5.2 The effect of a series capacitance

Figure 5.17 shows a voltage source of source resistance R_s and e.m.f. v_s coupled to its load R_L via a series capacitor C.

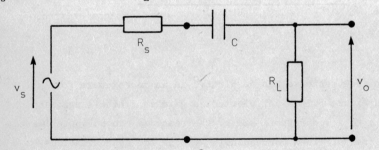

Figure 5.17

The output voltage may be written

$$v_o = \frac{R_L}{R_s + R_L + \frac{1}{j\omega c}} \cdot v_s$$

$$= \frac{\frac{R_L}{R_s + R_L}}{1 + \frac{1}{j\omega c(R_s + R_L)}} v_s \qquad 5.24$$

At frequencies at which the reactance of C is negligible

$$v_o = \frac{R_L}{R_s + R_L} \cdot v_s = v$$

Substituting this in equation (24)

$$v_o = \frac{v}{1 + \frac{1}{j\omega C(R_s + R_L)}}$$

or $\quad \dfrac{v_o}{v} = \dfrac{1}{1 + \dfrac{\omega_o}{j\omega}} = \dfrac{1}{1 - j\dfrac{\omega_o}{\omega}} \qquad 5.25$

where ω_o is defined as $\dfrac{1}{C(R_s + R_L)}$

Again, to study the magnitude and phase, this expression must be rationalised and converted into the polar form.

$$\frac{v_o}{v} = \frac{1}{1 - j\frac{\omega_o}{\omega}} = \frac{1 + j\frac{\omega_o}{\omega}}{1 + (\frac{\omega_o}{\omega})^2}$$

or $\quad \dfrac{v_o}{v} = \dfrac{\sqrt{1 + (\frac{\omega_o}{\omega})^2}}{1 + (\frac{\omega_o}{\omega})^2} \quad \underline{/\tan^{-1} \frac{\omega_o}{\omega}}$

$$= \frac{1}{\sqrt{1 + (\frac{\omega_o}{\omega})^2}} \quad \underline{/\tan^{-1} \frac{\omega_o}{\omega}} \qquad 5.26$$

We can now examine the form of the response.

(a) when $\omega \gg \omega_o$

If $\omega \gg \omega_o$, then the term $(\frac{\omega_o}{\omega})^2$ may be neglected when compared with unity.
The expression therefore becomes

$$\frac{v_o}{v} = \frac{1}{\sqrt{1}} \angle \tan^{-1} \frac{\omega_o}{\omega} = 1 \angle \tan^{-1} \frac{\omega_o}{\omega}$$

While this approximation holds, the output voltage is constant in magnitude, and in decibels

$$20 \log_{10} \frac{v_o}{v} = 20 \log_{10} 1 = 0 \text{ dB}$$

The output voltage leads the input voltage by a very small angle

$$\phi = \tan^{-1} \frac{\omega_o}{\omega}$$

(b) when $\omega = \omega_o$

At this particular frequency, the expression becomes

$$\frac{v_o}{v} = \frac{1}{\sqrt{1+1}} \angle \tan^{-1} - = \frac{1}{\sqrt{2}} \angle 45°$$

ω_o is, therefore, the lower cut off or corner frequency.
The gain is 3 dB down in comparison with that at mid range frequencies, and the phase angle has increased to $45°$.

(c) when $\omega \ll \omega_o$

At frequencies when $(\frac{\omega_o}{\omega})^2$ is large in comparison to unity, the expression may be written as

$$\frac{v_o}{v} = \frac{\omega}{\omega_o} \angle \tan^{-1} \frac{\omega_o}{\omega}$$

The voltage ratio is now directly proportional to frequency; doubling the frequency will also double the output voltage giving an increase of

$$20 \log_{10} 2 = 6 \text{ dB}$$

Amplifiers 169

While this approximation holds, therefore, the output increases with frequency at the rate of 6 dB per octave (20 dB per decade).

The output voltage now leads the input voltage by an angle greater than $45°$, the phase angle approaching $90°$ as the frequency reduces to 0 Hz.

The magnitude response can, as in the previous case, be shown as a Bode plot, approximated as two straight lines, corresponding to the two approximations

$$\omega \ll \omega_o \quad \text{and} \quad \omega \gg \omega_o$$

The response plot is shown in Figure 5.18.

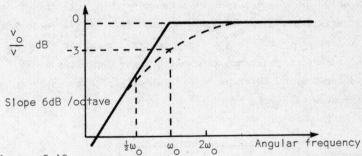

Figure 5.18

Notice that in the response plots of Figures 5.16 and 5.18, in each case 0 dB corresponds to the response in the mid-frequency range, i.e. to the response at frequencies at which the effect of the capacitive reactance is negligible. This is quite an arbitrary choice, and arises because the ordinate of the graph is $\frac{v_o}{v}$ expressed in decibels.

If the ordinate had been taken as $\frac{v_o}{v}$ and expressed in decibels, then the voltage ratio in the mid frequency range would have been

$$\frac{v_o}{v_s} = 20 \log_{10} \frac{R_L}{R_s + R_L} \text{ decibels}$$

and the complete response would have been lowered by this amount. To assist in sketching on to the Bode plot the final response, the following points should be noted:

170 Basic Electrical Engineering and Instrumentation for Engineers

(a) At the cut off frequency, the final response is 3 dB below the mid-range value,

(b) At $\omega = 2\omega_o$, the response is

$$20 \log_{10} \frac{1}{\sqrt{1 + (\frac{\omega_o}{\omega})^2}} = 20 \log_{10} \frac{1}{\sqrt{1.25}} \approx -1 \text{ dB}$$

In fact, for both the series and the shunt capacitor responses, there is a difference of very nearly 1 dB between the actual response and the approximated response at both $\omega = 2\omega_o$ and $\omega = 0.5\omega_o$.

Example

An amplifier has an open circuit voltage gain of 160, an input resistance of 20 kΩ and an output resistance of 2 Ω. It drives directly a resistive load of 15 Ω. The amplifier is driven from a source of resistance 5 kΩ via a series coupling capacitor of 0.2 μF. The amplifier input capacitance (including stray capacitance) is 500 pF. Determine the lower and upper cut-off frequencies and sketch the system response.

The amplifier is shown in Figure 5.19.

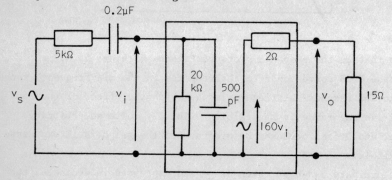

Figure 5.19

In the mid-frequency range, when the effect of the capacitors is negligible, then

$$\frac{v_i}{v_s} = \frac{20 \times 20^3}{20 \times 10^3 + 5 \times 10^3} = \frac{20}{25}$$

i.e. due to the input and source resistance

$$\text{voltage ratio} = 20 \log_{10} \frac{20}{25} = -1.94 \text{ dB}$$

The amplifier open circuit voltage gain is 160 or

$$20 \log_{10} 160 = 44.08 \text{ dB}$$

Due to the output and load resistances, there is a gain reduction given by

$$20 \log_{10} \frac{15}{2 + 15} = -1.09 \text{ dB}$$

The total mid-frequency gain is thus

$$-1.94 + 44.08 - 1.09 = 41.05 \text{ dB}$$

The lower cut-off frequency is given by

$$\omega = \frac{1}{C_c (R_s + R_I)}$$

$$= \frac{1}{\frac{0.2}{10^6}(20 \times 10^3 + 5 \times 10^3)} = 200 \text{ rad/sec}$$

$$\text{i.e. } f = \frac{200}{2\pi} = 31.8 \text{ Hz}$$

The upper cut-off frequency is given by

$$\omega = \frac{1}{C_s R}$$

where R is the value of R_s and R_I in parallel

$$R = \frac{5 \times 20}{5 + 20} \text{ k}\Omega = 4 \text{ k}\Omega$$

$$\omega = \frac{1}{\frac{500}{10^{12}} \cdot 4 \times 10^3} = \frac{10^6}{2} \text{ radians/sec}$$

$$\text{i.e. } f = \frac{10^6}{2} \times \frac{1}{2\pi} = 79.6 \times 10^3 \text{ Hz}$$

$$= 79.6 \text{ kHz}$$

The amplifier response is shown in Figure 5.20.

172 Basic Electrical Engineering and Instrumentation for Engineers

An amplifier frequency response can, in general, be divided into three sections. These are

(a) the mid-frequency range, in which the gain is substantially constant, and effects due to shunt and series capacitors are negligible,
(b) the high-frequency range, in which the shunt capacitance causes the gain to reduce as the frequency increases,
(c) the low-frequency range, in which series capacitance causes the gain to reduce as the frequency reduces.

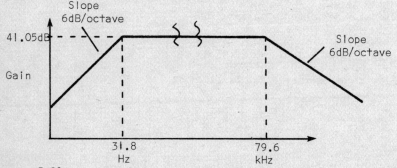

Figure 5.20

The frequencies which form the boundaries between the ranges are, as we have seen, determined by the circuit values, and it is not possible to give any general figures. Thus a frequency which would, for one particular amplifier system, be considered a low frequency, may quite easily fall in the high frequency section of the response of a different amplifier.

5.6 The application of feedback to amplifiers

In our work so far, we have considered the use of amplifiers which have fixed characteristics, i.e. as shown in Figure 5.6, the amplifier is specified by its input resistance R_i, its output resistance R_o and its open circuit gain A. In practice, these parameters are not constant, and would, in fact, vary with changes in ambient temperature, with changes in power supply voltages, or perhaps with age. These changes may be minimised by applying negative feedback to the amplifier. In a more general way, feedback, either positive or negative, can be used to modify the characteristics of an amplifier.

5.6.1 Effect on amplifier gain

Figure 5.21 shows a basic amplifier of gain A, to which feedback has been applied. In order to simplify the analysis, assume that the amplifier input resistance is sufficiently large, in comparison with the source resistance, that loading effects may be neglected. The feedback has been applied by connecting a potential divider across the amplifier output and tapping off a fraction β of the output voltage, this fraction being connected in series with the input signal in such polarity as to oppose the input signal. The arrows in Figure 5.21 show the effective directions of the signal voltages. A further assumption is that the loading effect of the feedback potential divider upon the output is negligible, i.e. the resistance of the potential divider is high in comparison with the amplifier output resistance.

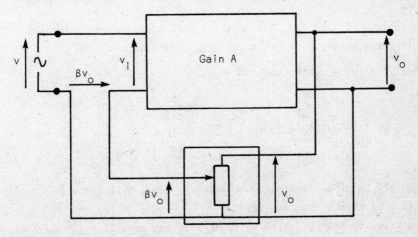

Figure 5.21

Since the feedback voltage opposes the input signal, the amplifier input voltage is

$$v_i = v - \beta v_o \qquad 5.27$$

The amplifier output voltage is then

$$v_o = A v_i$$

or $\quad v_o = Av - \beta A v_o \qquad\qquad 5.28$

The overall gain of the system is, therefore

$$A_f = \frac{v_o}{v} = \frac{A}{1 + \beta A} \qquad\qquad 5.29$$

The feedback, since it opposes the original input signal, is termed 'negative feedback'. Its effect is to reduce the gain from that of the basic amplifier, i.e. by the factor $1 + \beta A$.

If the original gain A was reasonably large, so that the factor βA is large compared to unity, then the overall gain becomes

$$A_f = \frac{v_o}{v} = \frac{A}{\beta A} = \frac{1}{\beta} \qquad\qquad 5.30$$

This is, of course, independent of the gain of the original amplifier and, in fact, the overall system is decided only by the feedback fraction β. Therefore, any changes in the gain of the amplifier, due to any cause, will not affect the gain of the overall system, provided that the product βA always is large compared with unity.

Example

An amplifier, of gain 200, has negative feedback applied by feeding back $\frac{1}{4}$ of the output voltage in series with its input.

$$\text{Amplifier gain, with feedback} = \frac{A}{1 + \beta A}$$

$$= \frac{200}{1 + \frac{1}{4}.200} = 3.92$$

If the amplifier gain should reduce by 20% to 160

$$\text{gain, with feedback} = \frac{160}{1 + \frac{1}{4}.160} = 3.90$$

The overall gain has, therefore, changed by about 0.5% for a 20% reduction in gain in the amplifier.

The sensitivity of the system to changes in the amplifier gain may be determined as follows:

The overall gain with feedback is

$$A_f = \frac{v_o}{v} = \frac{A}{1 + \beta A} \qquad\qquad 5.29$$

Differentiating with respect to A

$$\frac{dA_f}{dA} = \frac{(1 + \beta A) - \beta A}{(1 + \beta A)^2} = \frac{1}{(1 + \beta A)^2} \qquad 5.31$$

Combining equations (29) and (31)

$$\frac{dA_f}{A_f} = \frac{1}{1 + \beta A} \cdot \frac{dA}{A}$$

Thus the effect of a fractional change $\frac{dA}{A}$ in the gain of the amplifier is a fractional change of

$$\frac{\frac{dA}{A}}{1 + \beta A}$$

in the overall system.

One of the main causes of change in the gain of an amplifier is the reactive effects of series capacitance, and of shunt stray capacitance, as explained in section 5.5. The effects of these reactances upon the gain of the system will be reduced by negative feedback, giving a resultant increase in the bandwidth of the amplifier. Summarising the effects of negative feedback on the system gain,
(a) the overall gain of the system is reduced by the factor $(1 + \beta A)$,
(b) the sensitivity of the system gain to changes in the amplifier gain is reduced by the factor $(1 + \beta A)$.

5.6.2 Effect on the input resistance

The input resistance of any system is defined as the resistance R_i given by

$$R_i = \frac{v}{i}$$

where v is the input signal voltage applied to the system and i is the resulting signal current which flows into the system input. Figure 5.22 shows the input of the amplifier.

With no applied feedback, i.e. $\beta v_o = 0$, then

$$v = v_i$$

and the input resistance is

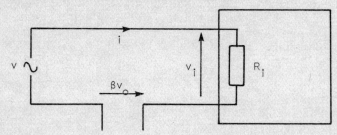

Figure 5.22

$$R_i = \frac{v_i}{i} = \frac{v}{i} \qquad 5.32$$

With applied negative feedback, however, the actual amplifier input v_i is much less than the system input v because

$$v_i = v - \beta v_o$$

For a given value of input voltage v, therefore, the signal current flowing into the amplifier is less with feedback than without. The effect of applying negative feedback in this manner is thus to raise the input resistance of the system.

Since $v_o = Av_i$

then $v_i = v - \beta A v_i$

or $v = v_i(1 + \beta A)$

The input resistance, with applied feedback, is therefore

$$R_{i_f} = \frac{v}{i} = \frac{v_i(1 + \beta A)}{i}$$

or $R_{i_f} = R_i(1 + \beta A)$ \qquad 5.33

The effect of negative feedback applied in this manner is to increase the input resistance by the factor $(1 + \beta A)$.

5.6.3 Effect on the output resistance

In section 5.3 it was shown that an amplifier output resistance is given by

$$R_o = \frac{\text{open-circuit voltage}}{\text{short-circuit current}}$$

The output circuit of the amplifier is shown in Figure 5.23. The assumption is still made that the loading effect of the feedback potentiometer on the amplifier output is negligible, i.e. the resistance of the potentiometer is large in comparison with R_o.

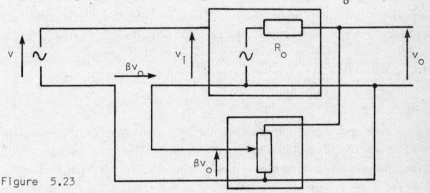

Figure 5.23

Without feedback, therefore, (i.e. with $\beta v_o = 0$)

open circuit output voltage $v_{oc} = Av_i = Av$

and

short circuit output current $i_{sc} = \dfrac{Av_i}{R_o} = \dfrac{Av}{R_o}$

output resistance $= \dfrac{v_{oc}}{i_{sc}} = \dfrac{Av}{\frac{Av}{R_o}} = R_o$ 5.34

With negative feedback applied,

open circuit output voltage $v_{oc} = Av_i$

$= A(v - \beta v_{oc})$

$= \dfrac{Av}{1 + \beta A}$

With the output short-circuited, in order to determine the short-circuit current, the output voltage is oviously zero. There is, therefore, no feedback voltage, and the feedback is removed from the system.

Thus the short-circuit output current is

$$i_{sc} = \frac{Av_i}{R_o} = \frac{Av}{R_o}$$

The output resistance with feedback is, therefore, given by

$$R_{o_f} = \frac{v_{oc}}{i_{sc}} = \frac{Av}{1 + \beta A} \cdot \frac{R_o}{Av} = \frac{R_o}{1 + \beta A} \qquad 5.35$$

The effect of negative feedback, applied in this manner, is thus to reduce the output resistance of the amplifier, by the factor $(1 + \beta A)$.

5.6.4 The effect of negative feedback upon signals generated within an amplifier

Ideally an amplifier should produce at its output an exact, amplified replica of the signal applied to its input. However, in all practical amplifiers, the output will contain, in addition to the amplified input signal, components at other frequencies. These unwanted output components may be considered as internally generated signals. They arise from many sources. In all circuits, for example, there exist signal components which are caused by the random motions of electric charges in the circuit. At low current levels, the particle nature of an electric current becomes more obvious, and there is a continuous random variation in the magnitude from instant to instant.

Because the magnitude of this variation is dependent upon temperature interfering signals, due to this effect, are described as 'thermal noise'. The noise energy is uniformly distributed over the whole of the frequency spectrum, and so produces an output at all frequencies within the amplifier frequency range.

Similar wideband noise signals are also produced in the active devices used in the amplifier. If the amplifier is an audio amplifier, designed to drive a loudspeaker, the noise signal would manifest itself as an audible background hiss.

A further source of interference is the 50 Hz main power system. This is very often used to produce the d.c. power supply for the amplifier circuit, and can affect the signal circuits by means of e.m.f's, electrostatically induced voltages in the higher impedance

Amplifiers 179

sections of the circuit, or electromagnetically induced currents in circuit loops. This interference would manifest itself in the amplifier output as a 50 Hz (or 100 Hz, depending upon the type of power supply) signal, which would give a deep hum note in the output loudspeaker.

These interfering signals form a lower limit to the allowable input signal to the amplifier, below which the signal to noise ratio is inadequate.

Internally generated noise of this sort normally causes difficulties only in the input stage of an amplifier, because after amplification in the input stage, the signal level is sufficiently large for noise generated in second amplifier stage to be negligible.

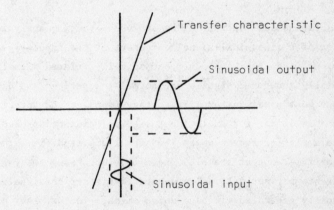

Figure 5.24

The amplitudes of the noise signals described are normally independent of the amplitude of the input signal, since the noise effects are properties of the amplifier circuit itself. There is, however, a further class of internally generated signals whose amplitude is a function of the amplitude of the signal itself. Figure 5.24 shows the input/output characteristic (or transfer characteristic) of an ideal amplifier. Because of the linear form of the characteristic, a sinusoidal input signal results in an amplified sinusoidal output signal. However, any departure from perfect linearity in the transfer characteristic will produce a non-sinusoidal output as is shown in exaggerated form in Figure 5.25.

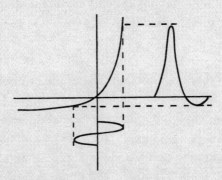

Figure 5.25

The non-sinusoidal output can be shown, by Fourier analysis, to consist of a sinusoidal signal component at the input signal frequency, together with other harmonically related signals. These internally generated signals are normally described as 'distortion'. Because most practical amplifier stage transfer characteristics are reasonably linear for small input signals, distortion occurs mainly in the amplifier stages where the signal amplitude is large, i.e. in the amplifier output stage. There is, therefore, the fundamental difference between internally generated distortion signals and internally generated wideband noise signals in that they are produced in different parts of the amplifier circuit.

Consider now the effect of applying negative feedback to an amplifier which produces some internally generated noise or distortion signal. Let the noise output from the amplifier without feedback be n. Let the noise output from the amplifier with feedback be n_f. With feedback, a fraction βn_f of the output noise is fed back into the input circuit, and is then amplified A times. The amplifier noise output is thus

$$n_f = n - \beta A n_f$$

since the applied feedback does not alter the amount of noise produced in the amplifier.

Thus
$$n_f = \frac{n}{1 + \beta A} \qquad 5.36$$

The output noise component is thus reduced by the same factor $(1 + \beta A)$. It must be carefully considered, however, whether this reduction in noise output provides a useful increase in the signal to noise ratio. In reducing the generated noise by the factor $(1 + \beta A)$, by the addition of negative feedback, we must also have produced a decrease in overall voltage gain by the same factor. The signal to noise ratio must, therefore, be the same with feedback as without. If, however, we can insert in the amplifier an extra amplifying stage which will increase the gain of the system without increasing the noise output, then the signal to noise ratio would be improved. Consider first, noise entering the system at the amplifier input, and including the thermal noise and transistor noise associated with the input amplifying stage. The application of negative feedback will cause a reduction in both the noise output and the signal output by the factor $(1 + \beta A)$.

If an extra amplifying stage is included in the amplifier to return the gain to its original value, this will affect equally the noise signal and the input signal. No improvement will be obtained in the signal to noise ratio.

Consider now, noise entering the system anywhere else except at the input stage, and including, of course, the distortion signals generated in the later stages of the amplifier. Negative feedback will, as before, reduce the output of noise, and of the signal, by $(1 + \beta A)$, but, in this case, the inclusion of an extra amplifying stage in the system, at any point earlier in the signal path than the noise is generated, will return the gain to its original value, without affecting the noise output. The overall signal to noise ratio will, therefore, be improved.

The following general conclusions may, therefore, be drawn:

1. The signal to noise ratio cannot be improved by feedback if the noise is entering the amplifier at the input or in the input stage.
2. The signal to noise ratio can be improved by feedback if an additional amplifying stage, free from noise, can be included in

the system prior to the point where the noise signal is being generated.

5.6.5 Positive feedback and amplifier stability

In the previous work in this section, we have shown that the effect of negative feedback upon the gain of the amplifier of Figure 5.21 is to give a gain, with feedback equal to

$$A_f = \frac{A}{1 + \beta A}$$

In Figure 5.21, the feedback voltage βA is produced by means of a potential divider, and is fed back in series with the input voltage in such a phase as to oppose the input voltage. If the phase of the feedback signal is reversed, then the overall gain will become

$$A_f = \frac{A}{1 - \beta A} \qquad 5.37$$

The overall gain of the system will, therefore, be increased by the feedback, which is now termed 'positive feedback' since the fed back signal and the input signal are now additive.

Example

An amplifier of gain 20 has 3.75% of its output voltage fed back in series with the input as positive feedback. Gain with positive feedback is

$$A_f = \frac{20}{1 - \frac{3.75}{100} \cdot 20}$$

$$= \frac{20}{1 - 0.75}$$

$$= 80$$

An interesting condition arises if the feedback factor is such as to make $\beta A = 1$.
Then

$$A_f = \frac{A}{1 - \beta A} = \infty$$

In fact, the feedback voltage is equal in amplitude and phase to the original input signal, and can, therefore, replace it. The amplifier thus produces its own input signal and is, therefore, an oscillator. This is inherently an unstable situation, and the output signal will increase until it is limited in amplitude by non-linearity in the oscillator circuit. The output will not in general be sinusoidal, and may take the form of a distorted sine wave, or even of a square or pulse waveform.

All oscillating systems thus require positive feedback, and in specially designed oscillators, for use as signal sources, the magnitude of the oscillation is carefully controlled to give the desired output waveform.

In practical amplifying systems, negative feedback is very often applied in order to produce the circuit gain and characteristics required. In considering the effects of negative feedback, it has been assumed that the feedback voltage is always in antiphase with the input signal voltage. However, we have also considered the effects of series and shunt capacitances, one effect of which is to cause the phase change in the amplifier to vary. The maximum phase change possible with one reactive element was shown to be $90°$ and this will occur at either zero or infinite frequency, depending upon the connection. However, practical amplifiers may contain several reactive components, and the phase change in the amplifier may be such that the feedback, designed to be negative, may at some frequency within the range of the amplifier become positive. The amplifier system will then oscillate at this frequency if the feedback magnitude is sufficiently great to make $\beta A = 1$. Phase or gain correcting networks are sometimes necessary to prevent instability of this sort.

5.7 Transistor amplifiers

In this chapter so far, we have considered the amplifier from the point of view of its external characteristics only, that is, from the point of view of the characteristics which may be determined by measurements made at its input and output terminals only.

Practical amplifiers consist of as few as one, or as many as several hundreds of active devices coupled together, and with the power supplies necessary to operate each device under its correct d.c. conditions as discussed in chapter 4.

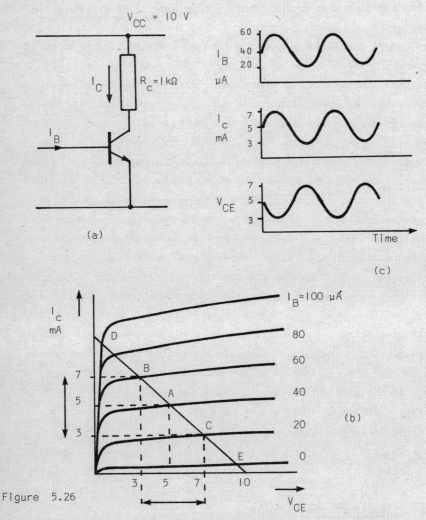

Figure 5.26

Figure 5.26(a) shows a single transistor with a collector load resistor, in the common emitter connection.

In Figure 5.26(b) typical common emitter characteristics are shown, with the load line for the collector resistor. The characteristics are typical of a transistor which would be used as a low level signal amplifier.

The d.c. conditions in the circuit are established by driving the transistor base with a steady direct current I_B; for the characteristic shown a current I_B = 40 µA is chosen. The circuit operating point is given by the intersection of the characteristic for a base current of 40 µA with the collector resistor load line, labelled B in the diagram. Under these static conditions, the collector voltage and current have the values 5 volt and 0.5 mA respectively. An alternating signal current may be applied to the circuit by adding to the steady base current, a sinusoidal current, a convenient value for the circuit being 20 µA peak. This will result in the base current varying between 20 µA and 60 µA in a sinusoidal manner. The circuit operating point will, therefore, move along the load line and will oscillate between the extreme limits marked A and C, these being the intersections of the load line with the 20 µA and 60 µA characterstics. The collector current swings from 0.3 mA to 0.7 mA, rising and falling in phase with the base current. The signal current gain of the circuit is thus

$$A_I = \frac{\text{change in load current}}{\text{change in input current}}$$

$$= \frac{7-3}{0.06-0.02}$$

$$= 100$$

The collector voltage falls as the collector current rises, due to the increased voltage drop across the load resistor. The signal voltage at the collector is, therefore, $180°$ out of phase with the input signal current. The signal voltage swing is between 3 and 7 volts, giving a peak sinusoidal output voltage of 2 volts. The input signal voltage can not be determined from these characteristics; however, since it will be the voltage change required to change the current in the base-emitter diode in a sinusoidal manner with a peak value of 20 µA about a mean value of 40 µA, an estimated voltage swing of about 0.2 V is reasonable. The circuit voltage gain may then be estimated as

$$A_V = \frac{\text{output voltage change}}{\text{input voltage change}}$$

$$= \frac{7-3}{0.2}$$

$$= 20$$

The following points should be noted.

In estimating the voltage and current values from the characteristics, equal swings in the values about the mean values have been assumed. Because, however, the characteristics are not exactly parallel, and not equally spaced, the circuit output voltage will not be exactly sinusoidal for a sinusoidal input current. The departure from a true sinusoidal signal will be greater for larger input signals. In particular, if the input current swing is sufficient to take the operating point to the points D or E, i.e. to bottom the transistor (point D), or to cut off the transistor (point E), then the collector voltage waveform, for a sinusoidal input current, will start to flatten at its peaks. It follows from this argument that, if such an amplifier stage is to be designed to allow the maximum sinusoidal output voltage possible, then the initial operating point must be positioned in the centre of the operating characteristics. The steady collector voltage will normally, therefore, be equal to $\frac{1}{2} V_{cc}$. The input signal to the circuit was specified as a sinusoidal base current. The impedance presented to the input signal generator is the forward biassed base-emitter diode, which is, of course, a non-linear impedance. For a sinusoidal current, therefore, the input voltage must be non-sinusoidal. In fact, the output voltage changes are reasonably linearly related to the input base current; the driving circuits should, therefore, be designed to produce a sinusoidal current.

It is also possible to derive for the transistor an equivalent circuit which may be used to give an estimation of the performance of the circuit under signal conditions. In fact there are many such small signal equivalent circuits, varying in complexity. It is, however, possible to obtain a reasonable estimate of circuit performance using only a very simple equivalent circuit, if its use

Amplifiers 187

is not extended beyond relatively low frequency operation.

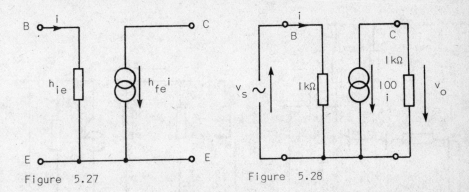

Figure 5.27   Figure 5.28

Figure 5.27 shows the simple equivalent circuit for a transistor connected in common emitter. The values of the parameters will obviously vary as the d.c. operating conditions are varied. The equivalent circuit is termed a 'small signal' equivalent circuit because the parameters are assumed to remain constant in value. It is one of the main functions of the d.c. design of a transistor circuit to set the conditions in the transistor so that the a.c. parameters have the optimum values for the purpose of the design. Further, the d.c. design should be such that the a.c. parameters remain as constant as possible, with changes in external conditions, such as temperature, etc.

The two parameters of the equivalent circuit are the resistor h_{ie}, which represents the impedance presented by the transistor to the source generator, and the current gain between the output and the input h_{fe}.

The value of h_{ie} is determined by the resistance of the forward biased base-emitter junction, together with the effective resistance of the narrow base region. A typical value for a transistor, similar to that represented by Figure 5.26, is about 1000Ω.

The current gain h_{fe} will be of the order of 100.

Figure 5.28 shows the equivalent circuit with a resistive load R_C of 1 kΩ, as in figure 5.26.

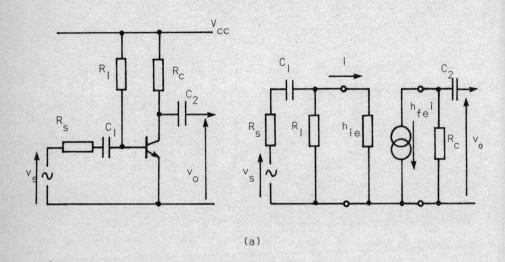

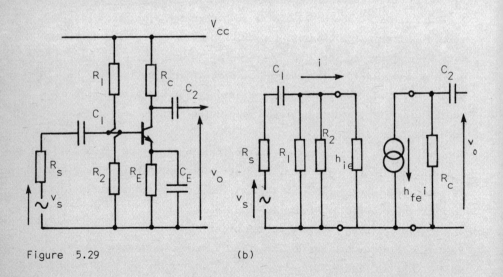

Figure 5.29 (b)

For an input current of i, the load resistor current is 100 i, giving the current gain as

$$A_i = 100$$

For an input voltage of v_s, the input current i

$$i = \frac{v_s}{1000}$$

The output voltage is then

$$v_o = h_{fe} \, i \, R_C$$

$$= 100 \times \frac{v_s}{1000} \times 10000$$

The voltage gain is therefore

$$A_v = \frac{v_o}{v_s} = 100$$

Figure 5.29 shows two versions of the single transistor amplifier, together with their equivalent small signal circuits. In Figure 5.29 a) the steady base bias current is provided via a resistor R_1 from the collector supply voltage V_{CC}. This method is not good, being susceptible to changes in the operating point conditions due to small variations in the transistor current gain or in the leakage currents, both of which are temperature sensitive. In the equivalent circuit, note that both the positive and negative voltage supply lines are effectively the same point. Although they are separated by a direct voltage V_{CC}, they are, in fact, shorted together to a.c. signals, via the very low a.c. impedance of the power supply. The bias resistor R_1 effectively is in parallel with the transistor input resistance h_{ie}. A better circuit is shown in Figure 5.29 b) in which the bias current is provided by fixing the d.c. potential of the transistor base by the potentiometer R_1 and R_2; the base and emitter currents are then determined by including a resistor R_E in series with the emitter. In order to remove the effect of R_E at signal frequencies, it is short circuited by a capacitor C_E. As long as C_E is sufficiently low in reactance, R_E and C_E need not appear on the signal equivalent circuit. The advantage of this circuit is that it is self compensating for small changes in d.c. conditions. Thus, if the emitter current tends to rise, due, perhaps to a temperature effect, then the emitter voltage will also rise. If the base voltage is adequately fixed by the potentiometer $R_1 R_2$, this will effectively reduce the base-emitter voltage, so reducing the emitter current again. Resistors R_1 and R_2 should, therefore, be as low in value as possible, consistent with not reducing the gain of the amplifier too much. In both circuits, the capacitors C_1 and C_2 prevent the external circuits from upsetting the d.c. conditions established in the transistor.

The emitter follower, shown schematically in Figure 5.30, together with its small signal equivalent circuit, is a widely used amplifier circuit. It also provides an interesting example of the use of an equivalent circuit in the determination of the properties of transistor circuits.

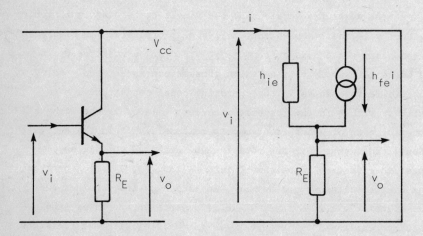

Figure 5.30

Because the circuit does not have a collector load resistor, the collector is connected directly to earth in the equivalent circuit. The circuit output voltage is taken from across the emitter resistor. Two basic equations may be written for the circuit

$$v_i = i h_{ie} + v_o \qquad 5.38$$
$$\text{and} \qquad v_o = i(1 + h_{fe}) R_E \qquad 5.39$$

If the current i is eliminated from these two equations, the circuit voltage gain is shown to be

$$A_v = \frac{R_E(1 + h_{fe})}{R_E(1 + h_{fe}) + h_{ie}} \qquad 5.40$$

With typical values of $h_{fe} = 100$ $h_{ie} = 1\ k\Omega$ and using $R_E = 1\ k\Omega$

$$A_v = 0.99$$

The circuit input resistance may also be determined, from the equations, to be

$$R_i = \frac{v_i}{i} = h_{ie} + R_E(1 + h_{fe}) \qquad 5.41$$

giving a value of

$$R_i = 102\ k\Omega$$

The output impedance of any circuit may be estimated by determining the ratio

$$R_o = \frac{\text{open circuit voltage}}{\text{short circuit current}} \qquad 5.42$$

where the open circuit voltage is the output voltage with no load connected across the output terminals, and the short circuit current is the current obtained in a short circuit placed across the output terminals. This method may actually be used in the laboratory if the d.c. conditions are not altered by the short circuit. The use of an a.c. short circuit (a capacitor) will normally allow the measurements to be made. The method is, however, very useful analytically.

Referring to Figure 5.30, the open circuit output voltage, with no load connected across v_o, is

$$v_{oc} = A_v v_i \qquad 5.43$$

The current in a short circuit placed across the output is

$$i_{sc} = \frac{v_i}{h_{ie}}(1 + h_{fe}) \qquad 5.44$$

The output resistance is then given by

$$R_o = \frac{v_{oc}}{i_{sc}}$$

$$= A_v \frac{h_{ie}}{1 + h_{fe}} \qquad 5.45$$

Substituting for A_v, from equation 5.40, we get

$$R_o = \frac{h_{ie}}{1 + h_{fe} + \frac{h_{ie}}{R_E}} \qquad 5.46$$

Typical values give

$$R_o = 9.8 \, \Omega$$

The emitter follower circuit has approximately unity gain, zero phase shift, a very high input resistance, and a very low output resistance. It finds great use as an impedance matching device, for example in instrument probes, or to couple effectively the signal from a high impedance device to a transmission line. Bias current for the base of the emitter follower is normally obtained via a resistor connected from the base to the voltage supply V_{CC}. If a very high input resistance is required, however, it must be borne in mind that this resistor is effectively in parallel with the circuit input, so giving a lower input resistance.

5.8 Operational amplifiers

The introduction of integrated circuit techniques in recent years changed the basic design philosophy of electronic circuits. In circuits constructed of discrete devices, i.e., separate transistors and diodes, the active devices constituted generally the most expensive components in the circuit. Consequently, the design techniques used concentrated on minimising the number of such devices used. Discrete resistors, and capacitors, were available cheaply and in a wide range of values, and the number used in a circuit was a secondary consideration.
An integrated circuit consists of a complete circuit fabricated on

a small chip of silicon; the circuit, which may contain 100 or more components, is accommodated on a chip, perhaps 0.3 mm thick and perhaps 2 mm square. Such integrated circuits (I.C's) are made in quantity batches. Active devices, all made at the same time, are very cheap and very small, a transistor occupying an area of as low as 0.1 mm^2. Resistors are fabricated by doping a strip of the silicon chip and making contacts to each end. Capacitors are made by reverse biassing a p-n junction on the chip, and utilising the depletion layer capacitance. Alternatively, silicon dioxide may be used as a dielectric between two conducting surfaces. Capacitors are, however, limited to very small values, a chip area of the order of 0.1 mm^2 being required to make a capacitor of 50 pF. Inductors are not realisable directly.

Integrated circuit designs thus use large numbers of transistors and diodes, but a small number of resistors and capacitors. Most circuits are directly coupled, because of the difficulty in making large value capacitors. Complex circuits are used to obtain the correct d.c. levels in the circuits.

The integrated circuit operational amplifier is a very high gain, d.c. coupled amplifier designed to have extremely low drift, a high input resistance and low output resistance. There are usually two signal input connections, the inverting and non-inverting inputs.

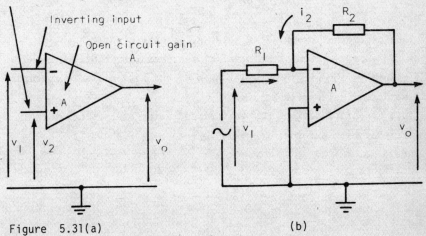

Figure 5.31(a)   (b)

A positive going signal applied to the non inverting input produces a positive going output; when applied to the inverting input it produces a negative going output. If identical signals are applied to both inputs, the output remains zero. The output is, in fact, proportional to the difference between the two inputs. Thus

$$v_o = A(v_2 - v_1) \qquad 5.47$$

Figure 5.31(a) shows the normal circuit symbol, while in Figure 5.31(b) the connections are shown for an inverting amplifier. The open circuit gain of the amplifier is very large. For any normal output voltage, the signal voltage at input, point E in the figure, is very small indeed, virtually at earth potential. Point E is said to be a 'virtual earth'. The circuit input current i_1 can, therefore be written

$$i_1 = \frac{v_1}{R_1}$$

Because the amplifier input resistance is very high, the current flowing into the input terminal is very nearly zero. Therefore,

$$i_1 = -i_2$$

or

$$\frac{v_1}{R_1} = -\frac{v_o}{R_2}$$

giving the circuit gain as

$$\frac{v_o}{v_1} = -\frac{R_2}{R_1} \qquad 5.48$$

As long as the basic amplifier gain, A, is sufficiently high, so that the approximations remain valid, the overall circuit gain is determined by the external resistances R_1 and R_2, and is independent of the actual value of the gain A.

Various circuit functions may be realised using operational amplifiers, with external feedback impedances, as shown in Figure 5.32.

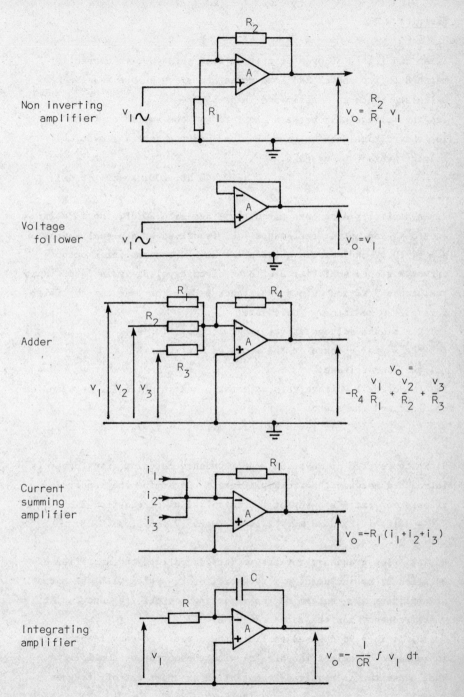

Figure 5.32

Examples

1. An amplifier, of open circuit voltage gain 400, is correctly matched to its source and load resistances, which are equal. Calculate, both as a ratio and in decibels
(a) the voltage gain between the load and the amplifier input,
(b) the voltage gain between the load and the source e.m.f.,
(c) the overall power gain.

                              (100, 40 dB, 200, 46 dB, 46 dB)

2. An amplifier A of open circuit voltage gain 20 dB, input resistance 5 kΩ and output resistance 1 kΩ is driven by a signal source of e.m.f. 10 mV and source resistance 1 kΩ. The amplifier output drives a second amplifier B of open circuit voltage gain 25 dB input resistance 3 kΩ and output resistance 8 Ω. The load for amplifier B is a 16 Ω resistor. Calculate
(a) the signal voltage at the input of amplifier A,
(b) the signal voltage at the input of amplifier B,
(c) the load voltage,
(d) the overall voltage gain as a ratio and in decibels, between the load and the source e.m.f.

                              (8.33 mV, 62.5 mV, 0.741 V, 74.1, 37.4 dB)

3. In the system of Question 2 an impedance matching transformer is introduced between the two amplifiers. Determine the required transformer ratio and repeat the calculations of question 2.

       (1 : 1.73, 8.33 mV, 72.14 mV, 0.855 V, 85.5, 38.6 dB)

4. (a) A low frequency cut-off is introduced into the amplifier of Question 2, by including a 1 µF capacitor in series with the output of amplifier A. Determine the resulting cut-off frequency. At what frequency has the gain reduced by 5%?

   (b) It is also desired to limit the upper frequency response by including a capacitor in parallel with the output of amplifier A. What capacitor is required to establish an upper cut-off frequency of 8 kHz?   (39.8 Hz, 121 Hz, 0.0265 µF)

5. An amplifier has an open circuit voltage gain of 10,000, an input resistance of 1 kΩ and a very low output resistance. Voltage negative feedback is applied by feeding back one fiftieth of its output voltage in series with its input. Calculate the resulting voltage gain and input resistance.

(49.75, 201 kΩ)

6. The amplifier of Question 5 reduces in gain to 8000, due to a faulty component. What is its effective gain with feedback?

(49.69)

7. An amplifier is required with a gain of 20, guaranteed to 0.5%. If production tolerances may produce a gain variation of 15%, what must be the value of β and of the amplifier gain without feedback for this specification to be met?

(0.0483, 600)

8. A transistor, connected in common emitter, has a 2 kΩ collector load and is biased by a single 100 kΩ resistor connected from its base to the collector power supply. The transistor parameters are h_{ie} = 1.25 kΩ, h_{fe} = 150. Determine the amplifier midfrequency gain. If the signal source is connected to the transistor base via a 1 μF capacitor, determine the cut off frequency.

(240, 129 Hz)

9. A transistor is connected in common emitter in the circuit of Figure 5.29. The collector load R_C = 2.2 kΩ, R_2 = 12 kΩ and R_1 = 100 kΩ. The emitter capacitor C_E is of negligible reactance. The transistor parameters are h_{fe} = 150, h_{ie} = 1.25 kΩ. The signal source is an e.m.f. of 10 mV with a source resistance of 600 Ω. What is the signal output voltage? The frequency is such that the capacitor C_1 and C_2 are of negligible reactance.

(1.719 V)

10. In the amplifier of Question 9, it is desired that the lower cut off frequency is 30 Hz. What should be the value of C_1? (3.08 μF)

CHAPTER SIX
Power Supplies

The majority of the direct current supplies required for electronic and control equipment is provided by means of rectifying equipment from the national alternating current mains supply. The term power supply in this context is used to refer to the complete circuitry which performs the conversion from a.c. to d.c., including the mains transformer which is normally used to isolate the d.c. supply from the a.c. mains, and which will usually also be required to alter the alternating line voltage. The term also includes any further circuitry which may be included to improve the operation of the supply, including ripple filters, voltage or current regulators or overvoltage or overcurrent protection. The direct current requirements vary over a wide range, from, say, 5 V at a current of 20 or 30 mA for small electronic circuits, to 80 or 100 volts at a current of several amperes for high power audio equipment. Welding equipment or motor controllers may have current requirements of several hundred amperes. The rectifying devices used in the majority of power supplies are silicon diodes, but thyristors are used to a large extent in controlled rectifier applications.

6.1 The half-wave rectifier with resistance load

The circuit of a half wave rectifier supplying a resistive load R is shown in Figure 6.1, together with the load voltage and current waveforms. For a sinusoidal alternating voltage supply, the load voltage and current waveforms are half sinusoidal in shape, corresponding in time with the half cycle of the supply voltage which foward biases the diode.

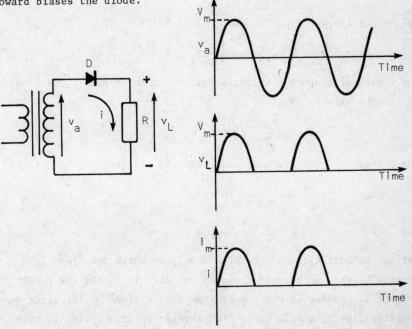

Figure 6.1

The peak voltage across the load resistor will be slightly smaller than the peak value of the applied voltage, because of the voltage drop in the diode (of the order of 1 volt). If, however, the diode voltage drop is neglected, the voltage across the load resistor may be expressed as

$$V_L = V_m \sin \omega t \qquad 0 < \omega t < \pi$$

$$V_L = 0 \qquad \pi < \omega t < 2\pi \qquad \qquad 6.1$$

A moving coil d.c. voltmeter connected across the load resistor will read the average value of the load voltage.

The d.c. voltage is, therefore,

$$V_{DC} = \frac{1}{2\pi} \int_0^\pi V_m \sin \omega t \, d\omega t + \frac{1}{2\pi} \int_\pi^{2\pi} 0 \, d\omega t$$

$$= \frac{1}{\pi} V_m$$

$$= 0.318 \, V_m \qquad \qquad 6.2$$

The direct current is

$$I_{DC} = \frac{V_{DC}}{R} = \frac{1}{\pi} \frac{V_m}{R} \qquad \qquad 6.3$$

For a sinusoidal supply voltage, the root mean square (r.m.s.) value of the voltage, V, is

$$V = \frac{V_m}{\sqrt{2}} \qquad \qquad 6.4$$

The direct load voltage may, therefore, be written

$$V_{DC} = \frac{1}{\pi} V_m$$

$$= \frac{\sqrt{2} \, V}{\pi} = 0.45 \, V \qquad \qquad 6.5$$

During the half cycle of the a.c. supply in which the diode is reverse biased, no current flows in the circuit. All the supply voltage is applied across the reverse biased diode. The diode must, therefore, be of sufficiently high reverse voltage rating to withstand safely the peak supply voltage.

The peak inverse voltage (P.I.V.) is defined as the maximum voltage appearing across the diode during the non-conducting period.

For the half wave rectifier, therefore,

$$\text{P.I.V. rating of rectifier} = V_m \qquad \qquad 6.6$$

The load waveforms show that whereas an ideal output would consist of a constant magnitude, steady voltage, or current, the output from a half wave rectifier consists of half sinusoidal pulses, and that, in fact, for 50% of the period, the voltage and current are actually zero. It can be shown that the output involves, in addition to the d.c. component of magnitude 0.318 V_m, the load fluctuation or ripple consisting of alternating components at frequencies which are harmonically related to the input supply frequency. Only the d.c. output is of use and, although in some applications the presence of such ripple may be tolerated, often ripple filtering circuits must be included to reduce the ripple magnitude to a low level. As an example, in a battery charging application, the efficiency of the circuit does not depend upon a smooth charging current, but depends rather upon the total amount of charge passed into the battery. A pulsating charging current is quite acceptable. In an audio equipment, however, the presence of excessive ripple components in the power supply voltage will result in signals at the ripple frequencies appearing in the system output. In the case of the half wave rectifier, the largest ripple component occurs at the power supply frequency; this would result in an audio equipment in a very low frequency (50 Hz) hum in the loudspeaker.

6.2 The full wave rectifier with resistance load

In the full wave rectifier circuit of Figure 6.2, a centre tapped transformer is used to provide effectively two secondary windings, which, together with separate rectifier diodes, act as two half wave rectifiers conducting during alternate half cycles of the a.c. supply and supplying a common load resistor. In the figure, when A is positive with respect to B, B is also positive with respect to C. Diode D_2 is, therefore, reverse biased but diode D_1 is conducting, the current flowing in the direction shown through the load resistor R. During the next half cycle of the a.c supply, diode D_1 is reverse biased, but diode D_2 conducts, and its current flows in the same direction through the load resistor.

As in the case of the half wave rectifier, the load voltage and current waveforms consist of half sinusoids, but as one occurs for each half cycle of the supply, the direct voltage and current values are double those of the half wave circuit.

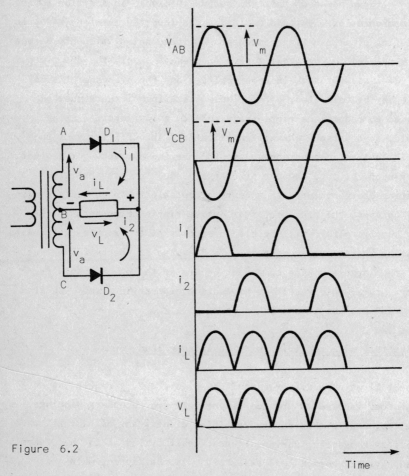

Figure 6.2

The load voltage may be expressed as

$$V_L = V_m \sin \omega t \quad 0 < \omega t < \pi$$

$$V_L = -V_m \sin \omega t \quad \pi < \omega t < 2\pi \qquad 6.7$$

The average value of the load voltage is then

$$V_{DC} = \frac{1}{2\pi} \int_0^\pi V_m \sin \omega t \, d\omega t + \frac{1}{2\pi} \int_\pi^{2\pi} - V_m \sin \omega t \, d\omega t$$

$$= \frac{2}{\pi} V_m$$

$$= 0.636 \, V_m \qquad 6.8$$

The direct current is

$$I_{DC} = \frac{V_{DC}}{R} = \frac{2}{\pi} \frac{V_m}{R} \qquad 6.9$$

In terms of the r.m.s. value of the supply voltage V,

$$V_{DC} = \frac{2}{\pi} \sqrt{2} \, V$$

$$= 0.90 \, V \qquad 6.10$$

It is important to note that each half of the transformer secondary winding is required to produce an alternating voltage of

$$v_a = V_m \sin \omega t$$

In other words, the alternating voltage between points A and C in the figure is twice this value.

We have seen that the peak value of the load voltage is equal to the peak value of the supply voltage, and both occur at the same time. The reverse voltage across the non-conducting rectifier diode is equal to the load voltage plus the voltage across half of the secondary winding. Because both of these voltages have peak values of V_m, the maximum reverse voltage is equal to $2 V_m$.

For the full wave rectifier, therefore,

$$\text{Peak Inverse Voltage (P.I.V.)} = 2 V_m \qquad 6.11$$

The load current and voltage waveforms, shown in Figure 6.2, consist again of repeated half sinusoids, both falling instantaneously to zero twice per cycle of the supply. The waveform may be shown, as in the half wave case, to consist of the direct value, together with harmonics of the supply frequency. In the full wave case, however, the lowest frequency component present is a component at twice the supply frequency. This makes the reduction of the magnitude of the ripple components using filtering techniques slightly easier than is the case with a half wave rectifier.

A disadvantage of the circuit of Figure 6.2 is the need for the centre tapped transformer, which is inefficiently utilised, since only half of the transformer is conducting at any given time. An alternative full wave rectifier circuit is shown in Figure 6.3 which does not require a centre tapped transformer, but does require four diode rectifiers instead of two. This is the bridge rectifier circuit. The transformer conducts during both half cycles of the

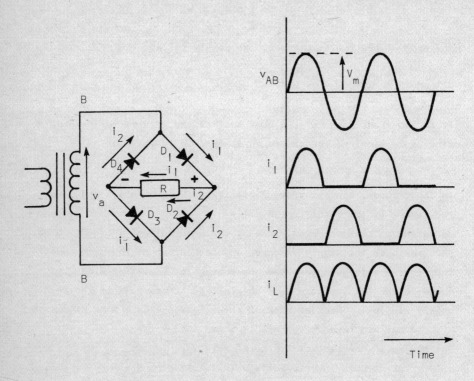

Figure 6.3

supply voltage, using two rectifiers in each cycle. Thus, when A is positive with respect to B, diodes D_1 and D_3 conduct, giving a half sinusoidal current pulse in the load R. During the succeeding half cycle, when B is positive with respect to A, diodes D_2 and D_4 conduct giving again a half sinusoidal current pulse in the load R, in the same direction as in the preceding half cycle. The load voltage and current waveforms are similar, therefore, to those of Figure 6.2. The reverse biased diodes are subjected to the sum of the voltage across the load resistor R, and the secondary voltage of the transformer, the maximum reverse voltage thus being $2 V_m$. However, since this voltage is shared between the two diodes, the peak inverse voltage of either diode is

$$\text{Peak Inverse Voltage} = V_m \qquad 6.12$$

In the forward conduction half cycle also, two diodes are in series together with the load R. The load voltage will, therefore, be less than the transformer voltage by two diode forward voltage drops. However, if the forward voltage drop is neglected, the direct output voltage is the same as that of the previous circuit, i.e.

$$V_{DC} = \frac{2}{\pi} V_m$$
$$= 0.636 \ V_m \qquad 6.13$$

6.3 Polyphase rectifiers

The full wave rectifier of Figure 6.2 may be considered as a two phase half wave rectifier circuit, each phase of the two phase supply conducting for one half of each cycle of the a.c. supply. The voltage produced by a single phase half wave rectifier has been shown to be

$$V_{DC} = 0.318 \ V_m \qquad 6.2$$

while that of a two phase half wave rectifier is

$$V_{DC} = 0.636 \ V_m \qquad 6.8$$

A further gain in rectified voltage is obtained if the number of phases is increased.

Figure 6.4 shows a three phase half wave rectifier fed from the star connected secondary windings of a three phase transformer. The figure also shows the voltage waveforms of the three phases of the transformer secondary, which are each sinusoidal and of peak value V_m volts. The three voltages are separated in phase by $\frac{2\pi}{3}$ radians (120°).

Consider the instant that the phase voltage v_{a1} is at its peak value, with A positive with respect to the star point O. Both of the other two phase voltages are such that B and C are negative with respect to the star point O. Diode D_1 is, therefore, conducting while diodes D_2 and D_3 are noth non-conducting. Phase voltage v_{a2} is, however, starting to rise positively while v_{a1} is starting to reduce. At the instant voltage v_{a2} exceeds voltage v_{a1}, diode D_2 commences to conduct raising the load voltage and reverse biasing D_1. Each diode,

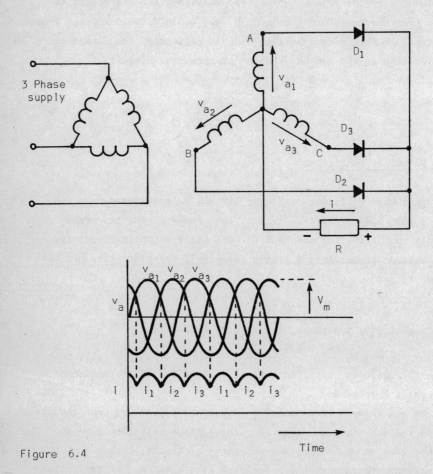

Figure 6.4

therefore, conducts for the fraction of the supply cycle that its phase voltage is in the forward biasing direction, and is greater in magnitude than both the other phase voltages. Each diode, therefore, conducts for one third of the supply cycle, giving a load current and voltage waveform as shown in Figure 6.4. It is easily seen that one advantage of increasing the number of rectifier phases is to decrease the magnitude of the ripple component of the rectified voltage. The lowest frequency ripple component is also increased to three times the frequency of the supply. It can be seen from Figure 6.4 that the voltage (or current) waveform consists of repeated identical segments of a sinusoidal curve, each segment being $\frac{2\pi}{3}$ radians in width. The equation to the waveform is best written in terms of a cosine function to give

$$v_L = V_m \cos \omega t \qquad -\frac{\pi}{3} < \omega t < +\frac{\pi}{3}$$

$$v_L = V_m \cos (\omega t - \frac{2\pi}{3}) \qquad \frac{\pi}{3} < \omega t < \pi$$

$$v_L = V_m \cos (\omega t - \frac{4\pi}{3}) \qquad \pi < \omega t < \frac{5\pi}{3} \qquad 6.14$$

Bearing in mind that the direct value of the rectified voltage is given by the average value of the voltage, and also that the average value of three identical consecutive segments is the same as the average of one of the segments, then

$$V_{DC} = \frac{3}{2\pi} \int_{-\pi/3}^{+\pi/3} V_m \cos \omega t \, d\omega t$$

$$= 0.827 \, V_m \qquad 6.15$$

In terms of the r.m.s. value of the supply voltage V, where

$$V = \frac{V_m}{\sqrt{2}} \qquad 6.4$$

then $\quad V_{DC} = 0.827 \times \sqrt{2} \, V$

$$= 1.17 \, V \qquad 6.16$$

In practice, because of the inductance of the circuit, due mainly to the transformer windings, the switch from one diode to the next diode does not take place instantaneously, and, in fact, a short overlap occurs during which time both diodes may conduct together. This tends to reduce slightly the value of the rectified output voltage. In power rectifier systems, the use of polyphase rectifiers is quite common, and rectifiers are built with a large number of phases. In general, with an m phase rectifier using m diodes, each diode will conduct for $\frac{2\pi}{m}$ radians per cycle. Neglecting losses the rectified output voltage may be written in general as

$$V_{DC} = \frac{m}{2\pi} \int_{-\frac{\pi}{m}}^{+\frac{\pi}{m}} V_m \cos \omega t \, d\omega t$$

$$V_{DC} = V_m \frac{\sin \frac{\pi}{m}}{\frac{\pi}{m}} \qquad 6.17$$

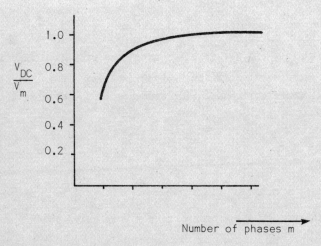

Figure 6.5

Figure 6.5 shows a graph of the direct rectified output voltage magnitude against m, the number of phases.

6.4 Ripple reduction

For many applications, especially those in which the power supply is required for electronic equipment, the ripple produced by the rectifier circuits described in the previous sections, is too large in magnitude. Additional ripple reducing circuits must be included.

6.4.1 The reservoir capacitor

Figure 6.6 shows the half wave rectifier circuit described in section 6.1, with the addition of a capacitor in parallel with the load. The effect of the added capacitor is to modify the output voltage waveform to that shown in the figure. During the period AB, the transformer supplies current, via the diode, to supply the load and also to charge the capacitor.

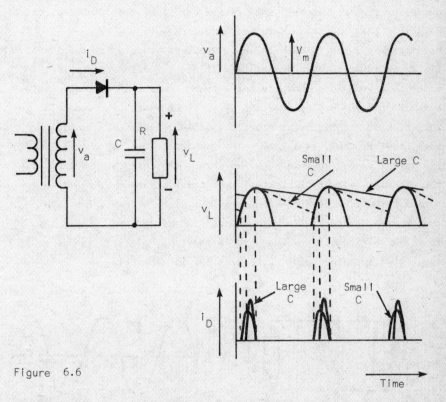

Figure 6.6

Because of the short charging time constant, the capacitor voltage (which is also the load voltage) follows almost exactly the waveform of the supply voltage. After its peak value, the supply voltage starts to reduce. If the capacitor is sufficiently large, its voltage does not fall as quickly as that of the supply. The diode, therefore, becomes reverse biased, and non-conducting. All the load current is now supplied from the capacitor; the capacitor voltage falls exponentially with a time constant determined by the value of the capacitor and by the effective resistance of the load. The diode remains non-conducting for the remainder of the supply cycle, and also during the next cycle, until the transformer voltage rises sufficiently to exceed the capacitor voltage and hence to forward bias the diode again. The action of the capacitor is very similar to that of a reservoir, in that it is charged by the supply, while the supply voltage is high; when the supply voltage is low, or negative, the current required by the load is suppled from the stored charge on the capacitor. This is the origin of the name used, the capacitor being termed a reservoir capacitor.

An important point to note is that, whereas without the capacitor the diode conducts for a complete half cycle, with a reservoir capacitor, the diode conducts for only a fraction of the half cycle. Further, the conducting fraction of the half cycle is reduced, if the value of the reservoir capacitor is increased. The capacitor acts, however, only as a reservoir of charge; all the charge flowing into the load must be supplied during the conducting period, via the diode. Increasing the capacitor value, therefore, shortens the period for which the diode conducts, but also increases the magnitude of the current pulse which flows. Care should be taken that the reservoir capacitor chosen for a particular design, is not so large that the diode peak current rating is exceeded.

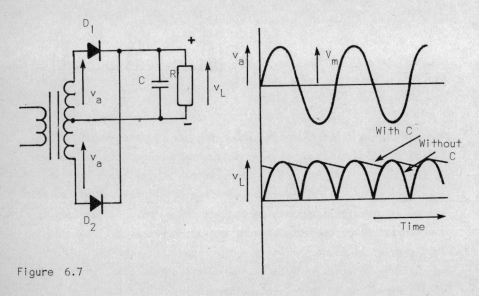

Figure 6.7

Figure 6.7 shows a full wave rectifier with an added reservoir capacitor, together with the rectified voltage waveform. It is interesting to note that if the load current taken from a power supply, which includes a reservoir capacitor, is reduced to zero, then the capacitor will charge up to the peak voltage of the supply. No further current will then flow in the rectifier diode. In practice this effect may occur if the power supply is disconnected from the circuit which it is supplying. Even if the a.c. supply is then switched off, the capacitor will retain the d.c. voltage across its terminals. Because the diode is reverse biased, the capacitor cannot discharge via the transformer secondary. If the reservoir capacitor is a good quality component, with high insulation resistance, the stored charge may be maintained for long periods. This dangerous condition is prevented in good designs by including, in parallel with the capacitor, a leakage resistor of sufficiently high value to not load the power supply excessively, but sufficiently low to discharge the capacitor within a reasonable time if the power supply is switched off after the load has been disconnected.

A measure of the effectiveness of a rectification system is given by the 'ripple factor' which is defined as

$$R = \text{ripple factor} = \frac{\text{r.m.s. value of ripple components}}{\text{average (d.c.) component}}$$

An estimation of the peak to peak ripple amplitude may be made by the approximation illustrated in Figure 6.8. For reasonably large values of reservoir capacitor, the direct output voltage will not fall excessively between successive supply voltage peaks.

In the figure the exponential capacitor discharge curve has been approximated by a linear discharge curve which has been extended in duration until the time of the next supply voltage peak. The ripple voltage waveform has thus been approximated to a triangular waveform. During the discharge period AB, the load current i_L is supplied by the capacitor.

We may write

$$i_L = C \frac{dv_L}{dt}$$

whence

$$\frac{dv_L}{dt} = \frac{1}{C} i_L \qquad \qquad 6.18$$

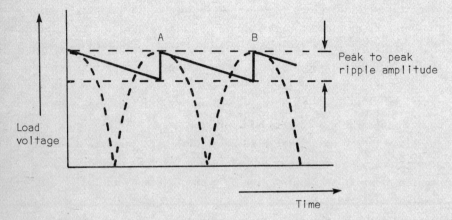

Figure 6.8

Example

Estimate the peak to peak ripple voltage for

(a) a half wave rectifier,

(b) a full wave rectifier

designed to produce from a 50 Hz supply a current of 2 amperes when used with a reservoir capacitor of 5000 µF.

Using the approximation, for a half wave rectifier, the duration of the discharge will be equal to the time of one full cycle of the a.c. supply

(a)   Discharge time = $\frac{1}{50}$ = 0.02 second

The rate of change of load voltage will be

$$\frac{dv_L}{dt} = \frac{1}{C} i_L \qquad\qquad 6.18$$

$$= \frac{10^6}{5000} \cdot 2 = 400 \text{ V/sec}$$

Fall in voltage during the discharge time is then

$$\delta V = \text{discharge time} \times \frac{dv_L}{dt}$$

$$= 0.02 \times 400$$

$$= 8 \text{ V}$$

Peak to peak ripple voltage = δV = 8 V

(b) For the full wave rectifier, the rate of fall of voltage will be the same as for the half wave rectifier.

$$\frac{dv_L}{dt} = 400 \text{ V/sec}$$

The discharge time will, however, be equal to one half cycle of the a.c. supply.

$$\text{Discharge time} = \frac{1}{2} \times \frac{1}{50} = 0.01 \text{ second}$$

Peak to peak ripple voltage is thus

$$\delta V = 0.01 \times 400$$
$$= 4 \text{ V}$$

This example shows that, for a given ripple voltage, a smaller reservoir capacitor may be used with a full wave rectifier then with a half wave rectifier.

When using the approximation of Figure 6.8 the ripple waveform is a triangular wave. The root mean square value of a triangular wave of peak to peak magnitude V_p is

$$\text{r.m.s. value} = \frac{1}{\sqrt{3}} V_p$$

Example

The full wave rectifier of the previous example produces an average direct output voltage of 40 V. Estimate the ripple factor.

Peak to peak ripple voltage is 4 V.

The r.m.s. value of a triangular waveform is

$$\text{r.m.s. value} = \frac{1}{2\sqrt{3}} \times \text{peak to peak value}$$
$$= \frac{1}{2\sqrt{3}} \times 4 = 1.15 \text{ V}$$

$$\text{Ripple factor} = R = \frac{1.15}{40} = .0289$$

The ripple factor is often expressed as a percentage,

i.e.   Ripple factor R = 2.89%

The use of a reservoir capacitor also has an effect on the 'peak inverse voltage' of a diode used in a half wave circuit. This was shown to be equal to V_m when used without a capacitor. The peak inverse voltage of the non-conducting diode is equal to the sum of the peak supply voltage and the peak direct voltage. As the reservoir capacitor maintains the direct voltage, during the non-conducting half cycle, the peak inverse voltage is increased to be

$$\text{peak inverse voltage} = 2 V_m \qquad 6.19$$

for the half wave circuit.

The peak inverse voltage for the full wave circuit is unchanged by the addition of a reservoir capacitor.

6.4.2 Ripple filters

For applications where the use of a reservoir capacitor does not reduce the ripple magnitude to a sufficiently low value, further ripple filtering circuits may be added.

A filter in its simplest form is in effect a potentiometer whose voltage output is a fraction of its input voltage, the fraction being designed to be different for signals of different frequencies. Figure 6.9(a) shows a potentiometer, consisting of two impedances Z_A and Z_B. The potentiometer output voltage is given by

$$v_o = \frac{Z_B}{Z_A + Z_B} v_i \qquad 6.20$$

Two conditions should be considered.

(a) $Z_B \gg Z_A$

If Z_B is large in comparison to Z_A then

$$\frac{v_o}{v_i} = \frac{Z_B}{Z_A + Z_B} \simeq \frac{Z_B}{Z_B} \simeq 1 \qquad 6.21$$

(b) $Z_A \gg Z_B$

If Z_A is large in comparison Z_B then

$$\frac{v_o}{v_i} = \frac{Z_B}{Z_A + Z_B} \simeq \frac{Z_B}{Z_A} \simeq 0 \qquad 6.22$$

A resistor capacitor ripple filter is shown in Figure 6.9(b). Considering the input voltage v_i to consist of two components, a direct voltage and an a.c. voltage, the effect of the filter on each com-

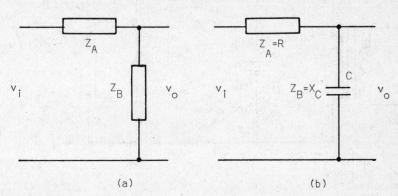

Figure 6.9

ponent may be considered separately.

Direct voltage component.   To the direct component, the impedance of the capacitor C is very high indeed, being virtually infinite.   The value of the resistor R is kept as low as possible.   Thus the direct output voltage is

$$v_{oDC} = \frac{\infty}{R + \infty} \times v_{iDC} \simeq v_{iDC}$$

The filter has little effect on the direct component.

Alternating voltage component.   To the alternating component, the impedance of the capacitor C is very low, a large value capacitor being chosen.   Thus the alternating output voltage is

$$v_{oAC} = \frac{X_C}{\sqrt{R^2 + X_C^2}} \times v_{iAC}$$

$$\simeq 0 \text{ if } X_C \ll R$$

where X_C is the reactance of the capacitor C and is equal to

$$X_C = \frac{1}{2\pi f C}$$

at a frequency of f Hz.

Figure 6.10 shows a half wave rectifier circuit, with a reservoir capacitor C_R and a resistor-capacitor ripple filter, $R_F - C_F$. As explained, the filter is required to reduce the ripple component to as low a value as possible.   Thus C_F is made as large as is con-

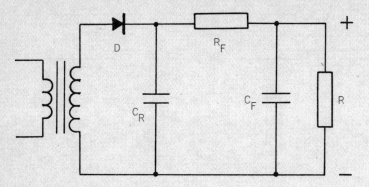

Figure 6.10

venient, while R_F should also be large. However, in order that the value of the direct voltage is not reduced by the flow of the load current through R_F, its value must be kept reasonably low. The value of R_F is thus a compromise between those two conflicting requirements. One solution to this problem is to replace the resistor R_F by an inductor L_F. The inductor, which would normally be wound upon a magnetic core to allow a large inductance value, would have a large impedance at the alternating ripple frequency while having a low resistance to the direct current. The use of an inductor, especially for high current power supplies, is a relatively expensive method, and the use of a resistor is more common.

6.5 Voltage multiplying rectifiers

Various circuits are used in which capacitors, charged via diode rectifiers, are connected in series in order to produce higher output voltages. Figure 6.11 shows one such circuit which operates as a voltage doubling rectifier.

When A is positive with respect to B, capacitor C_1 charges, with the polarity shown via the diode D_1. During the next half cycle, when B is positive with respect to A, capacitor C_2 charges with the polarity shown via the diode D_2. The load voltage, being taken from across the two capacitors connected in series, is the sum of the two voltages. With the load R disconnected, each capacitor would charge to the peak value of the voltage v_a; the direct output volt-

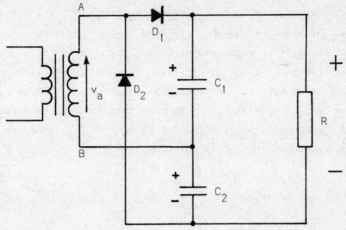

Figure 6.11

age would then be equal to $2 v_a$.

6.6 Power supply regulation

The main requirement of a power supply is that it shall produce its full load output current with as little change in output voltage as possible. The rectifier circuits discussed so far in this chapter all produce an output voltage which is related to the voltage of the a.c. supply to the transformer primary winding. Any variation in the supply voltage will, therefore, cause a variation in the direct output voltage. Further, due to the impedance of the transformer windings, to the voltage drop across a conducting non-ideal diode, and to the mode of operation of the reservoir capacitor, the direct output voltage is also reduced as the current drawn from a power supply is increased.

The regulation of a power supply may be defined generally as the amount that the output voltage (or current if it is designed as a constant current supply), changes when an external parameter changes. Thus 'Load Regulation' may be defined as

$$\text{Load regulation} = \frac{\text{No load output voltage} - \text{Full load output voltage}}{\text{Full load output voltage}} \times 100\% \qquad 6.23$$

Regulation may be similarly specified in terms of variation of the a.c. supply voltage or of ambient temperature changes. The effect of variations of load current upon the load voltage may also be specified in terms of an apparent internal resistance of the power supply. As the output load current taken from the supply is increased, the output voltage falls, and the power supply behaves as if it had an internal resistance. Suppose that a change of output current δI_L produces a change of output voltage δV_L. Then the internal resistance of the supply is given by

$$R_S = - \frac{\delta V_L}{\delta I_L} \qquad 6.24$$

Example

A 1 ampere power supply using a half wave diode rectifier circuit with a reservoir capacitor has a 'no load' voltage of 35 volts, and a full load voltage of 25 volts. Determine the regulation and its effective internal resistance.

$$\text{Regulation} = \frac{35 - 25}{25} \times 100 = 40\%$$

$$\text{Effective internal resistance } R_I = \frac{\delta V_L}{\delta I_L}$$

$$= \frac{35 - 25}{1}$$

whence
$$R_I = 10 \, \Omega$$

Although a load voltage/load current characteristic for a rectifier circuit would not, in general, exhibit a linear fall in voltage, the concept of an internal resistance is very useful in comparing different power supplies.

6.6.1 Regulated power supplies

For applications in which the regulation of a normal half or full wave rectifier power supply is not adequate (for TTL logic devices, for example, the direct supply voltage must be maintained within the range 5 V ± 0.25 V), a regulator circuit may be added to reduce the effective internal resistance of the supply. Two basic alternative regulator circuits are in common use and are illustrated in Figure 6.12.

Power Supplies   219

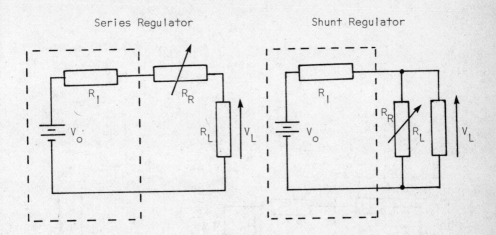

Figure 6.12

The power supply in the figure is represented by a battery of voltage equal to the no load voltage of the supply, together with a resistor representing its internal resistance. In the series regulator circuit, a variable resistor R_{REG} is included in series with the load. The value of the variable resistor is continually adjusted so that, at all values of load current, the output voltage is the same. In the shunt regulator circuit, a variable resistor R_{REG} is included in parallel with the load, its value being continually adjusted to maintain the total current taken from the supply, and hence the load voltage constant. A very common, simple regulator circuit makes use of a Zener diode, as a shunt regulator, as illustrated in Figure 6.13. The Zener diode, when reverse biased, has an almost constant voltage over a wide range of reverse currents. In the circuit shown the output load voltage V_L is equal to the Zener voltage and the no load voltage of the power supply V_o is dropped across the series resistance of the circuit, $R_I + R_S$, the resistor R_S being included to prevent the power dissipation in the diode exceeding its maximum rating, if the load R_L is disconnected. If the load current I_L is increased, the Zener current I_Z will decrease by an equivalent amount to maintain the total current, and hence the voltage drop across the series resistors constant. The stabilising effect of the Zener diode

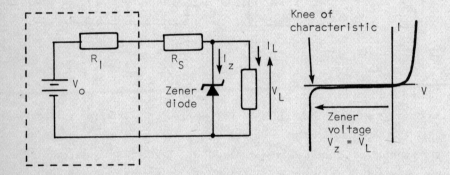

Figure 6.13

ceases when the load current I_L is increased, and the Zener current reduces to a very low value. The Zener diode operating point then moves round the knee of the characteristic. For more accurate stabilisation of the output voltage, regulating devices are included which incorporate an amplifier to increase the sensitivity of the device.

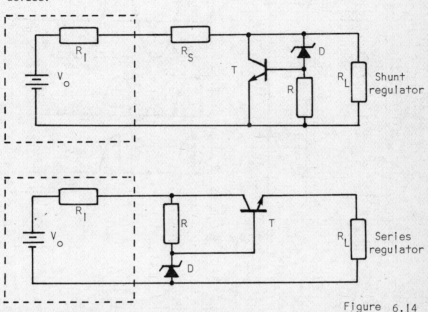

Figure 6.14

Figure 6.14 shows a series regulator and a shunt regulator, both using a transistor as the regulating device. In the shunt regulator circuit, if the load voltage increases, the transistor collector potential will rise positively with respect to the power supply negative line. A Zener diode regulator circuit is used to hold the potential of the transistor base constant with respect to the positive line; as the load voltage increases, therefore, the transistor base potential will rise positively with respect to its emitter. This, in turn, increases the current flowing via the transistor, increasing the voltage drop across the series resistors R_I and R_S, so counteracting the original rise in the load voltage. In the series regulator circuit, the difference between the power supply voltage and the load voltage is dropped across the series regulator transistor T. An increase of load current tends to cause the load voltage

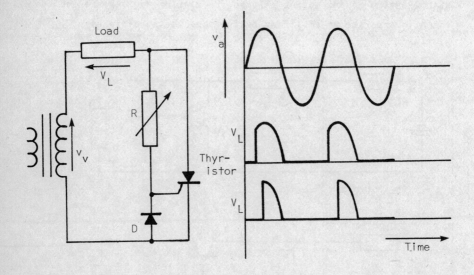

Figure 6.15

to fall. However, because the base of the transistor is fixed in potential by the Zener diode regulator circuit, as the load voltage starts to fall, the base emitter voltage of the transistor is increased, so increasing the transistor current and reducing the voltage drop across the transistor. This compensates for the original fall in load voltage.

The series regulator arrangement is more common, and the power dissipated in the transistor is normally smaller for a given power output. The disadvantage of the series connection is that, if the output of the regulated supply is short circuited, the full power supply voltage is applied directly across the regulator transistor, which will probably be destroyed. Series regulator circuits are, therefore, normally used in conjunction with protection circuits, which prevent the application of excessive load currents.

6.7 Controlled power rectification

For higher power systems, rectified a.c. mains supplies are often constructed utilising the thyristor as the rectifying element. The controlled conduction properties of the thyristor allow control of the power in the load.

A very simple method of obtaining controlled triggering of the thyristor is illustrated in Figure 6.15. The circuit shown is a half wave rectifier circuit in which the thyristor conducts during the half cycle in which its anode is positive with respect to its cathode. The thyristor differs from a simple diode rectifier, however, in that it does not conduct until a certain critical anode-cathode voltage is reached, as described in section 4.9.1. The critical voltage may be varied, being lowered by increasing the gate current, which is provided via the resistor R. Increasing the value of R will thus raise the critical anode-cathode voltage. Variation of the point of conduction can be achieved over the first quarter cycle of the a.c. mains supply, i.e. up to the instant of the peak voltage of the supply; a further decrease of gate current will then prevent the rectifier conducting altogether. Using this simple circuit, a reduction of the power to the load of up to 50% of that obtained using a normal half wave rectifier may be achieved. The diode is included in the circuit between the gate and cathode of the thyristor, to prevent the full voltage of the supply being applied in reverse direction between the cathode and gate during the non-conducting half cycle.

Conduction continues in the thyristor until the anode-cathode voltage is reduced almost to zero at the end of the conducting half cycle.

Examples

1. A half wave rectifier is fed from a transformer with a secondary voltage of 35 V r.m.s. Neglecting the voltage drop across the rectifier diode, determine
   (a) the average voltage output,
   (b) the r.m.s. voltage output,
   (c) the peak inverse voltage.

   (15.75 V, 24.75 V, 49.5 V)

2. If a reservoir capacitor is added to the rectifier of Question 1, determine
   (a) the peak voltage across the capacitor,
   (b) the peak inverse voltage,
   (c) the mean output voltage on no load,
   (d) the r.m.s. output voltage on no load.

   (49.5 V, 99 V, 49.5 V, 49.5 V)

3. A d.c. output of average value 300 V is required from a half-wave rectifier operating on a 230 V r.m.s. sinusoidal supply via a transformer. Neglecting rectifier voltage drop, determine
   (a) the transformer ratio,
   (b) the peak inverse voltage.

   (1 : 2.90, 942 V)

4. Repeat Question 3 for a full-wave rectifier.

   (1 : 1.45 + 1.45, 942 V)

5. A half-wave rectifier, with a 16 μF reservoir capacitor is required to supply an average current of 40 mA at an average voltage of 250 V. The mains supply is 240 V 50 Hz. Determine the required transformer ratio.

   (1 : 0.81)

6. A full-wave rectifier, operating from a 50 Hz supply has a peak output of 400 V. It is loaded with a resistor of 8 kΩ and is shunted by a reservoir capacitor. Calculate the capacitance for a ripple of 20 V peak to peak.

$$(25 \ \mu F)$$

7. A 50 Hz power supply consists of a transformer giving an output of 50 V r.m.s. to a bridge rectifier. The output current is 2 A and a reservoir capacitor of 2500 μF is used. Calculate the peak output voltage and estimate the peak to peak ripple.

$$(70.7 \ V, \ 8 \ V)$$

8. A stabilised voltage source consists of a 5.6 volt Zener diode which is connected via a 220 Ω limiting resistor R_s across a d.c. supply of voltage 10 V ± 10%.
(a) What is the maximum load current which can be taken from the stabilised source if the Zener diode current must not fall below 1 mA?
(b) If the maximum permissible power dissipation of the diode is 250 mW, is there a minimum value of load current which must be taken from the voltage source?

(14.5 mA, No, power dissipation in the diode with no load is 138 mW)

CHAPTER SEVEN

Transformers

The transformer is the electrical equivalent to the mechanical gear box. The gear box transmits mechanical power at different speed and torque levels whilst the transformer "transforms" electrical power from one voltage and current level to another. Both devices work with very little power loss, the efficiency of electrical transformers being of the order of 98%.

The alternating current transmission and distribution systems have developed largely because of the transformer. Electrical power is generated at 11,000 volts, transformed up to 33,000 or 132,000 volts and transmitted to the centres of power demand, where it is transformed down to a voltage level of 415 or 240 volts and distributed for industrial or domestic use.

7.1 Transformer Action

Consider the simplest form of single-phase transformer shown in figure 7.1 in which there are two coils wound on a ferrous core. Usually the coil which is connected to the supply is called the

PRIMARY and that which is connected to the load the SECONDARY. Transformers are completely reversible, i.e. power can flow in either direction and transformation of voltage and current can be up or down.

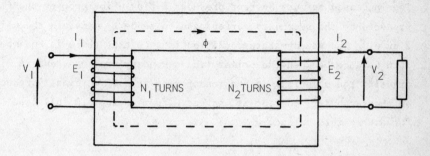

Figure 7.1

Let a sinusoidal alternating voltage V_1 be applied to the primary winding. A current I_1 will flow in the primary winding and a flux ϕ will be produced in the core. If it is assumed that the flux links both windings perfectly, i.e. there is no flux "lost", then by Faraday's Law, equation 1.15,

$$E_1 = N_1 \frac{d\phi}{dt}$$

and

$$E_2 = N_2 \frac{d\phi}{dt}$$

Therefore

$$\frac{E_1}{E_2} = \frac{N_1}{N_2} \qquad 7.1$$

The ratio N_1 to N_2 is termed the TRANSFORMATION RATIO of the transformer. The difference between the applied and induced primary voltages is the primary winding impedance volt drop $I_1 Z_1$ which for most practical transformers is very small, so that $V_1 \approx E_1$. This is also the case in the secondary winding and $V_2 \approx E_2$. Therefore

$$\frac{V_1}{V_2} \approx \frac{N_1}{N_2} \qquad 7.2$$

When a load is connected to the transformer a current I_2 will flow in the secondary winding producing a flux which, by Lenz's law, tends to oppose the main flux ϕ. If the main flux were reduced then E_1 would be reduced and the primary current I_1 would increase according to the equation,

$$\overline{V}_1 = \overline{E}_1 + \overline{I}_1 \overline{Z}_1 \qquad 7.3$$

The increased primary current produces a flux which opposes the flux produced by the secondary current and so tends to maintain the main flux ϕ. The steady-state condition is such that there is an ampere-turn balance between the primary and secondary windings, and in practice the main flux in the transformer is only marginally reduced as the secondary current is increased from no-load to full load. For ampere-turn balance

$$I_1 N_1 = I_2 N_2$$

i.e.
$$\frac{I_1}{I_2} = \frac{N_2}{N_1} \qquad 7.4$$

7.2 The Equivalent Circuit of a Transformer

When dealing with electrical devices such as transformers and machines, an attempt is usually made to devise a simple EQUIVALENT circuit consisting of R's, L's and C's, to act as an "aide-memoire" into the operation of the device and to facilitate analytical solutions. The quantities to be modelled in a transformer are,

    (i) The transformation ratio of the transformer.
    (ii) The copper losses in the windings.
    (iii) Useful and leakage fluxes.
    (iv) Iron losses in the core.

7.2.1 The Ideal Transformer

The ideal transformer, shown in figure 7.2, consists of two coils which simply model the transformation ratio of the transformer. The coils possess no resistance or inductance, and thus the ideal transformer can be characterised by the simple transformation equations,

$$\frac{V_1}{V_2} = \frac{N_1}{N_2} = \frac{I_2}{I_1} \qquad 7.5$$

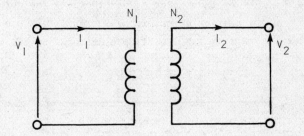

Figure 7.2

7.2.2 Copper Losses

Both the primary and secondary coils of an actual transformer possess resistance so that when current flows there is a copper loss ($I^2 R$ loss) associated with both coils. These winding resistances can be represented in the equivalent circuit by resistances included in series with the ideal transformer.

7.2.3 Useful and Leakage Fluxes

The useful or main flux is that flux which links both coils, as shown in figure 7.3. Considering sinusoidal operation of the transformer and assuming that the instantaneous value of the useful flux may be written as

$$\phi = \hat{\phi} \sin \omega t$$

then the induced e.m.f. is,

$$e_1 = N_1 \frac{d\phi}{dt} = \omega N_1 \hat{\phi} \cos \omega t$$

Thus the flux is proportional to the e.m.f. and lags it by $90°$. The current producing this useful flux is called the MAGNETISING CURRENT, and is designated I_m. It is in phase with the flux and therefore lags the induced voltage E_1 by $90°$.

230 Basic Electrical Engineering and Instrumentation for Engineers

The effect of this magnetising current can be included in the
equivalent circuit by connecting an inductance, called the
MAGNETISING INDUCTANCE in parallel with the primary coil of the
ideal transformer. An inductance is used because the current in an
inductance lags the voltage across it by $90°$ and it is connected in
parallel because the flux is directly proportional to the e.m.f.

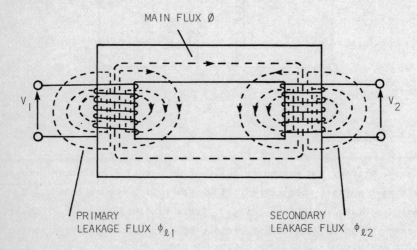

Figure 7.3

The LEAKAGE FLUX is that flux which links only one winding and
produces no magnetic coupling between the coils. When the transformer is operating on load the primary and secondary currents
produce fluxes which are mutually opposing, causing leakage flux
paths as shown in figure 7.3. The leakage flux path is largely in
air and thus the relationship between the leakage fluxes and the
current producing them is linear.

$\phi_{\ell 1}$ is proportional to and in phase with I_1

$\phi_{\ell 2}$ is proportional to and in phase with I_2

The induced e.m.f's produced by the leakage fluxes are proportional
to and leading them by $90°$, similar again to the conditions existing
in an inductor. The effect of the leakage fluxes can therefore be
included in the equivalent circuit by including an inductance in

series with each coil of the ideal transformer.

7.2.4 Iron Losses

It was shown in sections 1.3.6 and 1.3.7 that iron cores which are subjected to time varying fluxes exhibit hysteresis and eddy current losses. Therefore, when a sinusoidal voltage is applied to the primary winding, hysteresis and eddy current losses are present in the transformer core. These losses are termed the IRON or CORE losses of the transformer. For a particular transformer, the iron losses are constant for a fixed value of supply voltage, therefore they can be represented in the equivalent circuit by a resistance connected in parallel with the primary coil of the ideal transformer. The equivalent circuit of a practical transformer is obtained by combining the effects outlined above, and is shown in figure 7.4.

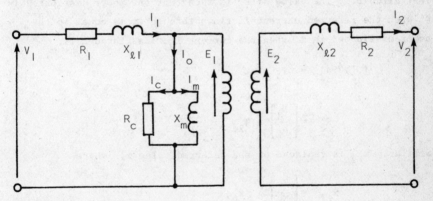

Figure 7.4

R_1 R_2 Primary and secondary winding resistances

$x_{\ell 1}$ $x_{\ell 2}$ "   "   "   leakage reactances

X_m Magnetising reactance

R_c Resistance to represent core losses

I_o No load current

I_m Magnetising current

I_c Core loss current

7.2.5 Referred Values

The ideal transformer can be eliminated from the equivalent circuit if the secondary parameters are "referred" to the primary side. The secondary current I_2 is replaced by the referred current I_2' flowing in the primary side of the equivalent circuit, where

$$I_2' = \frac{N_2}{N_1} I_2 \qquad 7.6$$

The secondary voltage V_2 is replaced by the referred voltage V_2', where

$$V_2' = \frac{N_1}{N_2} V_2 \qquad 7.7$$

The secondary winding resistance R_2 is replaced by the referred secondary resistance R_2' connected in the primary side of the equivalent circuit. The value of R_2' is such that the power dissipated in R_2' when the referred current I_2' flows through it is equal to the power dissipated in R_2 when the current I_2 flows through it.

$$(I_2')^2 R_2' = I_2^2 R_2$$

$$R_2' = \left[\frac{N_1}{N_2}\right]^2 R_2 \qquad 7.8$$

Similarly $x_{\ell 2}$ is replaced by the referred value $x_{\ell 2}'$ where

$$x_{\ell 2}' = \left[\frac{N_1}{N_2}\right]^2 x_{\ell 2} \qquad 7.9$$

The resulting equivalent circuit is shown in figure 7.5.

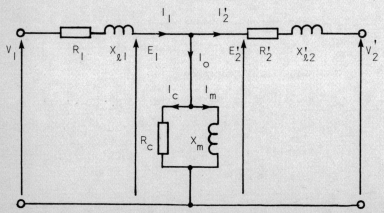

Figure 7.5

7.2.6 Approximate Equivalent Circuit

In a practical transformer the no-load current is only a small percentage of the full-load current. Power transformers have a no-load/full load current ratio of about 1/20, therefore, when considering transformers on load (say, above ¾ full-load), the parallel branch of the equivalent circuit is neglected as it has little effect on the overall performance and the approximate equivalent circuit of figure 7.6 can be used.

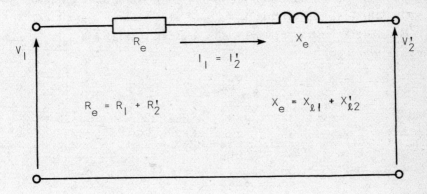

Figure 7.6

7.3 Transformer tests

7.3.1 Short Circuit Test

The secondary winding of the transformer is short circuited and the primary winding is supplied from a variable voltage source with a reduced voltage. An ammeter, voltmeter, and wattmeter are connected in the primary circuit and the primary voltage is increased until the rated (full load) current flows in the windings, when readings of voltage current and power are taken. Let these readings be $V_{s.c.}$, $I_{s.c.}$, and $P_{s.c.}$, then considering the equivalent circuit of figure 7.6 with the secondary short circuited

$$z_e = \frac{V_{s.c.}}{I_{s.c.}} \quad , \quad R_e = \frac{P_{s.c.}}{I_{s.c.}^2}$$

$$X_e = \sqrt{Z_e^2 - R_e^2} \qquad 7.10$$

7.3.2 Open Circuit Test

Full rated voltage is applied to the primary winding and readings of primary voltage current and power are taken with the secondary winding open circuited. Let these readings be $V_{o.c.}$, $I_{o.c.}$, and $P_{o.c.}$, then considering the equivalent circuit shown in figure 7.4 and noting that on open circuit $I_2 = 0$ and also that in a practical transformer R_c and X_m are very much greater than R_1 and $x_{\ell 1}$. R_c and X_m are given by,

$$R_c = \frac{V_{o.c.}}{I_c} \quad \text{and} \quad X_m = \frac{V_{o.c.}}{I_m} \qquad 7.11$$

where $I_c = I_{o.c.} \cos\theta_{o.c.}$ and $I_m = I_{o.c.} \sin\theta_{o.c.}$ and $\theta_{o.c.}$ is the open circuit power factor angle.
Also $\cos\theta_{o.c.} = \frac{P_{o.c.}}{V_{o.c.} I_{o.c.}}$ hence $I_c = \frac{P_{o.c.}}{V_{o.c.}}$
and, therefore,

$$R_c = \frac{V_{o.c.}^2}{P_{o.c.}}$$

a result which may have been written down at sight.

7.4 Phasor Diagram of a Transformer on Load

The phasor diagram shown in figure 7.7 is based on the equivalent circuit shown in figure 7.5.

7.5 The Voltage Equation of a Transformer

The relationship between the voltage, flux, turns and frequency can be obtained directly from Faraday's Law. On the primary side,

$$e_1 = N_1 \frac{d\phi}{dt}$$

Figure 7.7

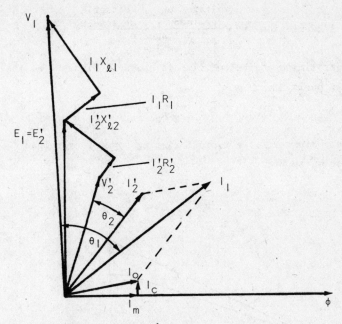

Let the sinusoidal flux be $\hat{\phi} \sin \omega t$

thus $e_1 = N_1 \dfrac{d(\hat{\phi} \sin \omega t)}{dt}$

$e_1 = 2\pi f N_1 \hat{\phi} \cos \omega t$

The r.m.s. value of induced e.m.f. is

$E_1 = \dfrac{2\pi f N_1 \hat{\phi}}{\sqrt{2}}$

$E_1 = 4.44 f N_1 \hat{\phi}$ 7.12

Similarly

$E_2 = 4.44 f N_2 \hat{\phi}$ 7.13

7.6 Voltage Regulation

As in most other voltage sources the terminal voltage of the transformer drops when the load current drawn from it is increased. The difference between the rated no-load output voltage and the output voltage on load is termed the regulation, and the percentage regulation of the transformer is defined as,

$$\% \text{ Regulation} = \frac{\text{Voltage on no load} - \text{Voltage on load}}{\text{Voltage on no load}} \times 100 \qquad 7.14$$

Using the approximate equivalent circuit shown in figure 7.6, the voltage equation is

$$\overline{V'_2} = \overline{V_1} - \overline{I'_2 Z_e} \qquad 7.15$$

and the phasor diagram for a typical lagging power factor load current is shown in figure 7.8.

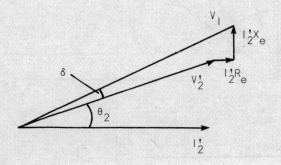

Figure 7.8

Now in a practical transformer the voltage drops $I'_2 R_e$ and $I'_2 X_e$ are small compared to V'_2 and V_1, therefore the angle δ is negligibly small and thus

$$V_1 = V'_2 + I'_2 R_e \cos \theta_2 + I'_2 X_e \sin \theta_2 \qquad 7.16$$

and the percentage regulation is given by,

$$\% \text{ Reg.} = \frac{V_1 - V'_2}{V_1} \times 100$$

$$\% \text{ Reg.} = \frac{100}{V_1} \{I'_2 R_e \cos \theta_2 + I'_2 X_e \sin \theta_2\} \qquad 7.17$$

7.7 Efficiency of the Transformer (η)

The efficiency of any device is defined as the ratio of the power output to the power input, the difference between the output and input being the inherent power loss in the device, which in the case of the transformer is made up of two components, i.e. copper loss and iron loss. Therefore the efficiency of the transformer is given by

$$\eta = \frac{\text{output}}{\text{input}} = \frac{\text{output}}{\text{output + copper losses} + F_e \text{ losses}}$$

$$\eta = \frac{V_2 I_2 \cos \theta_2}{V_2 I_2 \cos \theta_2 + I_2^2 R_e + F_e} \qquad 7.18$$

(F_e represent the constant iron losses)

In equation 7.18 there are two variables, the load current I_2 and the power factor θ_2. Considering operation at a constant power factor, equation 7.18 may be rewritten as

$$\eta = \frac{V_2 \cos \theta_2}{V_2 \cos \theta_2 + I_2 R_e + \frac{F_e}{I_2}} \qquad 7.19$$

differentiating the denominator of equation 7.19 with respect to I_2, and equating to zero gives,

$$\frac{d(\text{den})}{d I_2} = R_e - \frac{F_e}{I_2^2} = 0$$

$$I_2^2 R_e = F_e$$

The second differential shows that this is the minimum condition for the denominator, and therefore the maximum efficiency of a transformer occurs when the copper loss is equal to the iron loss. The variation of the efficiency and the losses with the load current is shown in figure 7.9a. If the load current is now assumed constant, then equation 7.18 becomes

$$\eta = \frac{V_2 I_2}{V_2 I_2 + \dfrac{I_2^2 R_e + F_e}{\cos \theta_2}} \qquad 7.20$$

It is easy to see from equation 7.20 that as the power factor decreases from unity the efficiency will decrease, as shown in figure 7.9b.

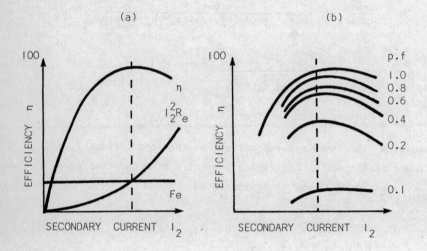

Figure 7.9

7.8 Auto-transformers

The auto-transformer utilises only one winding, the secondary voltage and current being obtained from a tapping point along the winding as shown in figure 7.10. The auto-transformer shown in figure 7.10 is equivalent to a conventional two winding step down transformer; however a considerable saving in copper has been effected since in the two winding transformer the windings were carrying a total current of $I_1 + I_2$ and in the auto-transformer the single winding carries only $I_2 - I_1$.

Figure 7.10

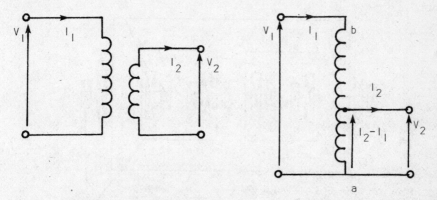

Auto-transformers find a wide application when a variable voltage is required; the tapping is made into a sliding contact and a secondary voltage of zero, when the contact is at "a", to V_1, when the contact is at "b", can be obtained. The main disadvantage of the auto-transformer is that the primary and secondary windings are not electrically isolated from each other, a safety feature which is essential on all distribution transformers.

7.9 Three-phase Transformers

A three-phase transformer can be obtained by
(a) suitably connecting three single-phase units, or
(b) using a three-limbed core as shown in figure 7.11.
The latter is far more common, particularly in distribution transformers, because it is less costly.

Electrically the three-phase transformer is made up of six coils, three primary and three secondary, which may be connected in either star or delta, giving the possibility of twelve different transformer connections. It is important, therefore, to know exactly how a three-phase transformer is connected especially when two or more transformers are to operate in parallel. The different transformer connections are arranged in four main groups according to the phase displacement between the primary and secondary line voltages, as

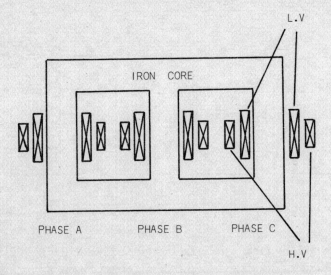

Figure 7.11   Three-Phase Core Type Transformer

shown in the table of figure 7.12. For successful parallel operation of three-phase transformers it is essential that they belong to the same main group and that their voltage ratios are equal.

Figure 7.12

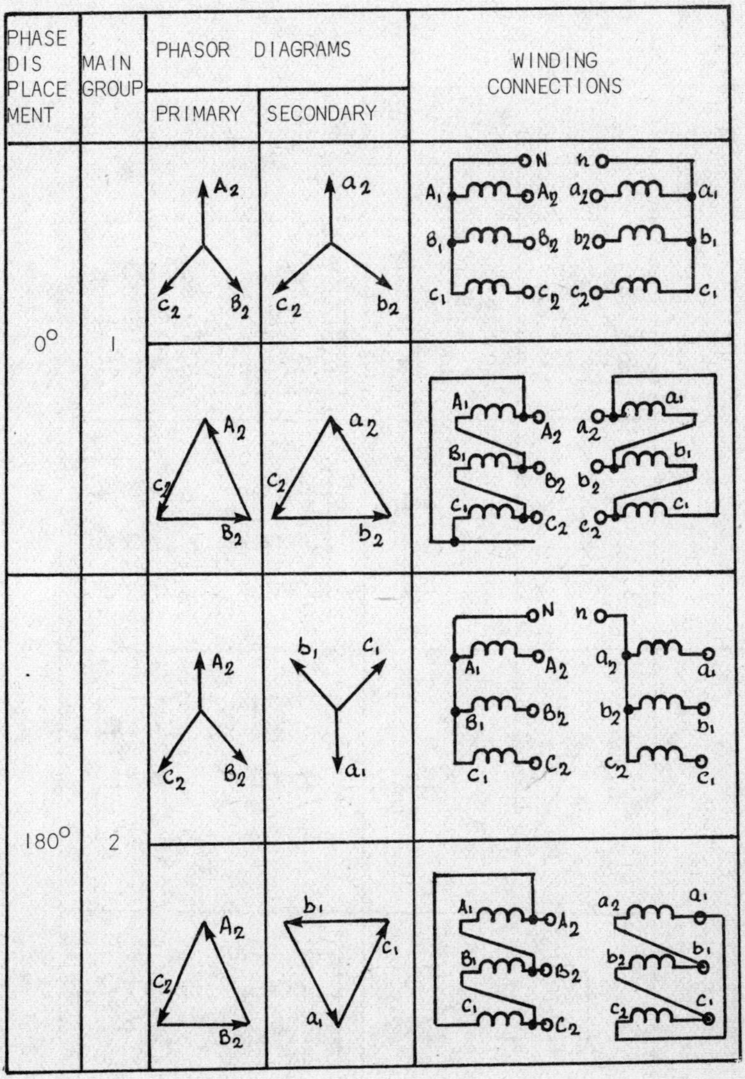

PHASE DISPLACEMENT	MAIN GROUP	PHASOR DIAGRAMS		WINDING CONNECTIONS
		PRIMARY	SECONDARY	
$0°$	1	Y (A_2, B_2, C_2)	Y (a_2, b_2, c_2)	Star-star winding connection
		Δ (A_2, B_2, C_2)	Δ (a_2, b_2, c_2)	Delta-delta winding connection
$180°$	2	Y (A_2, B_2, C_2)	Y inverted (a_1, b_1, c_1)	Star-star winding connection
		Δ (A_2, B_2, C_2)	Δ inverted (a_1, b_1, c_1)	Delta-delta winding connection

Figure 7.12(cont.)

PHASE DISPLACEMENT	MAIN GROUP	PHASOR DIAGRAMS		WINDING CONNECTIONS
		PRIMARY	SECONDARY	
−30°	3			
30°	4			

Examples

1. Tabulate as many applications of transformers as possible.
A welding transformer is required to operate from the 240 V, 50 Hz mains and the output voltage required is 10 V. If the maximum flux in the transformer core is not to exceed 1 mW, calculate the number of turns required for the primary and secondary windings.
What other factors would be taken into account in the design of such a transformer?

(1080, 45)

2. The test results on a 5 kVA, 6600/240 V, 50 Hz single phase transformer were,

Open Circuit Test, on the h.v. side
6600 V, 20 W, 0.04 A
Short Circuit Test, on the h.v. side
76 V, 23.1 W, Full load current

Calculate the parameters of the equivalent circuit, assuming $R_1 = R_2^1$, and $x_{\ell 1} = x_{\ell 2}^1$.

($R_c = 2.18$ MΩ, $X_m = 165$ kΩ, $R_1 = R_2^1 = 20\Omega$, $x_{\ell 1} = x_{\ell 2}^1 = 45.83$ Ω)

3. A 110 kVA, 11000/240 V, 50 Hz, single phase transformer gave the following test results on short circuit. With the secondary short circuited a voltage of 500 V on the primary produced full load current in the transformer, the power supplied being 3000 W. Calculate (a) the percentage regulation, (b) the secondary terminal voltage, at full load current for power factors of unity, 0.8 lagging and 0.8 leading.

(2.73%, 233.45 V, 4.36%, 229.54 V, 0%, 240 V)

4. A 100 kVA, 10000/1000 V, 50 Hz transformer gave the following test results. With the low voltage winding open circuited, the input power to the high voltage side was 1500 W. With the low voltage winding short circuited, the power supplied to the transformer with full load current flowing was 2680 W.

Calculate the efficiency of the transformer at (a) full load unity power factor, (b) 50% full load at 0.8 power factor lagging. At what load does the maximum efficiency occur?

$$(0.96, 0.948, \tfrac{3}{4} \text{ Full Load})$$

5. A 110 kVA, 11000/2200 V, two winding transformer may be connected to form an auto transformer. Calculate the voltage ratios available and the new rating of the auto transformer.

$$(13200/2200, 132 \text{ kVA}, 13200/11000, 660 \text{ kVA})$$

6. A transformer has a maximum efficiency of 0.98 which occurs at 15 kVA and unity power factor. During one day it is loaded as follows:

        12 hours − 4 kW at 0.5 power factor
        8 hours − 12 kW at 0.8 power factor
        4 hours − 9 kW at 0.9 power factor

Calculate the "all day" efficiency, i.e. the ratio of $\dfrac{\text{kWhr output}}{\text{kWhr input}}$

$$(0.965)$$

CHAPTER EIGHT
Digital Systems

8.1 Analogue and digital signals

In the work so far, we have been concerned with signals in an analogue form. Let us consider carefully what this means. We can take, as an example, the problem of producing a signal which will convey information regarding the depth of water in a tank. Problems such as this are quite common in practice and require the construction of a measurement system which, in some cases, must operate for long periods without attention, and with great reliability. One solution to the problem is illustrated in Figure 8.1. The figure, which does not attempt to show the mechanical details of the measurement system, represents the water tank in which is mounted a potentiometer, the sliding contact being connected to a float, and which, therefore, alters its position as the water level changes. Obviously such a construction would require great mechanical refinement in order to produce a reliable system. Let is assume that at the maximum allowed water depth the float and slider just reach the top of the potentiometer travel. The output voltage, V_o, will then equal the supply voltage V. If also, at the lowest levels allowed, the float and slider have fallen to the bottom of the potentiometer,

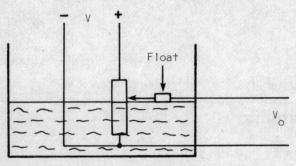

Figure 8.1

the output voltage, V_o, will equal zero. At any fraction, x, of the full water depth, the float and slider will rise to the same fraction x of the travel up the potentiometer and the output voltage will be

$$V_o = x\ V$$

The output voltage is thus proportional to the depth of water above the lowest allowed level, and can be used as an indication of the depth. The output voltage is said to be an analogue of the water depth, and in an ideal system would vary in exactly the same way as the water depth. The relationship between the water depth and its ideal voltage analogue is shown in Figure 8.2.

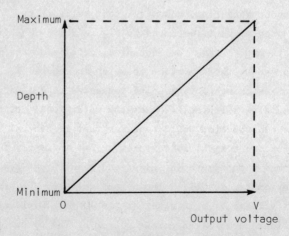

Figure 8.2

Many examples occur of the use of an analogue quantity to represent a physical variable. Thus in a motor car speedometer, the car speed is represented by the angular rotation of the speedometer pointer, while in an audio system, the intensity of the sound waves produced by the performers is converted by the microphone into a voltage analogue for amplification, recording and other processing.

However, in life, systems are never ideal, and there is a limit to the accuracy of the proportional relationship between the variable and its analogue.

Returning to the water depth system, we can point to various sources of inaccuracy. Thus, because of friction between the slider and the potentiometer wire, very small slow changes in water level may occur without the slider moving; the slider would, therefore, tend to respond to a slowly changing water depth in a series of small jerks. Further, because of corrosion, etc., on the surface of the potentiometer wire, the contact resistance between the contact and the wire will vary at different positions of the slider. Because of stray electromagnetic fields in the vicinity of the tank, perhaps due to adjacent mains electricity or other cables, various small induced alternating signals would also be present at the system output. These, and many other inherent factors, result in a 'noise' voltage at the output which forms a limit to the smallest detectable change in water depth. There is, therefore, at any water depth, a range of uncertainty, within which the true water depth is not known. This is illustrated in Figure 8.3.

The amplitude of the noise voltage is, to some extent, independent of the value of the voltage V applied to the potentiometer. Some improvement in the depth uncertainty can, therefore, be obtained by using an increased supply voltage. A signal amplifier could also be used, especially if the depth indicator is to be mounted in a remote place, connected via lengthy lines to the data processing equipment. The use of such long lines will cause a reduction (attenuation) in the signal level reaching the end of the line, making the signal more easily affected by further noise voltages induced in the lines themselves. It must be remembered that the noise voltage production process is an inherent fact of life, which occurs everywhere, and

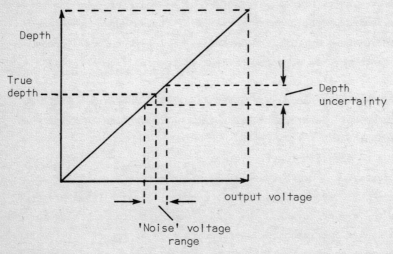

Figure 8.3

which must be taken account of by good design techniques which minimise its effect. Further, once the minimum noise voltage has been achieved, by good design, it must be ensured that the amplitude of any signal used is always sufficiently greater than the amplitude of the noise, to give the required system accuracy.

Summarising, an analogue signal is a signal which varies in exactly the same manner as the variable which is to be measured, and which can take any value within the allowable signal range. There is, therefore, theoretically, an infinite number of values which an analogue signal can take within an allowed signal range. In practice, as we have seen, there is, in fact, the smallest discernable change in signal value which is fixed by the unavoidable noise signals induced into the system.

A different solution to the water depth problem can be arranged by replacing the tank potentiometer with a multisegment switch, as illustrated in Figure 8.4. The slider, moving with the float, now connects the supply voltage to one only of the eight output lines shown. Thus, in the figure, a voltage V will exist between line 5 and earth, while all the other lines will have, theoretically, zero voltage. Of course, there will be the unavoidable noise voltage on all the lines, but in order to detect which segment of the switch

Digital Systems 249

the float is connecting, we need only to be able to distinguish the presence of the signal voltage V, and this can be achieved even with noise voltages almost as large as the supply voltage V itself. This solution to the problem is a digital solution, and although the system proposed is a relatively crude one, it does illustrate the fundamental difference between analogue and digital systems. That is, that whereas in the analogue system, we must be able to detect small changes in the analogue output voltage, in the digital system we are only interested in detecting the presence, or absence, of the voltage V.

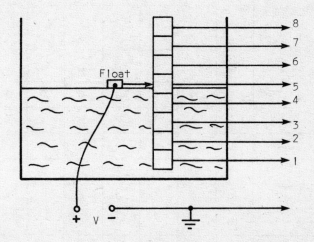

Figure 8.4

In fact, since the magnitude of the voltage V contains no information, (it must, of course, be larger than the interfering noise voltages), it is common practice to denote the presence of the voltage by the code 1 and the absence of the voltage by the code 0. The operation of the system can then be discussed without reference to the size of the voltage in any way. In fact, digital systems are quite common in which the signal is not a voltage but an air pressure or water pressure. The connecting wires are then, of course, pipes. However, whatever the nature of the actual signal, the use of the two codes 0 and 1 represents the absence or presence of the signal quantity.

The use of an eight segment switch in the water tank only allows the division of the full depth range into eight levels. The amplitude range is said to be 'quantised'. In place of the depth uncertainty in the analogue system, due to noise, we have introduced a depth uncertainty, due to the quantising error. Of course, we can reduce the quantising error by using a switch with more contacts, but a limit to this improvement will be provided again by sticking friction between the slider and the switch contacts.

Having decided how many contact segments are to be used, we can draw

DEPTH RANGE ABOVE MINIMUM LEVEL	SIGNAL ON LINES							
	Line 1	Line 2	Line 3	Line 4	Line 5	Line 6	Line 7	Line 8
$\frac{7}{8} - 1$	0	0	0	0	0	0	0	1
$\frac{3}{4} - \frac{7}{8}$	0	0	0	0	0	0	1	0
$\frac{5}{8} - \frac{3}{4}$	0	0	0	0	0	1	0	0
$\frac{1}{2} - \frac{5}{8}$	0	0	0	0	1	0	0	0
$\frac{3}{8} - \frac{1}{2}$	0	0	0	1	0	0	0	0
$\frac{1}{4} - \frac{3}{8}$	0	0	1	0	0	0	0	0
$\frac{1}{8} - \frac{1}{4}$	0	1	0	0	0	0	0	0
$0 - \frac{1}{8}$	1	0	0	0	0	0	0	0

Figure 8.5

up a table which lists the signals available on the output lines at each quantising level. This table is shown in Figure 8.5. We can see that the depth range, from one half to five eighths of the full range depth above the zero level, is represented by the 'BINARY WORD'

0 0 0 0 1 0 0 0

This is said to be a 'binary' word because the signal on each line can only take one of the two binary values 0 or 1. Each line signal is represented by one 'binary digit' or bit. The word is, therefore, an eight bit word. The process of assigning a binary word to

represent each possible output value is termed 'encoding'.
One obvious drawback to the system described is the fact that 8
signal lines are required to transmit the depth information. A
system in which each bit of the code word is transmitted at the same
instant is said to be a parallel system, and requires as many signal
channels as there are bits in the complete word. A parallel data
system is obviously the fastest way to convey the data, since the
time required to transmit the complete word is the same as the time
required to transmit one bit. An alternative, cheaper method, is to
convert the parallel signal into a 'serial' signal, i.e. the individual bits are transmitted one at a time, in time sequence, down a
single channel. Although cheaper, the serial system is slower,
because the time required to transmit the complete word is equal to
n times the time required to transmit each bit, where n is the
number of bits in the word. There is, therefore, a choice between
speed and cost. If it is decided to use a serial transmission
method with the water depth transducer, than a parallel to serial
data converter would be required, the output from the transducer
described being inherently a parallel signal.

Figure 8.6

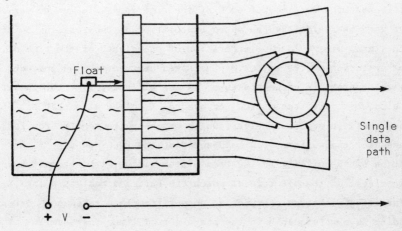

Figure 8.6 shows one method by which the serial data signal can be
produced. The eight output lines from the transducer are connected
to the eight contacts on a rotary switch which is motor driven. As
the rotating contact performs one complete revolution, it contacts in

turn each of the eight lines, starting at line 1, and ending at line
eight. The signal voltages on the eight lines, (i.e. either 0 volt
or V volt), appear in sequence upon the single output data line.
Figure 8.7 shows, for each of the eight possible depth ranges, the
resulting serial code word produced by the depth transducer.

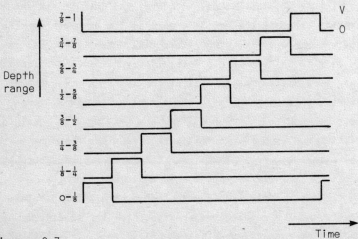

Figure 8.7

Notice that the duration of each of the time intervals in the serial
code word is decided by the speed of rotation of the rotating switch.
Each signal bit would have a duration of one-eighth of the time of
one revolution of the switch. The speed of transmission in such
digital systems is often spoken of as the bit rate, i.e. the number
of bits which are transmitted per second. Notice also that, in
order that the signal receiver, which may be quite remote from the
transducer/transmitter, can make sense of the received signal, it
must be synchronised with the transmitter timing, i.e. it must have
a knowledge of the bit rate at which the serial signal is being
transmitted. Let us consider an output from the transducer corresponding to a water depth in the range $\frac{7}{8}$ to 1 times full scale. The
data word transmitted would be

0 0 0 0 0 0 0 1

The word would exist on the data line as a 0 signal for seven
consecutive bit periods, followed by a 1 signal for one bit period.

Unless the receiver is aware of the length of a bit period, it will be unable to determine the number of 0 signals intended. For this reason digital systems are often timed by a system clock, which produces a regular sequence of pulses which are used to synchronise events. An additional data line is often used to distribute the clock signal throughout the system.

One other point which is important in serial systems, is a knowledge, by the receiver, of the order in which the word is being transmitted. Consider again the transducer transmitting an output corresponding to a water depth in the $\frac{7}{8}$ to 1 times full scale range. If the direction of rotation of the rotary switch is now reversed, the transmitted serial word will change to

$$1\ 0\ 0\ 0\ 0\ 0\ 0\ 0$$

giving an ambiguity between the upper and lower depth ranges. We can thus see that, for the transmitted word to be meaningful to the receiver, the system as a whole must be aware of

    (a) the transmitted bit duration, and

    (b) the order of transmission of bits.

The inverse of the bit duration is known as the bit frequency, or bit rate, and is equal to the clock frequency of the system. For the eight bit system described, a total of eight bit periods would be required, in order to transmit one word. This finite word time would form a limit to the number of readings of water depth which could be made in any given time interval. A water depth transducer is, of course, inherently a relatively slowly operating system, and the speed of operation would be limited more by the mass, stiffness, etc. of the float potentiometer, than by the speed of the data transmission system. However, if the serial digital data system is used with a very fast physical variable providing the data input, the ultimate limit to the speed will probably be provided by the time required to convert the value of the data variable into a digital word. A digital data system, because of the time required to transmit each coded data value, is inherently only capable of transmitting sample values of the data input. The basic principle is illustrated in Figure 8.8, which shows an analogue data signal being sampled, the time required to obtain a sample being t_s. Once the

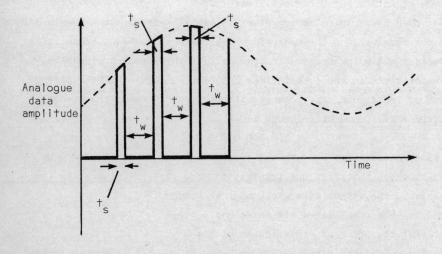

Figure 8.8

sample value has been obtained, the value can be converted into a digital word form, this taking the time t_w. The total conversion process thus takes a time

$$t_c = t_s + t_w$$

During the conversion time the analogue signal will, in general, change its value due to its normal variation. The conversion process is thus subject to a sampling error. It is obviously impossible to represent exactly a continuously varying quantity by fixed amplitude samples of finite duration.

Returning to the analogue system, is there a similar limit to the speed of obtaining readings? Remember that the output signal is a voltage whose magnitude is proportional to the water depth. Obviously again, the water level system is inherently a slowly operating system. However, if an analogue transducer is used with a very fast physical variable, the ultimate limit to the speed may again be provided by the speed of response of the data transmission system. In an analogue system this means the speed at which the signal voltage on the data line may be changed. This in turn will be decided by the properties of the line, i.e. the resistance, inductance and capacitance of the lines themselves. Because, however, in a digital system each transmitted digital word may consist of several changes of signal voltage, the digital system is inherently a slower method. In choosing to use a digital system, we have traded speed of response for relatively better immunity from electrical interference.

8.2 Encoding the amplitude samples - number systems and codes

In the simple digital encoder described, eight bits were used to represent the eight amplitude segments into which the full range of the depth recorder was divided. The use of eight bits to represent eight sample values is very inefficient.

Bearing in mind that each binary digit can take one of two values, i.e. 0 or 1, then a digital word of n bits can represent 2^n different values. The power of 2 table shown in Figure 8.9 indicates the range capability of digital words of various bit lengths.

No. of bits in word n	Range capability $= 2^n$
1	2
2	4
3	8
4	16
5	32
6	64
7	128
8	256
9	512
10	1024
11	2048
12	4096

Figure 8.9

It can be seen that a 10 bit word has a capability of $2^{10} = 1024$ different amplitude samples, which will allow an accuracy of slightly better than 1 in 1000, i.e. of approximately 0.1%. The eight amplitude samples of the water depth transducer could have been represented by a three bit word, since

$$2^3 = 8$$

Figure 8.10 shows the eight different values which may be represented by a 3 bit word. Listed in the order shown, the words form a code known as simple binary code. For use with the water depth transducer, the lowest segment would, perhaps, be represented by the binary word 000, while the highest segment, corresponding to the depth range $\frac{7}{8}$ to 1 times full scale could be represented by the binary word 111.

The use of a three bit word code would allow a considerable increase in the speed of transmission of the digitally coded data. However, the encoder used, i.e. the device which actually produces the digitally coded output word would, of necessity, be slightly more complex.

The simple binary code can be expressed in a general form as

$$N = a_0 2^0 + a_1 2^1 + a_2 2^2 + a_3 2^3 + \ldots a_n 2^n \qquad 8.1$$

in which the constants $a_0 \ldots a_n$ can take either of the values 0 or 1.

It is seen that the word is formed as the sum of a series of powers of 2 and is commonly written with the highest power of 2 at the left hand side, following the method used normally for decimal numbers. Thus the binary equivalent of decimal 15 is written as

$$1\ 1\ 1\ 1$$

i.e.

$$N = 1 \times 2^0 + 1 \times 2^1 + 1 \times 2^2 + 1 \times 2^3$$
$$= 15$$

The left hand bit is the 'most significant bit', since it involves the highest power of two in the word, while the right hand bit is the least significant bit.

DECIMAL EQUIVALENT	3 BIT BINARY WORD		
0	0	0	0
1	0	0	1
2	0	1	0
3	0	1	1
4	1	0	0
5	1	0	1
6	1	1	0
7	1	1	1

Figure 8.10

As with decimal numbers, it is possible to deal with numbers of magnitude less than unity by using a binary point.
Thus the binary word
$$1\ 0\ 1\ 1\ .\ 1\ 0\ 1$$
is understood as (starting with the MSB)
$$N = 1 \times 2^3 + 0 \times 2^2 + 1 \times 2^1 + 1 \times 2^0 + 1 \times 2^{-1} + 0 \times 2^{-2} + 0 \times 2^{-3}$$

$$= 8 + 0 + 2 + 1 + 0.5 + 0 + 0.125$$

$$= 11.625 \text{ decimal}$$

Although decimal and binary number systems are in quite common use, it is, of course, possible to develop number systems using any base. In general, the decimal equivalent of a number in a system of base r is
$$N = a_0 r^0 + a_1 r^1 + a_2 r^2 + \ldots\ldots a_n r^n$$
in which the constants $a_0 \ldots\ldots a_n$ may take any of the values 0 up to

$(r - 1)$.

It is good practice, when dealing with numbers in different bases, to indicate the base applicable to the number as a subscript. Thus

$$1010_2 = 10_{10} = 12_8$$

The great interest in the binary system, as far as digital system engineers are concerned, lies in the fact that only two levels, or characters, are used.

In the decimal system, the ten characters 0 to 9 occur. In the binary system, however, only the characters 0 and 1 occur, and these are conveniently characterised by the presence or absence of a voltage, or the opening or closing of a switch.

In this country we have, over the years, used in our normal life a combination of number systems of different bases. For example, in addition to the decimal system, used for numerical calculations, we have employed the base 12, for shilling and pence, 12 for feet and inches, 3 for feet and yards, 20 for pounds and shillings, and so on. However, notice that even in this jumble of systems, we are not consistent, since, for example, in counting pence from 0 to 11, we have persisted in using decimal characters. The two numbers 10 and 11 are decimal numbers, which is why, of course, they each require two digits to write down.

Digital system engineers use, in addition to the binary system, the octal system (base eight), and the hexadecimal system (base 16). In writing down numbers in the hexadecimal system, the characters 0 to 9 are used together with six other characters. Any symbols, of course, may be used, but commonly the first six letters of the alphabet are employed.

The hexadecimal code, and its decimal equivalent, are listed in Figure 8.11.

It is interesting to note that one reason, often offered as an advantage of the decimal system over other systems, is that arithmetic is simpler in the decimal system, this assumption being based upon the fact that in the decimal system multiplication by ten is accomplished by moving the decimal point one place to the right. It is, however, true that multiplication by ten in a number system of any base is accomplished by moving the point one place to the right.

8.2.1 Conversion between bases

(a) Binary to decimal conversion
This process has already been considered. However, to recapitulate, the decimal equivalent is obtained by summing the decimal equivalent of each power of two.

HEXADECIMAL VALUE	DECIMAL EQUIVALENT
0	0
1	1
2	2
3	3
4	4
5	5
6	6
7	7
8	8
9	9
A	10
B	11
C	12
D	13
E	14
F	15

Figure 8.11

Example
Convert 110110_2 to its decimal equivalent.
110110_2 = 0 x 2^0 + 1 x 2^1 + 1 x 2^2 + 0 x 2^3 + 1 x 2^4 + 1 x 2^5
       =   0   +   2   +   4   +   0   +   16   +   32
       = 54_{10}

(b) Decimal to binary conversion

This is best performed by a process of repeated division by 2.

Example

Convert 185_{10} to its binary equivalent

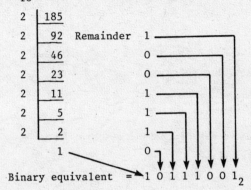

Binary equivalent = 10111001_2

(c) Binary to octal conversion

This conversion is often used if a relatively long binary number is to be remembered, even for a very short time, for example, in order to enter the number by keyboard into a digital system. The conversion is very easy, because of the special relationship which exists between the bases 2 and 8 (i.e. $8 = 2^3$).

The binary word is written with the bits separated slightly into groups of three bits. Note that if the word does not divide exactly into groups of three, zeros may be added at the 'most significant bit' end to complete the groups. Each group of three is now replaced by its octal equivalent, as given in Figure 8.10. (Note that the octal and decimal equivalents of 3 bit binary numbers are identical.)

Example

Convert the binary word

$$110000011111_2$$

into its octal equivalent.

```
110 000 011 111
 6   0   3   7
```

Octal equivalent = 6037_8

(d) Binary to hexadecimal conversion

A similar relationship exists between the bases 2 and 16 ($2^4 = 16$).

In this case the binary word is split into groups of four.

Example

Convert the binary word

$$0100111101_2$$

into its hexadecimal equivalent.

The binary word has ten bits. To divide it into groups of four bits, two zeros must be added at the left hand (MSB) end.

```
        0001  0011  1101
          1     3     D
```

Hexadecimal equivalent = $13D_{16}$

Remember that the sixteen hexadecimal digits are the numbers 0 to 9 and the six letters A to F.

8.2.2 Arithmetic using the binary number system

The rules applied in performing simple arithmetic operations using the binary system are the same as those familiar from decimal arithmetic. However, any difficulties which arise from an inability to remember the 2 to 10 times tables in decimal arithmetic do not exist in binary arithmetic, as, of course, only the characters 0 and 1 exist.

(a) Addition

Remember, in performing additions, that 1 + 1 = 10, which gives a 0 in the sum and a 1 to carry into the next higher column.

Example

Perform the addition 101011 + 1001

```
      101011     THE AUGEND
  +     1001     THE ADDEND
      110100     THE SUM
```

(b) Subtraction

In subtraction, where a borrow must be made from the next higher column, the borrow has the value of two.

Example

Perform the subtraction 101101 - 11010

```
    101101       THE MINUEND
     11010       THE SUBTRAHEND
     10011       THE DIFFERENCE
```

(c) Multiplication

In multiplication, each partial product formed consists of 0 or 1 times the multiplicand. The partial products are, therefore, equal to zero, or to the multiplicand shifted left by the required number of places.

Example

Perform the multiplication 110101 x 101

```
        110101       THE MULTIPLICAND
    x      101       THE MULTIPLIER
        110101
        000000       THE PARTIAL PRODUCTS
        110101
     100001001       THE PRODUCT
```

(d) Division

In division, the method is the same as for decimal long division. However, the divisor will divide into each section of the dividend either 0 times or 1 times.

Example

Divide 100011 by 101

```
                 111       THE QUOTIENT
THE
DIVISOR   101  100011      THE DIVIDEND
               101
               00111
                 101
                 0101
                  101
                  000      THE REMAINDER
```

8.2.3 Negative numbers

Two methods are in common use whereby both positive and negative numbers may be represented in a binary code. Both methods are illustrated, for 8-bit codes, in Figure 8.12.

SIGN + MAGNITUDE METHOD

	b_7	b_6	b_5	b_4	b_3	b_2	b_1	b_0	
by SIGN BIT	0	1	1	1	1	1	1	1	$+127_{10}$
	0	1	1	1	1	1	1	0	$+126_{10}$
by = 0 +IVE				·					
by = 1 -IVE				·					
				·					
	0	0	0	0	0	0	0	1	+1
	0	0	0	0	0	0	0	0	+0
	1	0	0	0	0	0	0	0	−0
	1	0	0	0	0	0	0	1	−1
				·					
				·					
				·					
	1	1	1	1	1	1	1	1	-127_{10}

TWOS COMPLEMENT METHOD

	b_7	b_6	b_5	b_4	b_3	b_2	b_1	b_0	
by SIGN BIT	0	1	1	1	1	1	1	1	$+127_{10}$
	0	1	1	1	1	1	1	0	
b_7 = 0 +IVE				·					
b_7 = 1 -IVE				·					
				·					
	0	0	0	0	0	0	0	1	+1
	0	0	0	0	0	0	0	0	0
	1	1	1	1	1	1	1	1	−1
	1	1	1	1	1	1	1	0	−2
				·					
				·					
				·					
	1	0	0	0	0	0	0	0	−128

Figure 8.12 Representation of positive and negative numbers

In both methods, the same conventions are assumed:-

   a) The MSB of the binary number is used as a sign indicator.

   b) The MSB value is to be 0 for a positive number and 1 for a negative number.

In the SIGN + MAGNITUDE method, the bits b_0 to b_6 give the magnitude of the number coded in simple binary. Although this is an obvious and simple method it has disadvantages.

If the code for +1 is added to the code for -1, then we obtain:-

```
+1        0 0 0 0 0 0 0 1
-1        1 0 0 0 0 0 0 1
-2        1 0 0 0 0 0 1 0
```

i.e. it appears that the result is -2.

In fact because the sign + magnitude code is not a weighted code (see Section 8.2) it is difficult to develop rules which allow arithmetic to be performed.

This difficulty is overcome by the use of the TWO'S COMPLEMENT method.

For positive numbers, both methods are identical. However, the code for -1 is obtained by subtracting 1 from the code for 0. This method is continued, with each succeeding negative number being obtained by subtracting 1 from the preceding number.

This removes the difficulty in the SIGN + MAGNITUDE method of having two codes to represent zero. The number range of the 8-bit TWO'S Complement is therefore -128 to +127.

Using the same example again

```
+1              0 0 0 0 0 0 0 1
-1              1 1 1 1 1 1 1 1
         1  ←   0 0 0 0 0 0 0 0
         carry
```

then, as long as a 'carry' is allowed from the 8-bit number into the non-existent ninth bit, the result obtained within the 8 bits is correct.

The two's complement method allows simple arithmetic because it is a weighted code; the weightings of the bits b_0 to b_6 are the simple binary weightings of 2^0 to 2^6. The weighting of the sign bit is 2^{-7}.

A simple rule for writing down the 2's complement of a binary number is as follows:-
'Write down the same binary number starting at the LSB (right hand) end, until and including the first logical 1 value. After the first 1, invert all values.

Example

Write down the two's complement of the 16-bit binary number
$$0\ 1\ 1\ 0\ 1\ 0\ 1\ 1\ 0\ 1\ 1\ 0\ 1\ 0\ 0\ 0$$
The two's complement is
$$1\ 0\ 0\ 1\ 0\ 1\ 0\ 0\ 1\ 0\ 0\ 1\ 1\ 0\ 0\ 0$$

8.2.4 Other possible binary codes

The simple binary code is not the only three bit binary code which could be used for encoding the amplitude levels. In fact, any arrangement of the three bit characters would be usable provided that each amplitude level has a unique coding, and, of course, also provided that a knowledge of the code used is available at both the transmitter and the receiver.

The simple binary code is an example of a 'weighted' code, by which it is meant that each bit in the code has a definite weighting assigned to it. Thus, each bit in the simple binary code has a weighting which is two times the weighting of the bit on its right hand side. The choice of code used will, however, affect the design of the encoder, and perhaps of the data transmission circuitry, and certain codes may have definite advantages, in cost or in reliability. A code which is often used in such applications is shown in Figure 8.13. This code, known as Gray binary code, is listed in the figure in 4 bit form. If required in 3 bit form, for example, to code the eight levels of the water depth transducer, then the most significant bit will not be required. Code listings such as this are best remembered by looking for the pattern which exists in the code. Thus the least significant bit pattern consists of a sequence of four zeros and four ones, but starting with two zeros. The number of zeros and ones in the pattern increases for each bit as a power of two.

Digital Systems 267

EQUIVALENT DECIMAL VALUE	4 BIT GRAY CODE
0	0 0 0 0
1	0 0 0 1
2	0 0 1 1
3	0 0 1 0
4	0 1 1 0
5	0 1 1 1
6	0 1 0 1
7	0 1 0 0
8	1 1 0 0
9	1 1 0 1
10	1 1 1 1
11	1 1 1 0
12	1 0 1 0
13	1 0 1 1
14	1 0 0 1
15	1 0 0 0

Figure 8.13

The Gray binary code is an example of a special class of codes which are cyclic and which also change, between adjacent code characters in only one bit position. The use of this type of code gives advantages in certain types of transducer.
Figure 8.14 shows a shaft position encoder which uses an array of light sensitive detectors to produce the electrical output, each detector having its own output line. The digitally coded output is thus available in parallel form. The light sensitive detectors are illuminated by an equivalent set of lamps which are directed through a transparent cylinder. An encoding pattern is formed photographically upon the cylinder, obscuring certain detectors in each position of the cylinder. The use of a four bit code, giving only $2^4 = 16$ different code patterns would, of course only allow an angular discrimination of 360/16 = 22.5°.

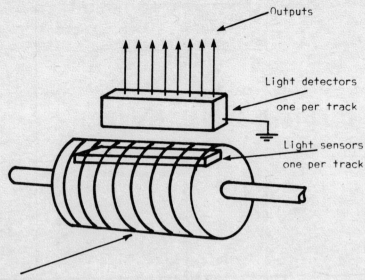

Figure 8.14

However, the use of a ten bit code, requiring 10 lamps and detectors, will allow an angular discrimination of 360°/1024, i.e. of approximately one third of one degree. What are the advantages of the GRAY code over simple binary code for such an application? Consider first the method of operation of the encoder. In any position of the shaft certain of the light detectors will be illuminated. As the shaft is turned the output word will change to a new value. If all the bits do not change to their new values at exactly the same instant, a set of false output words will occur before the output settles to give the new correct value. Examination of the GRAY code shows that between each adjacent code, in only one bit is a change valid. Multiple output changes, which would thus occur in a practical design, may, therefore, be ignored by the system. The code also is cyclic, in that there is only one bit difference between the first and last code characters.
Comparison with the equivalent change in the simple binary system will show that in returning from the last to the first code

character all bits in the code are required to change.
A further common form of binary code is the Binary Coded Decimal
(B.C.D.) code. This code is often used where the decimal form
of the coded number is to be retained. The decimal characters
0 to 9 are coded using simple binary code. These characters are
then used to replace the decimal characters. The decimal number
346_{10} would then be coded as

$$0011 \quad 0100 \quad 0110$$
$$3 \quad\quad 4 \quad\quad 6$$

Four binary bits are required for each decade in the decimal number.
The code is, therefore, less efficient in the use of bits than
is the simple binary, for twelve bits are required to represent
numbers up to 999_{10}.
Taking a specific example, to represent the decimal number 346
in simple binary,

i.e. $346_{10} = 11011010_2$

eight bits are required, while 12 bits are required if binary
coded decimal representation is used.

8.3 Mathematical treatment of coded signals

Boolean algebra

The original work of the mathematician George Boole, describing
propositions whose result is either true or false, has been applied
to signals in which the variable can take one of two values.
The Boolean algebra is now extensively applied to binary coded
signals and forms a major tool for engineers employed in digital
system design. The basic rules and identities are easily described,
with reference to the opening or closing of sets of switch contacts.
Thus the two binary values 0 and 1 can be used to represent a
switch open (non conducting) or closed (conducting), or the absence
or presence of a voltage, or the absence or presence of a current.
The interpretation can be extended to include the absence or presence
of all the necessary conditions for the success of a particular
process.

270 Basic Electrical Engineering and Instrumentation for Engineers

8.3.1   Boolean variables, operations and identities

Let us consider the simple circuit of Figure 8.15

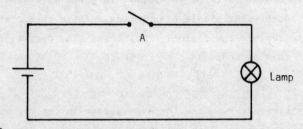

Figure 8.15

This consists of a battery, a switch A and a lamp. If we denote
the action of lighting the lamp by the Boolean variable F, which
will take the value 1 if the lamp is lit and 0 if the lamp is
unlit, then we can write

$$F = A \qquad (8.2)$$

where A is another Boolean variable, which takes the value 0 if
the switch is open, and the value 1 if the switch is closed.
An extension of this method defines the complement or negation
of a variable. Thus the negation of the variable A is written
as $\bar{A}$ (sometimes A'). If the variable A has the binary value 0
then its complement $\bar{A}$ has the value 1 and vice versa.

Two basic operations must also be clearly defined. Consider the
circuit of Figure 8.16, in which the operation of the lamp is
controlled by two switches, A and B, in series. Again, denoting
the lighting of the lamp by the variable F, then we can write

$$F = A \text{ AND } B$$

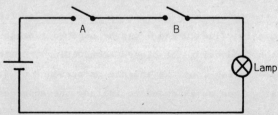

Figure 8.16

It is clear that both switches must be closed for the lamp to light. The logical operation AND is denoted in Boolean equations by a dot, although in common practice the dot is often omitted. Thus the equation would be written

$$F = A.B$$

or more commonly

$$F = AB \qquad 8.3$$

When a Boolean equation, written in this manner is read, it must be read as F equals A AND B.

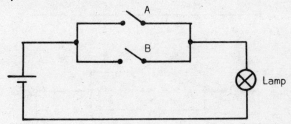

Figure 8.17

An alternative operation is denoted by the electrical circuit of Figure 8.17, which shows a lamp controlled by two switches A and B connected in parallel. In this case, closure of either of the switches will result in the lamp lighting. The circuit is a realisation of the logical OR function and the logical equation is

$$F = A \text{ OR } B$$

The OR function is usually written

$$F = A + B \qquad 8.4$$

where the + sign denotes the OR function. Again it is necessary when reading the equation to read the + sign as OR and to make sure that no confusion with the arithmetical version of the + sign occurs. The OR and the AND are the basic logical functions and quite complex switching circuits may be represented in Boolean form using these functions, together with the system variables in their normal and complemented forms.

It is also common practice to use brackets to link together sections of equations which are to be ANDed together. Thus the equation

$$F = A + BC$$

represents the circuit of Figure 8.18(a), while the equation
$$F = A(B + C)$$
represents the circuit of Figure 8.18(b).

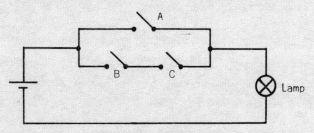

Figure 8.18(a)

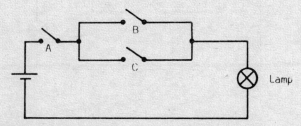

Figure 8.18(b)

In Figure 8.18(a) either A OR both B AND C are required to be closed for the lamp to operate, while in Figure 8.18(b) switches A AND either B OR C are to be closed.

Using these basic functions, we can now specify some basic identities

$$A + 0 = A \qquad (1)$$
$$A + 1 = 1 \qquad (2)$$
$$A \cdot 0 = 0 \qquad (3)$$
$$A \cdot 1 = A \qquad (4)$$
$$A + A = A \qquad (5)$$
$$A \cdot A = A \qquad (6)$$
$$A + \bar{A} = 1 \qquad (7)$$
$$A \cdot \bar{A} = 0 \qquad (8)$$
$$\bar{\bar{A}} = A \qquad (9) \qquad 8.5$$

The validity of these identities may be checked by checking their truth for all possible values of the variables. It also helps to visualise the meaning by drawing the equivalent electrical circuit, remembering that the value 0 represents an open circuit, while 1 represents a short circuit. The equivalent electrical circuit for each identity is shown in Figure 8.19.

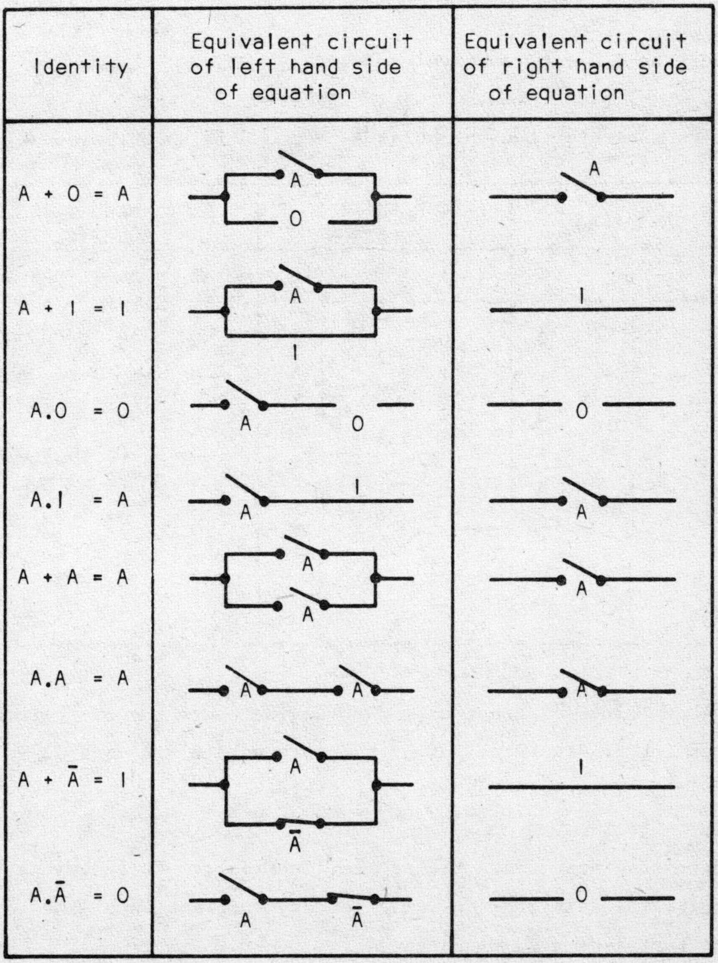

Identity	Equivalent circuit of left hand side of equation	Equivalent circuit of right hand side of equation
$A + 0 = A$		
$A + 1 = 1$		
$A.0 = 0$		
$A.1 = A$		
$A + A = A$		
$A.A = A$		
$A + \bar{A} = 1$		
$A.\bar{A} = 0$		

Figure 8.19

For identities involving a larger number of variables, it becomes useful to list the possible values of the variables in an ordered manner to facilitate checking each possible combination. Such

an ordered listing of all possible values of the system variables
is called a TRUTH TABLE. For any logical equation, therefore,
the equality may be checked by drawing its truth table. It is
very convenient, when listing the combinations of variables, to
do so according to the simple binary code.

Example

Consider the Boolean equation

$$A(B + C) = AB + AC \qquad 8.6$$

The truth table for the variables A, B and C is shown below.

VARIABLES			LEFT HAND SIDE OF EQUATION		RIGHT HAND SIDE OF EQUATION		
A	B	C	(B + C)	A(B+C)	AB	AC	AB+AC
0	0	0	0	0	0	0	0
0	0	1	1	0	0	0	0
0	1	0	1	0	0	0	0
0	1	1	1	0	0	0	0
1	0	0	0	0	0	0	0
1	0	1	1	1	0	1	1
1	1	0	1	1	1	0	1
1	1	1	1	1	1	1	1

For the two sides of the equation to be identical the two columns
representing the two sides, i.e. the columns A(B + C) and AB +
AC should be identical for all combinations of variables. This
is seen to be true.

Two functions whose truth tables are identical, i.e. functions
which have the same Boolean value for any combination of the input
variables are identical functions.

This example also illustrates that the technique of removal of
brackets by 'multiplying out', familiar from simple arithmetic,
is also applicable to Boolean equations. There are, however,
steps which are valid in the manipulation of Boolean equation,
which have no counterpart in simple arithmetic.

Example

Use a truth table to prove the validity of the Boolean equation

$$A + BC = (A + B)(A + C) \qquad 8.7$$

The two columns headed $A + BC$ and $(A + B)(A + C)$ are identical, and the validity of the equation is proved.

Comparison of the equations proved in these two examples is interesting

$$A(B + C) = AB + AC$$
$$A + BC = (A + B)(A + C)$$

VARIABLES			LEFT HAND SIDE OF EQUATION		RIGHT HAND SIDE OF EQUATION		
A	B	C	BC	A + BC	A + B	A + C	(A+B)(A+C)
0	0	0	0	0	0	0	0
0	0	1	0	0	0	1	0
0	1	0	0	0	1	0	0
0	1	1	1	1	1	1	1
1	0	0	0	1	1	1	1
1	0	1	0	1	1	1	1
1	1	0	0	1	1	1	1
1	1	1	1	1	1	1	1

These two equations illustrate the dual nature of the two Boolean operations, the logical AND and the logical OR. The second equation is obtained from the first, by replacing each AND operation with an OR operation, and also by replacing each OR operation with an AND operation. The removal of the brackets by multiplying out in the first equation also has its dual in the second equation. Thus the right hand side may be obtained by 'anding out' the left hand side of the equation, i.e. by ANDING together the logical OR of the variable A with each of the variables in the second term in turn. This operation has no counterpart in simple arithmetic, but the use of the dual nature of the logical operations AND and OR in Boolean arithmetic is of great value. Although the use of the Boolean operation may seem strange at first, familiarity is soon obtained with a little practice in the manipulation of equations.

One further very useful manipulation tool will be introduced at this point. This is the theorem due to De Morgan. This may be stated in equation form as

$$\overline{A + B} = \overline{A}\,\overline{B} \qquad 8.8$$

$$\overline{AB} = \overline{A} + \overline{B} \qquad 8.9$$

The truth of these two forms of De Morgan's Theorem should be checked using the truth table method.

In words, the theorem states that to negate a logical expression of two or more variables, each variable should be negated, and the logical operation changed (i.e. AND becomes OR and vice versa). The method will be illustrated with one or two examples.

Example

Negate the function $F = A\overline{B}$

Applying De Morgan's theorem directly

$$\overline{F} = \overline{A\overline{B}} = \overline{A} + B$$

Example

Negate the function $F = AB + CD$

Here De Morgan's theorem must be applied twice.

First consider AB and CD to be single variables.

Then

$$\overline{F} = \overline{AB + CD} = \overline{AB} \cdot \overline{CD}$$

Applying De Morgan a second time

$$\overline{F} = \overline{AB} \cdot \overline{CD} = (\overline{A} + \overline{B}).(\overline{C} + \overline{D})$$

Example

Negate the function $F = A + \overline{B} + \overline{CD}$

Applying De Morgan's theorem

$$\overline{F} = \overline{A + \overline{B} + \overline{CD}}$$
$$= \overline{A} \cdot B \cdot \overline{\overline{CD}}$$

The theorem may be applied yet again to the two variable term $\overline{CD}$ to give

$$\overline{F} = \overline{A} \cdot B \, (C + \overline{D})$$

8.4 Electronic gates

The Boolean arithmetic has been explained in terms of the interconnection of various switch contacts. In modern digital systems, however, the use of moving mechanical contacts would result in systems with a very slow maximum speed of operation. As a result of the requirement for faster operation, electronic switches, or gates as they are usually termed, have been developed which are capable of operating in times as short as several nanoseconds. These have enabled the design and production of the very fast and complex digital systems of today, such as the digital computer. Over recent years, various technologies have been used in the manufacture of electronic gates, resulting in the availability of several series or families of gates, the members of each family being compatible with each other in terms of the required power supplies, and the voltage or current levels at the inputs and outputs of the gate. Simple interconnections between arrays of gates will, therefore, quite easily allow the realisation of very large switching systems. Although it will be instructive to consider briefly the principles of one or two of the various gate families, the internal construction of each gate may be very complex. When using the gates the mechanism of their internal operation is of little interest to the systems designer, whose aims are affected only by their external characteristics. Thus the user will wish to know the specified power supply voltages, the time required for each gate to operate correctly, the FAN OUT, i.e. the number of similar gates which may be safely driven from the output of the gate, and the FAN IN, i.e. the number of similar gate outputs which may be safely connected to the gate input. The gates are available in integrated circuit form in units which may contain up to eight separate gate circuits in one package. Each individual gate circuit may be an extremely complex design whose price is very low only because the units are made and sold in enormous quantities. The design and construction of the gates is, therefore, a matter for the large semiconductor manufacturing companies only. With the advent of the large scale integration (LSI) techniques, it is commonplace for

278 Basic Electrical Engineering and Instrumentation for Engineers

complex arrays of gates to be produced and sold on one chip, so providing the designer with large segments of systems which he may interconnect to realise the complete system.

8.4.1 Positive and Negative Logic

Before proceeding to examine various gate circuits, we must decide the meaning of the terms logic 0 and logic 1 when applied to actual circuits. In the previous sections we have discussed the logic operations in a general mathematical way, without reference to the actual voltage or current levels represented by the logic 0 or 1. When looking at actual gate circuits, however, the 0 and 1 levels are specified in terms of the actual voltage or current which is required at the gate input in order to cause the gate to switch correctly.

Switching systems may be described in terms of either positive or negative logic, depending whether the logic 1 level is more positive or more negative than the logic 0 level.

Figure 8.20(a) shows several typical logic voltage levels, each of which is a positive logic system, while Figure 8.20(b) shows

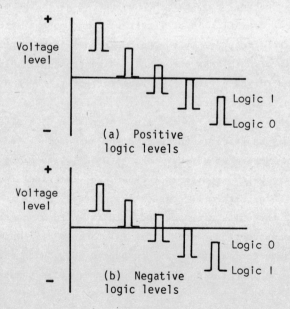

Figure 8.20

possible negative logic voltage levels. Notice that in a positive
logic system both the 0 and 1 logic voltages may be negative with
respect to the system earth or zero line; however, the 1 level will
be less negative (more positive) than the 0 level.

Instead of the terms logic 1 or logic 0, the terms HIGH and LOW
are often used. Thus, in changing from the logic 0 to the logic
1 level, a gate output is said to go from LOW to HIGH. Further,
if the normal resting state of the output is in the LOW state,
and if transition to the HIGH state initiates some further step
in the overall system, the gate is said to have an active HIGH
output. With the large scale integration (LSI) circuits using
the MOS technology, which are now widely available, and which
will be described in a later section, the active signal level
is often the low state.

8.4.2 Diode logic

The earliest form of electronic gate used the unidirectional conducting property of a diode to produce two distinct circuit conditions.
Figure 8.21 shows a positive logic diode AND gate with three inputs.

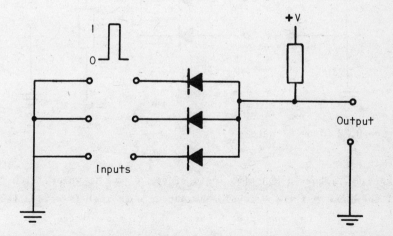

Figure 8.21 Diode AND gate

Its action may be explained as follows:

With all inputs at or near the earth potential, all the input diodes will be held in the conducting state, with the total current flowing to earth, via the diodes being limited by the resistor R. The output voltage is thus held at a potential more positive than the logic 0 level of the input by a voltage equal to potential across the diodes in the conducting condition. This potential is normally of the order of 0.5 to 0.7 volt. If one input is taken to the HIGH or logic 1 state, the diode connected to that input will become non-conducting, while the output potential will still be held in the low state by the remaining diodes, which are still conducting. Thus, only if all the inputs are taken to the logic 1 level, will the output level rise positively to the logic 1 level. The circuit thus performs the logical AND function for positive logic inputs.

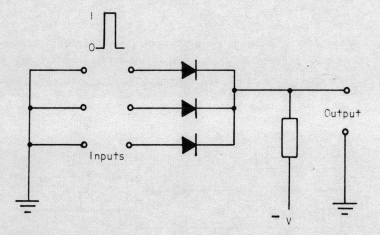

Figure 8.22 Diode OR gate

Figure 8.22 shows a positive logic diode OR gate in which, with all the inputs in the 0 level, the output potential is held slightly less positive than the input 0 level, due to the voltage drop across the diodes which are all conducting. If any input is raised in potential to the logic 1 level, the output potential will also rise to a value slightly less positive than the input logic 1

level, all the other input diodes becoming reverse biassed. The
gate thus performs the logical OR function.

One interesting point can be noted from Figure 8.23, which shows
the diode AND gate of Figure 8.21 with a negative logic input.
In other words the logic 0 level of Figure 8.21 is called the
logic 1 level in Figure 8.23. Note that the actual signal voltages
are not altered, only the logic level name assigned to it.

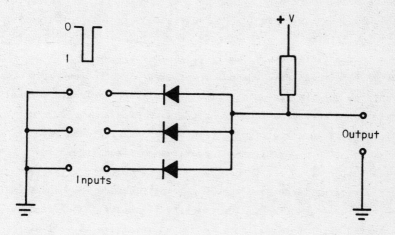

Figure 8.23 Positive logic diode AND gate - negative logic inputs

In Figure 8.23, if all the inputs are at logic 0 (which in a negative
logic system means they are all at their most positive value),
then the output potential will also be at its most positive value,
i.e. at logic 0. If, however, any input is taken negative to
its logic 1 value, the output voltage will also be taken negative
to its logic 1 value. The gate now is functioning as an OR gate.
Thus, the function performed by any circuit depends also upon
the definition of the logic levels assigned at its inputs and
outputs.

What then are the disadvantages of the simple diode gates described?
The main disadvantage derives from the use only of passive devices,
i.e. devices which have no power gain. Without gain, the signal
power at the output of the gate must be less than that at its

input, due to the unavoidable power loss in the gate. As can
be observed from a consideration of the method of operation of
the gate, the low and high voltage levels are not well defined,
and vary between input and output by almost one volt, due to the
voltage drop in the conducting diode. This problem is made rapidly
worse if similar gates are cascaded in order to perform more complex
Boolean functions, a process which is obviously necessary in any
but the most simple digital system. The result is that in practice
the low and high voltage levels rapidly converge until there is
little discrimination between them, resulting in very unreliable
operation. A satisfactory family of gates would, in practice,
have well defined logic 0 and 1 levels, whereby the manufacturer
is able to specify reasonably fine limits about the nominal levels
within which correct switching of the gate is guaranteed. Another
obvious disadvantage of the gates described is the need for the
provision of both positive and negative power supplies.

The inclusion of an active device in the circuit, to give satisfactory power gain, leads also to a major modification in the switching
theory already described. The majority of amplifying devices,
in their simpler forms, also give signal inversion as well as
amplification. Figure 8.24 shows a natural extension to the diode
logic gate, in which a common emitter mode transistor amplifier

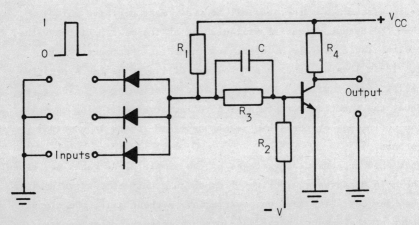

Figure 8.24 Positive logic AND gate with inverting amplifier

is added to give power gain. The amplifier also gives signal
inversion. In the figure, the AND gate is formed by the diodes
and the resistor R_1. The transistor inverting amplifier is coupled
to the AND gate with the resistors R_2 and R_3, the negative supply
voltage being included to ensure that the transistor is completely
non-conducting unless the AND gate output is in the HIGH state.
The capacitor C is included to improve the speed of operation
of the gate.

Let us now examine the logical function performed by an AND gate,
followed by an inverting amplifier. Considering a two input system
for simplicity, the truth table is as follows. The function performed is the NOT AND or NAND function, which is defined, for
a three input gate as

$$F = \overline{A.B.C.} \qquad 8.10$$

The output from a positive logic NAND gate is thus HIGH unless
all the inputs are HIGH, when the output is LOW.

INPUTS		OUTPUT FROM AND GATE	INVERTED OUTPUT FROM AND GATE
A	B		
0	0	0	1
0	1	0	1
1	0	0	1
1	1	1	0

The addition of a similar inverting amplifier to the OR gate results
in the NOT OR or NOR function. The NOR function is defined, for
a three input gate, as

$$F = \overline{A + B + C} \qquad 8.11$$

The truth table for a two input NOR function is as shown.

INPUTS		A + B	OUTPUT
A	B		$\overline{A + B}$
0	0	0	1
0	1	1	0
1	0	1	0
1	1	1	0

The output from a positive logic NOR gate is thus LOW unless all
its inputs are low, when the output is HIGH.

8.4.3 More complex logic families

The four logic functions already described, the AND, the OR, the
NAND and the NOR, form the basis of the logic families commonly
in use at the present time. The circuits of the gates are, however,
more complex than those shown in Figures 8.21 to 8.24.

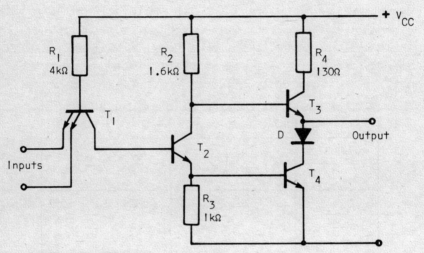

Figure 8.25 TTL NAND gate

Figure 8.25 shows the circuit of a TTL (transistor-transistor
logic) NAND gate. This family of gates employs a specially produced
multiple emitter transistor for the inputs. The operation of
the circuit is as follows. With all the inputs (emitters of transistor T_1) in the logic 1 state, the current in resistor R_1 flows
through the base-collector diodes which are forward biassed.
The resulting current flows into the base of transistor T_2,
saturating the transistor. Transistors T_3 and T_4 form an output
buffer stage, providing the gate with a very low output impedance
in both the LOW and HIGH states. The voltage developed by the
emitter current of T_2 across the resistor R_3 causes transistor

T_4 to be saturated, while the low voltage at T_2 collector ensures that transistor T_3 is switched off, i.e. into its high impedance state.

When the input at any emitter of transistor T_1 is taken into the LOW (logic 0) state, transistor T_1 is switched into the conducting state. Provided the input voltage is sufficiently low the collector voltage of T_1 will also be sufficiently low to prevent transistors T_2 and T_4 from conducting. The collector of T_2 will hence be at a high potential, ensuring that T_3 is in the saturated low impedance state. The output potential thus switches between the LOW and the HIGH states. When any input emitter of T_1 is taken into the LOW state, the driving circuit must be capable of accepting (sinking) the emitter current. This means that the emitter current must be able to pass to earth, via the driving stage output impedance. The gate described is typical of the 7400 family of TTL gates which are a standard in use at the present time. These gates operate from a single 5 V power supply, and have a logical 0 output impedance of 12 ohms, and a logical 1 output impedance of 70 ohms for the standard gates. The gates change state as the changing input voltage passes through a threshold value of approximately 1.4 volts. The output voltage is typically 3.3 volts in the logical 1 state, and 0.2 volt in the logical 0 state. The output can, therefore, tolerate typically 1.9 volts of negative going noise in the 1 state, and 1.4 volts of positive going noise in the 0 state, before falsely triggering the gate it is driving. Switching time for these gates is of the order of 10 nanoseconds. The gates are available in dual in line (DIL) packages, with either 14 or 16 pins. Depending upon the number of pins required per gate, one or more gates may be provided in one package. Thus, allowing 2 pins for the power supply connections, 4 - 2 input NAND gates (3 pins each) can be provided on a 14 pin package. Alternatively, one 8 input NAND can be provided on a 14 pin package. This family of TTL gates is produced by several different manufacturers. The standard gates in this family have a FAN OUT of the order of 10, i.e. the output will drive correctly the input of 10 similar gates.

The TTL gates are an example of saturated transistor logic in that the transistors used therein are used in either the saturated or the cut off states. Transition between those two states is arranged to be as rapid as possible. This is the most efficient method when looked at from the power dissipation point of view, because the cut off and the saturated states are the states in which the power dissipation in the device is a minimum. However, from the point of view of speed of operation, an improvement may be made by preventing the transistors from switching between these two extreme levels. The operating voltage swing is restricted to a very small level, so minimising the amount of charge which must be supplied to, or extracted from the device. Emitter coupled logic gates use this technique to provide a very fast system (propagation delay times typically of the order of 2 nanoseconds or less), at the expense of increased power dissipation. Use of this family of gates is, therefore, normally restricted to the application requiring the high switching speed.

The emitter coupled logic gate circuit is based upon the circuit of the differential amplifier shown in Figure 8.26. The circuit is completely symmetrical so that, with equal input and reference voltages, the potentials at the two transistor collectors are

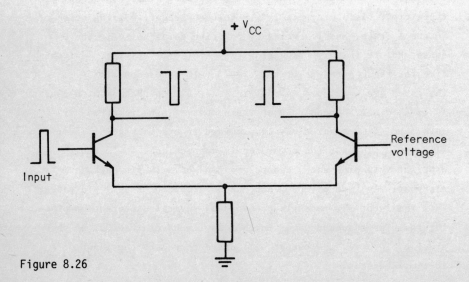

Figure 8.26

equal. When used as a differential amplifier, this is the normal
mode of operation of the circuit. If, however, the input voltage
is raised above that of the reference, the collector voltage of
the input transistor will increase. If the input voltage is reduced
below the reference voltage, the opposite effects occur. The
current in the emitter resistor remains virtually constant. The
application of a step input voltage change will, therefore, result
in voltage steps at the transistor collectors, the changes at
the collectors being in antiphase.

To allow multiple inputs to the circuit, several input transistors
are connected in parallel. The gate input, therefore, performs
the OR function, and the collector voltages provide both the OR
and the NOR (= $\overline{OR}$) outputs.

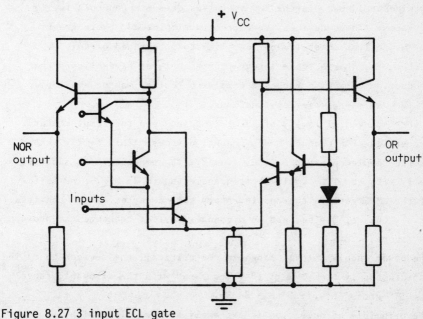

Figure 8.27 3 input ECL gate

In Figure 8.27, a typical 3 input ECL gate is shown. This circuit
is basically the same as that of Figure 8.26, with added refinements
to improve its operation. Three input transistors are operated
in parallel, while the output from each collector is taken via

an emitter follower circuit, whose functions are to give a reduced
output impedance, so giving an improved FAN OUT, and also to assist
in making the output voltage levels equal to the input levels,
this, of course being necessary if the gates are to be cascaded.
The reference voltage input is also provided with an emitter
follower circuit, together with a diode whose function is to improve
the temperature stability of the gate.

In a typical commercially available gate, the supply voltage is
5.2 volts, while the logic 0 and logic 1 voltages are 2.95 volts and
3.70 volts, a logic voltage swing of 0.75 volt. Other names sometimes given to this type of gate are 'current steering logic'
(CSL) or 'current mode logic' (CML).

A more recent development, which is now readily available and
in widespread use, is the MOS (metal-oxide-semiconductor) family
of gates. These devices are extremely simple, and their great
advantage lies in their applicability to integrated circuit
techniques. Large scale integration techniques have allowed the
production of a wide range of complex circuits, making available
a wide variety of digital functions.

A very successful family of gates is based upon the complementary
properties of p channel and n channel devices. This family, known
as the complementary MOS (cMOS) family, is competing most success-
fully with the TTL family at frequencies up to 25 MHz. The basis
of the cMOS gate is the complementary pair inverter, which consists
of a p channel device, and an n channel device, connected as shown
in Figure 8.28.

The power supply voltage range is not critical, the devices operating
satisfactorily from 3 V to 18 V (compared with the allowable range
for TTL gates of 4.75 V to 5.25 V).

The principle of operation is extremely simple. If a high input
voltage is applied to the inverter, the n channel device turns
on, giving a low impedance from the output to earth. The p channel
device is off giving a high impedance to V_{DD} the supply voltage.
For a low input voltage the opposite conditions apply; the output
voltage is virtually at the V_{DD} value.

Digital Systems   289

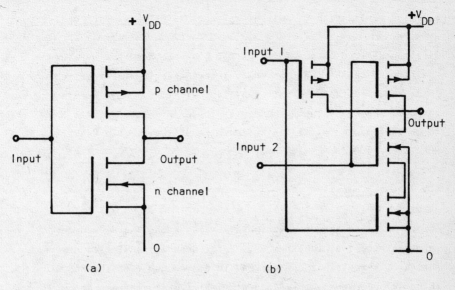

Figure 8.28 (a) cMOS invertor (b) 2 input cMOS NAND gate

The main advantage of the CMOS family is the low power dissipation. The d.c. current drain for a CMOS inverter is in the low nanoampere range. The noise immunity of the family is of the order of 30% of the value of the supply voltage V_{DD}.

The devices are available in dual in line (DIL) packages of up to 40 pins, although the smaller gate circuits are supplied in the common 14 or 16 pin forms.

The relative characteristics of the 3 families of gates are summarised in the table.

LOGIC TYPE	RELATIVE COST	TYPICAL PROPAGATION DELAY nsec	POWER DISSIPATION mW	SPEED-POWER PRODUCT pJoules	NOISE MARGIN volts	FAN OUT
TTL	MEDIUM	6	20	120	1.0	10
ECL	HIGH	2	150	300	0.4	25
cMOS	VERY LOW	15	0.0001	0.005	$0.3\ V_{DD}$	50

The product
          speed (nanoseconds) x power dissipation (mW)
with the unit of picojoules, is taken as a Figure of Merit for a gate system.

290 Basic Electrical Engineering and Instrumentation for Engineers

8.5 The realisation of switching functions using simple logic gates

Once the desired output function for a system has been obtained, the circuit can be constructed by interconnecting the required gates selected from the gate families described in the previous section. Depending upon the family chosen, the design will have to be implemented mainly in the more readily available NAND or NOR logic.

8.5.1 Gate symbols

In drawing gate interconnection diagrams, symbols are used to represent the individual gates. Unfortunately, a single universal series of symbols has not emerged in common use with the result that, in order to be able to deal with literature from various sources, an engineer must become familiar with several systems. The system which appears most likely to become universally adopted is the American Military Specification (AM.MIL.SPEC), being used by the majority of large digital system manufacturers. This system, although more complex to understand initially, has definite advantages when dealing with large scale systems, and is the system used in logic diagrams in this book. Figure 8.29 summarises the most common systems.

8.5.2 AND/OR logic

Let us consider the realisation of the Boolean expression
$$F = AB + CD$$
using simple AND and OR gates. The term AB may be realised directly, using a two input AND gate, as can the second term CD. The outputs from the two AND gates will then form the inputs to a two input OR gate whose output is the required function. This realisation is shown in Figure 8.30. The function is thus directly realisable using AND gates as the input gates followed by OR gates as the output gate. The Boolean expression is said to be in AND/OR form.

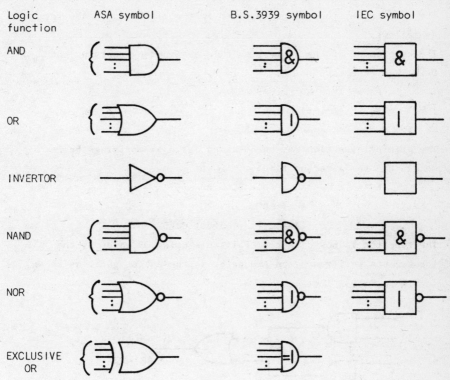

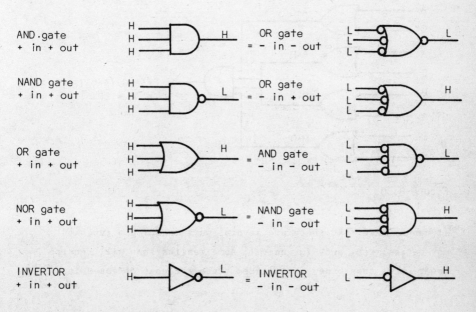

American Military Spec. The same gate will perform different functions depending upon the logic definition at the inputs and the outputs.

Figure 8.29 Gate symbols

The same expression may be obtained in the different OR/AND form as follows.

The complement of the expression F is obtained by applying the De Morgan theorem.

$$\overline{F} = \overline{AB + CD} = \overline{AB} \cdot \overline{CD}$$
$$= (\overline{A} + \overline{B})(\overline{C} + \overline{D})$$
$$= \overline{AC} + \overline{AD} + \overline{BC} + \overline{BD}$$

The original function may be obtained again by applying the De Morgan theorem a second time

$$\overline{\overline{F}} = F = \overline{\overline{AC} + \overline{AD} + \overline{BC} + \overline{BD}}$$
$$= \overline{\overline{AC}} \cdot \overline{\overline{AD}} \cdot \overline{\overline{BC}} \cdot \overline{\overline{BD}}$$
$$= (A + C)(A + D)(B + C)(B + D)$$

The function is now directly realisable using OR gates as the input gates followed by an AND gate, as the output gate, as shown in Figure 8.30.

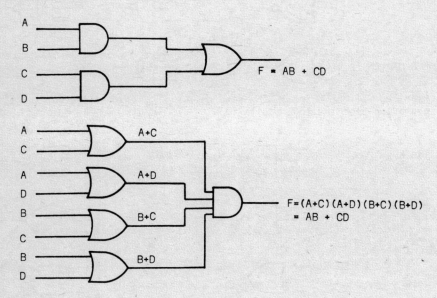

Figure 8.30

It can be seen that there are several ways in which a function may be realised, and, in general, some realisations will require fewer gates than others. In fact, since the cost of gates is to a great extent dependent upon the number of pin connections, the number of gate inputs used in a realisation is a good guide to the relative cost of different circuits. To realise the function $F = AB + CD$ will require 6 gate inputs for the AND/OR realisation, but 12 gate inputs for the OR/AND realisation.

The process of simplification or reduction of Boolean expressions to different forms is often termed minimisation; the property of the function which is being minimised will, however, depend upon the actual design problem.

<u>Example</u>

Realise the function

$$F = A\bar{B} + \bar{A}B$$

This is directly realisable in AND/OR form as shown in Figure 8.31(a). The function can be converted to the OR/AND form by the application of De Morgan

$$\bar{F} = \overline{A\bar{B} + \bar{A}B}$$

$$= \overline{A\bar{B}} \,.\, \overline{\bar{A}B} = (\bar{A} + B)(A + \bar{B})$$

This may be simplified by multiplying out the brackets.

$$\bar{F} = A\bar{A} + \bar{A}\bar{B} + AB + B\bar{B}$$

This may be simplified, remembering that

$$A\bar{A} = 0 \text{ and } B\bar{B} = 0$$

$$\bar{F} = \bar{A}\bar{B} + AB$$

Since F represents the lamp being lit, then the complement $\bar{F}$ must represent the lamp being unlit. The lamp is unlit if neither of the switches A and B are operated, or, alternatively, of both switches are operated.

The original function may now be obtained by a further application of De Morgan

$$F = \bar{\bar{F}} = \overline{\bar{A}\bar{B} + AB} = \overline{\bar{A}\bar{B}} \,.\, \overline{AB}$$

$$= (A + B)(\bar{A} + \bar{B})$$

The function can now be realised directly in OR/AND form, as shown in Figure 8.31(b).

In these realisations, the inputs A and B are required in both the normal (A) and the complemented ($\bar{A}$) forms. Invertors would, therefore, have to be included if the complemented variables are

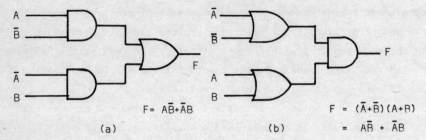

Figure 8.31

not directly available from the system inputs. The two realisations, in this case, are seen to require the same number of gate inputs. As previously mentioned, the function

$$F = \bar{A}B + A\bar{B}$$

is frequently used in digital systems, and is known as the EXCLUSIVE OR function. It is so commonly used that it is made available as a gate in its own right in most logic gate families. The symbol representing the 'exclusive or' gate is also shown in Figure 8.29.

Example

Simplify the function

$$F = A + \bar{A}B$$

The method of solution of this simplification is not obvious, but the example is included as this expression is often met in more complex problems.

$$F = A + \bar{A}B = A(B + \bar{B}) + \bar{A}B$$

The addition of the factor $B + \bar{B}$ does not change the expression because

$$B + \bar{B} = 1$$

and

$$A \cdot 1 = A$$

The new form of the expression may be expanded to give

$$A + \bar{A}B = AB + A\bar{B} + \bar{A}B$$

$$= AB + A\bar{B} + \bar{A}B + AB$$

The addition of the extra term AB does not change the expression because

$$AB + AB = AB$$

The expression may now be factorised to give

$$F = A + \bar{A}B = AB + A\bar{B} + \bar{A}B + AB$$
$$= A(B + \bar{B}) + B(A + \bar{A})$$
$$= A + B$$

Thus

$$F = A + \bar{A}B = A + B \qquad 8.12$$

This identity is a useful simplifying tool for use in more complex problems.

8.6 The Karnaugh Map

The Karnaugh map provides an alternative method for listing the output value of a Boolean expression for all possible combinations of the Boolean input variables. In this, the Karnaugh map acts in a very similar manner to the truth table in that expressions that are identical have identical Karnaugh maps also.
The map method, however, also forms the basis of a method for simplifying Boolean expressions in a systematic manner, so removing many of the difficulties which are met in simplifying Boolean expressions by the algebraic methods previously described.

8.6.1 Plotting Boolean expressions on a Karnaugh Map

A Karnaugh map consists of a set of boxes, each box representing one possible combination of the Boolean input variables. The box is marked with either a 1 or a 0 to indicate the value of the Boolean expression for the particular combination of the inputs that the box represents.
For one variable, (A), therefore, only two boxes are required, one to represent the input value A, and one to represent the value $\bar{A}$. For two variables (A and B), there are four possible combinations of variables ($\bar{A}\bar{B}$, $\bar{A}B$, $A\bar{B}$, AB). In general, for n input variables, the map will require 2^n boxes. There are, therefore, exactly as many boxes as there are entries in the equivalent truth table.

296 Basic Electrical Engineering and Instrumentation for Engineers

| $\bar{A}$ | A |

| $\overline{ABC}$ | $\overline{A}B\overline{C}$ | $AB\overline{C}$ | $A\overline{B}\overline{C}$ |
| $\overline{AB}C$ | $\overline{A}BC$ | ABC | $A\overline{B}C$ |

| $\overline{AB}$ | $A\overline{B}$ |
| $\overline{A}B$ | AB |

$\overline{ABCD}$	$\overline{A}B\overline{CD}$	$AB\overline{CD}$	$A\overline{B}\overline{CD}$
$\overline{ABC}D$	$\overline{A}BC\overline{D}$	$AB\overline{C}D$	$A\overline{B}\overline{C}D$
$\overline{AB}CD$	$\overline{A}BCD$	$ABCD$	$A\overline{B}CD$
$\overline{A}B\overline{C}D$	$\overline{A}BC\overline{D}$	$AB C\overline{D}$	$A\overline{B}C\overline{D}$

Figure 8.32

Figure 8.32 shows the Karnaugh map arrangement of the boxes for 1, 2, 3 and 4 input variables.

The maps can be extended for greater numbers of variables, but owing to the difficulty in actually drawing the maps in a useful way for greater than seven variables, the method is rarely used for expressions of this size.

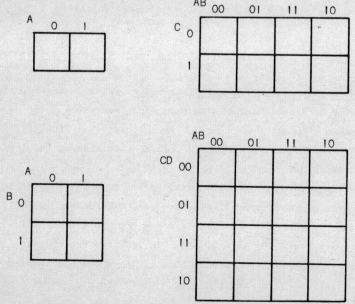

Figure 8.33 Normal labelling of Karnaugh maps

In the figure, the maps have been labelled in each box with the
particular combination of variables represented by the box. The
usual method for labelling the maps is shown in Figure 8.33, with
the individual rows and columns marked at the top or left hand
side of the map. Comparison of Figures 8.32 and 8.33 will show
that the two methods of labelling give the same result.

The maps are used by placing a 1 in each box for which the combination of variables makes the expression take the logical value
1. It is not the usual practice to enter a 0 into the remaining
boxes, although, of course, these represent the combinations of
the input variables which make the expression take the value 0.

Example

Plot on a Karnaugh map, the Boolean expression

$$F = A\bar{B}\bar{C}\bar{D} + ABCD + A\bar{B}\bar{C}D$$

The solution is shown in Figure 8.34.

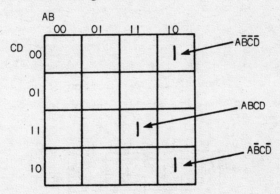

Figure 8.34

Example

Plot on a Karnaugh map the Boolean expression

$$F = \bar{A}\bar{B}\bar{C}\bar{D} + \bar{A}B\bar{C}D + A\bar{B}\bar{C}D + A\bar{B}CD + \bar{A}BCD$$

The solution is shown in Figure 8.35.

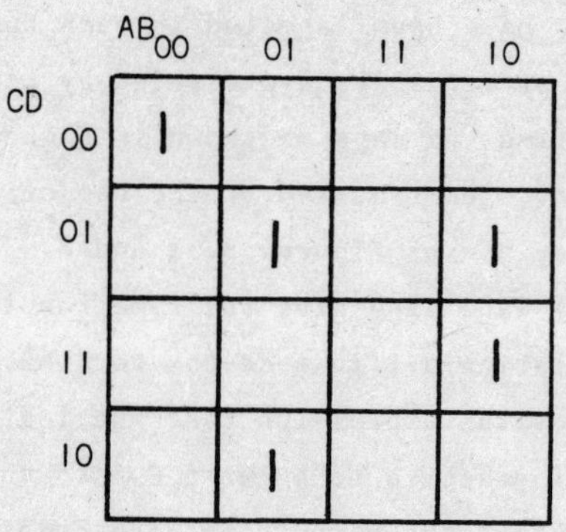

Figure 8.35

Example

Plot on a Karnaugh map, the Boolean expression

$$F = \overline{AB}\overline{C}D + BC + \overline{C}\overline{D}$$

This expression is again a 4 variable expression. However, not all the variables are included in each term. For example, the term BC indicates that all squares in which both B and C are included must contain a 1.

The solution to this example is given in Figure 8.36, from which it can be seen that there are in all 4 squares whose labels contain

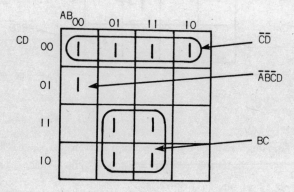

Figure 8.36

the variables B and C together. Similarly, the term $\overline{C}\overline{D}$ also represents 4 squares on the map, while the term $\overline{AB}\overline{C}D$ represents only one square. It can be concluded that, in a 4 variable problem, a term containing 4 variables represents only one square, a term containing 3 variables represents two squares, a term containing 2 variables represents four squares, while a term containing only one variable represents 8 squares.

Example

Plot on a Karnaugh map, the Boolean expression

$$F = \bar{A} + \bar{C}\bar{D} + A\bar{B}\bar{C}$$

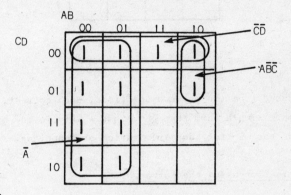

Figure 8.37

The solution to this example is shown in Figure 8.37. The three terms of the function consist of a group of 8, a group of 4 and a group of two squares. However, note that the terms overlap each other on the map. This is quite reasonable, since there are, for example, two squares on the map which contain both $\bar{C}\bar{D}$ and $\bar{A}$. These are the two squares in which $\bar{C}\bar{D}$ and $\bar{A}$ overlap.

Example

Determine the Boolean function, represented by the Karnaugh map of Figure 8.38.

The map consists of 4 squares containing 1's.
The function may be read off as

$$F = \bar{A}\bar{B}\bar{C}D + AB\bar{C}D + ABCD + \bar{A}BCD$$

or, alternatively, by grouping together two 1's to give a 3 variable term

$$F + \bar{A}\bar{B}\bar{C}D + AB\bar{C}D + ACD$$

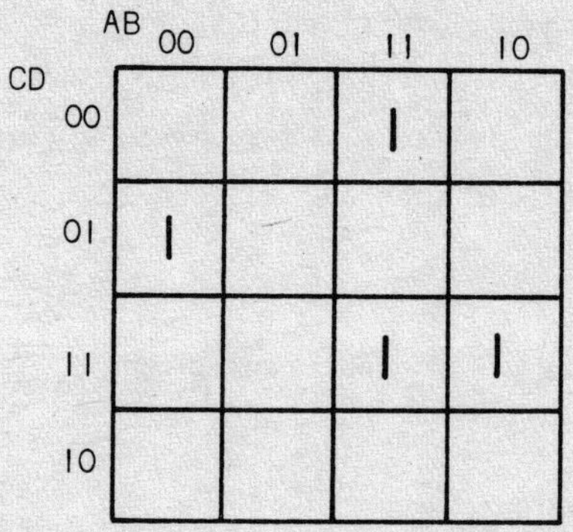

	AB 00	01	11	10
CD 00			1	
01	1			
11			1	1
10				

Figure 8.38

We have produced from the map two solutions, both correct, one being in a more simple form than the other. This is the basis of the use of the map for the minimisation of functions.

8.6.2 <u>Minimisation using the Karnaugh map</u>

Minimisation, using a Karnaugh map, is based upon the Boolean identity

$$A + \bar{A} = 1 \qquad 8.13$$

The function $F = ABCD + ABC\bar{D}$
may be minimised as follows

$$F = ABCD + ABC\bar{D} = ABC(D + \bar{D})$$
$$= ABC$$

This function of two terms is plotted on a Karnaugh map in Figure 8.39.

The minimisation is possible because it is possible to read the function from the map as either

      (a) 2 single square terms,

or   (b) 1 double square term.

This property of the Karnaugh map is possible because of the method used to label the squares.

The method used is chosen because squares next to each other in the same row or column, differ in their labelling by only one

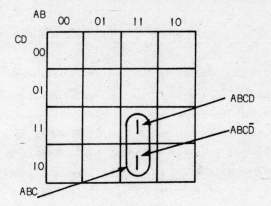

Figure 8.39

variable. Thus, in Figure 8.39, the two adjacent 1's are ABCD, and ABC$\bar{\text{D}}$, and differ only in the value of the variable D. The rows and columns of the map have been labelled according to the GRAY binary notation, which is, of course, a binary code in which adjacent characters differ only in one variable.
Consider an extension of this principle.

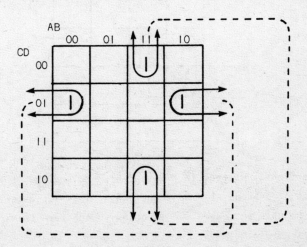

Figure 8.40

Figure 8.40 shows a Karnaugh map of the function
$$F = AB\bar{C}\bar{D} + AB C\bar{D} + \bar{A}\bar{B}CD + A\bar{B}CD$$

which minimises as

$$F = AB\bar{D}(\bar{C} + C) + \bar{B}\bar{C}D(A + \bar{A})$$
$$= AB\bar{D} + \bar{B}\bar{C}D$$

The four terms thus reduce to two, three variable terms.

From this example we see that the idea of adjacent squares is extended so that the top edge of the map is the same as the bottom edge of the map, as if the map had been rolled round a cylinder. Similarly, between the left and right hand edges there is also the same relationship as if the map had been rolled round a vertical cylinder. It is obviously difficult to visualise the map rolled round two cylinders at the same time. The map is, however, in effect a closed surface. For ease of drawing, it is, therefore, drawn in square form.

Example

Minimise, using the Karnaugh map, the function

$$F = \bar{A}\bar{B}\bar{C}D + \bar{A}\bar{B}CD + \bar{A}B\bar{C}D + \bar{A}BCD$$

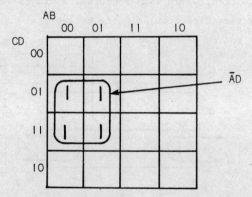

Figure 8.41

The function is plotted in Figure 8.41. We see that the function consists of a group of 4 adjacent ones. This suggests that we should be able to reduce the function to one term only.

Using algebraic methods

$$\begin{aligned}
F &= \bar{A}\bar{B}\bar{C}D + \bar{A}\bar{B}CD + \bar{A}B\bar{C}D + \bar{A}BCD \\
&= \bar{A}\bar{B}D(C + \bar{C}) + \bar{A}BD(C + \bar{C}) \\
&= \bar{A}\bar{B}D + \bar{A}BD \\
&= \bar{A}D(B + \bar{B}) \\
&= \bar{A}D
\end{aligned}$$

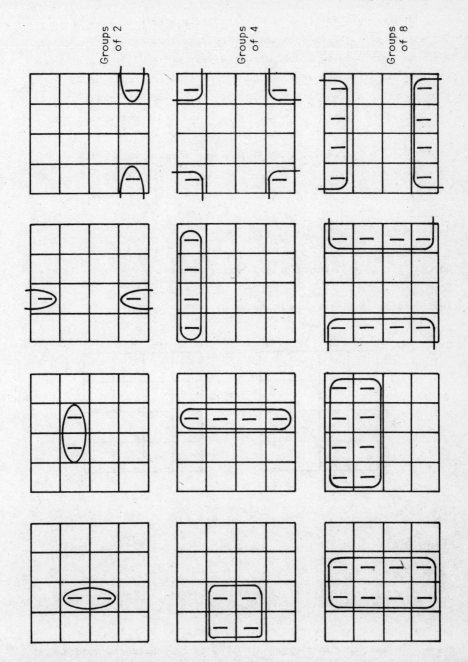

Figure 8.42 Possible groupings for 4 variable function

Minimisation on a Karnaugh map is thus only a task of looking for groups of adjacent ones which can be grouped together. Notice that ones can only be grouped together in powers of 2. On a 4 variable map of 16 squares, we can join together groups of 2, 4 or 8 ones. A group of 16 would, of course, mean that all squares are ones which would mean that the function is itself always equal to 1.

Figure 8.42 summarises the possible types of groupings for a 4 variable function.

Finally, let us extend the method to a 5 variable function.

Example

Minimise the function

$$F = ABCDE + ABCD\overline{E}$$

Obviously this will minimise algebraically to

$$F = ABCD(E + \overline{E}) = ABCD$$

The function is plotted in Figure 8.43.

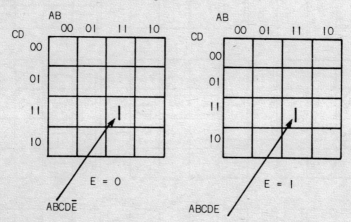

Figure 8.43

For a 5 variable problem we need extra squares; in fact we need $2^5 = 32$ squares.

The map consists of two sets of 16 squares, as shown in Figure 8.43. The two ones in the map are also to be considered adjacent. This can be better visualised in the 3 dimensional drawing of Figure 8.44, in which the two 4 variable maps have been positioned,

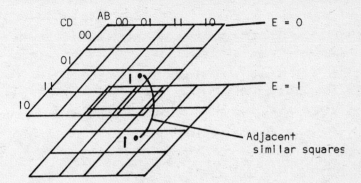

Figure 8.44

one on top of the other. Similar squares on each map are now adjacent to each other.

Many actual problems can, however, be solved, using up to 6 variables. Minimisation, using algebraic methods, can be quite awkward to apply in some cases. The Karnaugh map method will, with a little practice reduce the task to one of trivial proportions.

Example

Minimise the Boolean function

$$F = \overline{A}\overline{C}\overline{D} + \overline{A}B\overline{C}D + BCD + BC\overline{D}$$

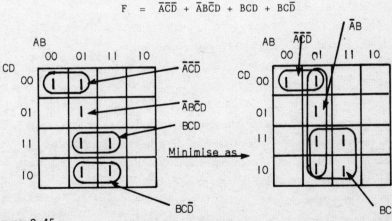

Figure 8.45

The function is plotted in Figure 8.45 and minimises to

$$F = \overline{A}\overline{C}\overline{D} + \overline{A}B + BC$$

Note, in this example, that it is perfectly correct, in grouping

together adjacent ones, to use a one in more than one grouping.
A circuit to realise this function would thus require

    1   3 input   AND   gate
    2   2 input   AND   gate
    1   3 input   OR   gate

Inverters may again be required if the negated forms of the input variables are not already available.

8.6.3 'Don't care' conditions

In certain problems, involving a number of input variables, certain combinations of the variables may never occur. If a certain input combination is known to never occur, the network output may be allowed to take any output value for this combination of the inputs. In certain cases, this technique can allow simpler network realisations. Such an input variable combination is described as a 'dont't care' condition. The technique is, perhaps, best explained by means of an example.

Example
Minimise the Boolean function

$$F = \bar{A}\bar{B}\bar{C}D + AB\bar{C}D + \bar{A}BCD \qquad 8.14$$

The input combination ABCD will never occur.
A Boolean function, including 'don't care' conditions, is often written

$$F = \bar{A}\bar{B}\bar{C}D + AB\bar{C}D + \bar{A}BCD + \{ABCD\}_{\text{x or } \phi} \qquad 8.15$$

where the terms included in the bracket are the 'don't care' terms. The bracket usually has the symbol x or ϕ against it. When the function is plotted on a Karnaugh map, the 'don't care' terms are again marked with either a x or with the ϕ symbol.
A plot of the function is shown in Figure 8.46.
If the 'don't care' term is ignored, the minimisation will require two two-term loops, giving

$$F = B\bar{C}D + \bar{A}BD$$

However, if the network output is allowed to be 1 for the 'don't care' term (which, of course, will never occur), a minimisation can be used of one four term loop, giving

$$F = BD$$

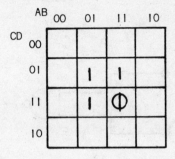

Figure 8.46

This technique can produce significant savings in the realisation of functions.

8.7 Realisations using NAND or NOR gates

We have already considered the problem of the realisation of a Boolean function, using only AND or OR gates. Any Boolean function may be realised in this way.
The majority of gates, available in the standard logic gate families, are, however, of the NAND or NOR type, and some method must be obtained which will enable the realisation of any Boolean function, using these more complex gates, with their inherent inverting action. Let us consider first whether it is indeed possible to realise all Boolean functions using the NAND/NOR gates.
Figure 8.47 shows two circuits, 8.47(a) representing a 2 input NAND gate driving into a single input NAND gate, while 8.47(b) represents a 2 input NAND gate with a single input NAND gate in series with each input.
Consider first the circuit of Figure 8.47(a).
The two input NAND gate has an output given by

$$F_1 = \overline{AB}$$

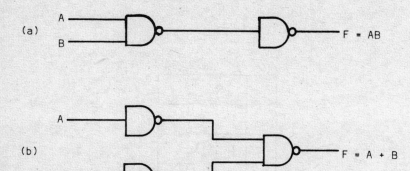

Figure 8.47

This provides the input to the single input NAND gate whose output is

$$F = \overline{F_1} = \overline{\overline{AB}} = AB$$

A single input NAND gate operates simply as an inverter. The two NAND gates of this circuit thus perform together the logical AND function.

Consider now the circuit of Figure 8.47(b).

Each input is connected to a single input NAND gate; the outputs of these gates are thus $F_1 = \bar{A}$ and $F_2 = \bar{B}$. The outputs of the two inverters form the inputs to the two input NAND whose output function is, therefore,

$$F = \overline{F_1 F_2} = \overline{\bar{A}\bar{B}} = A + B$$

The circuit of Figure 8.47(b) will thus perform the logical OR function.

Because we can use NAND gates to perform both the OR and the AND function it is perfectly possible to realise any Boolean function using NAND gates alone.

Similarly, it can be shown that the same is true using NOR gates alone. Figure 8.48 shows the NOR gate realisation of the simple AND and OR functions.

However, realisations performed by replacing AND or OR gates directly with the circuits of Figure 8.47 and 8.48 would be very inefficient indeed.

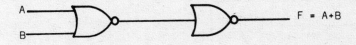

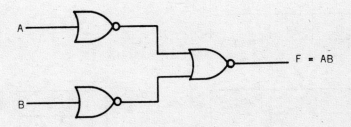

Figure 8.48

A method is required which will allow a simple derivation of NAND or NOR gate realisations.

Let us examine the action of a NAND gate more closely.

Figure 8.49(a) shows the truth table for a 2 input positive logic NAND gate, with the inputs and outputs marked as H (high) or L (low) rather than as 1 or 0.

If this same gate was used in a negative logic system, in which the level H would represent a 1 and the level L would represent a 0, (for both inputs and outputs), then the output would be at logic 1 only when both the inputs were at logic 0. This is true of the NOR function. We can conclude that the function performed by the gate is determined not only by the action of the gate, but also by the way in which the logic levels at its terminals are defined.

Figure 8.49(b) to (d) summarises the action of the positive input NAND gate with all possible definitions of the logic levels. This shows that, provided we can define, at will, the levels at the gate terminals, we can use the gate to perform any of the four basic logic functions.

(a)

Inputs		Output
A	B	F
L	L	H
L	H	H
H	L	H
H	H	L

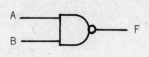

(b)

Inputs		Output
A	B	F
0	0	1
0	1	1
1	0	1
1	1	0

Input logic positive
Output logic positive
Function   NAND

(c)

Inputs		Output
A	B	F
1	1	0
1	0	0
0	1	0
0	0	1

Input logic negative
Output logic negative
Function   NOR

(d)

Inputs		Output
A	B	F
0	0	0
0	1	0
1	0	0
1	1	1

Input logic positive
Output logic negative
Function   AND

(e)

Inputs		Output
A	B	F
1	1	1
1	0	1
0	1	1
0	0	0

Input logic negative
Output logic positive
Function   OR

Figure 8.49 The action of a positive logic NAND gate

Digital Systems   311

Example

Realise the exclusive OR function using NAND gates only. The function $F = A\bar{B} + \bar{A}B$ was realised in simple AND/OR gate form in section 8.5.2. The AND/OR gate network produced is repeated in Figure 8.50(a).

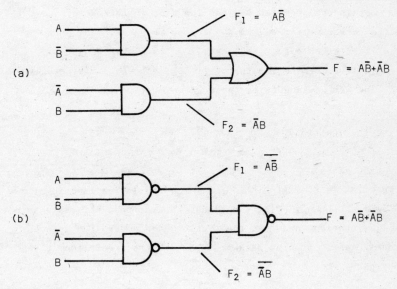

Figure 8.50

The output OR gate in this figure may be replaced by a NAND gate directly if we can define the logic levels at its inputs as negative logic. We can also use NAND gates to replace the input AND gates because, having defined the logic at their outputs as negative logic, NAND gates will actually perform the AND function. Figure 8.50(b) shows the resulting NAND gate realisation. Bear in mind, however, that as we have defined the input and output logic as positive logic, but the logic between the gates as negative, the Boolean function obtained at points between the gates will be the logical inversion of that obtained at the same point in the circuit of Figure 8.50(a).

It is common practice to call the output gate in such a network the LEVEL 1 gate, the gates which drive the LEVEL 1 gates are

called the LEVEL 2 gates and so on, working backwards towards
the input. The circuits of Figure 8.50 would thus be described
as two level networks.

A simple rule may now be stated for obtaining a NAND realisation
of a Boolean function.

RULE. Obtain the required Boolean function in AND/OR form, and
draw the network required, using AND and OR gates. The output
gate will be an OR gate. Replace all gates with NAND gates.
The logic level at the inputs to all ODD level gates is negative
logic, and all new inputs to these gates must be inverted.

Example

Realise the function

$$F = AB = C(D + E)$$

using NAND gates.

The function is directly in AND/OR form, and is realised in simple
gate form as shown in Figure 8.51(a). Notice that this is a three
level realisation, and that the type of gate alternates between
OR and AND as we proceed backwards through the levels from the
output gate.

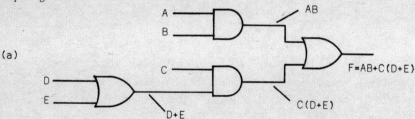

(a)

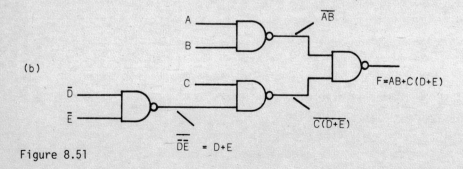

(b)

Figure 8.51

(a)

Inputs		Output
A	B	F
L	L	H
L	H	L
H	L	L
H	H	L

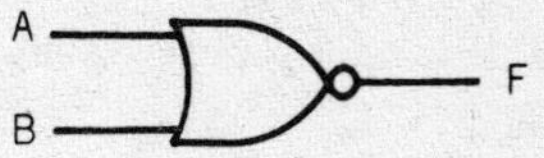

(b)

Inputs		Output
A	B	F
0	0	1
0	1	0
1	0	0
1	1	0

Input   logic positive
Output logic positive
Function   NOR

(c)

Inputs		Output
A	B	F
1	1	0
1	0	1
0	1	1
0	0	1

Input   logic negative
Output logic negative
Function   NAND

(d)

Inputs		Output
A	B	F
0	0	0
0	1	1
1	0	1
1	1	1

Input   logic positive
Output logic negative
Function   OR

Inputs		Output
A	B	F
1	1	1
1	0	0
0	1	0
0	0	0

Input   logic negative
Output logic positive
Function   AND

Figure 8.52 The action of a positive logic NOR gate

Because the output gate is an OR gate we can replace all the gates with NAND gates giving the final realisation shown in Figure 8.51(b). At the level 3 gate, however, because at the inputs to all ODD levels the logic is inverted, the input variables required are $\bar{D}$ and $\bar{E}$.

A consideration of the dual nature of the logic functions NAND and NOR will lead us to expect that a similar rule can be derived for realising functions using NOR gates only.

Figure 8.52 examines the action of a positive logic NOR gate for all the possible definitions of the logic levels at its terminals. Again, by a suitable choice of the logic level definitions, we can make the NOR gate perform any of the four simple logical operations. In particular, with a positive logic output and negative logic input, the NOR gate will perform the AND function. With negative logic output, and positive logic input, the function performed is the logical OR. NOR gates can be used, therefore, to replace directly the gates in a simple OR/AND network, provided that the output gate is an AND gate. Again, care must be taken with inputs to ODD level gates, because the logic definition at such inputs is inverted.

Example

Realise the exclusive OR function, using NOR gates only.
The exclusive OR function is

$$F = A\bar{B} + \bar{A}B$$

As shown in section 8.5.2, this AND/OR function can be converted into the OR/AND form of the same function to give

$$F = (A + B)(\bar{A} + \bar{B})$$

A simple gate realisation of this function is given in Figure 8.53(a). The gates can then be all replaced directly by NOR gates to give the solution of Figure 8.53(b).

The rule for obtaining a NOR gate realisation of a Boolean function may be stated as:

RULE. Obtain the required Boolean function in OR/AND form and realise the function using AND and OR gates. The output gate will be an AND gate. Replace all gates with NOR gates. The logic levels at the inputs to all ODD levels are inverted.

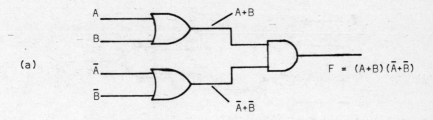

(a)

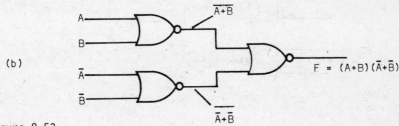

(b)

Figure 8.53

Example

Realise the function

$$F = AC + \bar{A}BC$$

in both NAND and NOR forms.

The function is already in AND/OR form, and may be realised using NAND gates to give a two level solution as shown in Figure 8.53(a). However, the function may be minimised (either algebraically, using the identity $X + \bar{X}Y = X + Y$, or, alternatively, using the Karnaugh map method), to give the result

$$F = AC + BC$$

This will allow a two level NAND realisation, with fewer gate inputs, as shown in Figure 8.54(b).

To obtain the NOR realisations, the function may be factorised

$$F = C(A + B)$$

This form of the function allows the NOR realisation using only 4 gate inputs, although, of course, it may be necessary to include an inverter to produce the inverted input $\bar{C}$ for the input to the level 1 gate. The NOR realisation is shown in Figure 8.54(c).

316 Basic Electrical Engineering and Instrumentation for Engineers

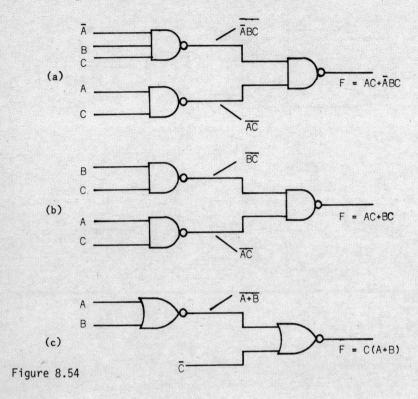

Figure 8.54

8.8 The use of the American Military Specification gate symbols

A difficulty often arises when dealing with relatively large digital system designs in deciding the function for which a particular gate is included. Figure 8.55 shows a NAND gate realisation of

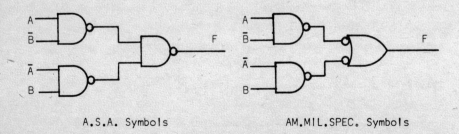

Figure 8.55

the exclusive OR function, using two different symbol schemes. Using the A.S.A. symbols, no indication is given of the fact that the output gate is performing basically an OR function upon the outputs of the other two NAND gates, which are included to perform basically an AND function upon their inputs. In trying to trace a signal through a large network drawn in this way, great difficulty would be met.

The AM.MIL.SPEC system obviates this problem by maintaining the characteristic AND gate shape for gates used basically in an AND function, and the characteristic curved OR shape for gates used in a basic OR function. The system thus allows identical gates, even gates on the same chip, to be represented using different symbols. In the system, an input or output marked with a small circle is taken as 'active LOW', while a connection not so marked is taken as 'active HIGH'. In Figure 8.55, therefore, the symbol shapes tell that the two input gates are performing an AND function upon their inputs, and that if both A and $\bar{B}$ are HIGH, then the top gate output will go LOW, while if both $\bar{A}$ and B are HIGH, the lower gate will go LOW. The output gate shape tells us that it is performing an OR function, and that if either of its inputs goes LOW, its output will go HIGH. A little practice with this system soon brings familiarity, and its use is recommended.

Example

Determine the method of operation of the network of Figure 8.56. The network consists of three similar segments, together with some output gating. Each segment consists of two input AND gates whose outputs drive an OR gate. Because one of the input AND gates is driven via inverters, one AND gate will go to active LOW if the inputs A_1 and A_2 are both HIGH, and the other AND gate will go to active LOW if the inputs A_1 and A_2 are both LOW. If either of these events occurs, the OR gate output will go HIGH. The OR gate outputs will thus indicate, by going HIGH, if the inputs A_1 and A_2 are in the same state.

The three OR gate outputs, together with the output from the fourth CONTROL input OR gate, drive an AND gate. The output will thus

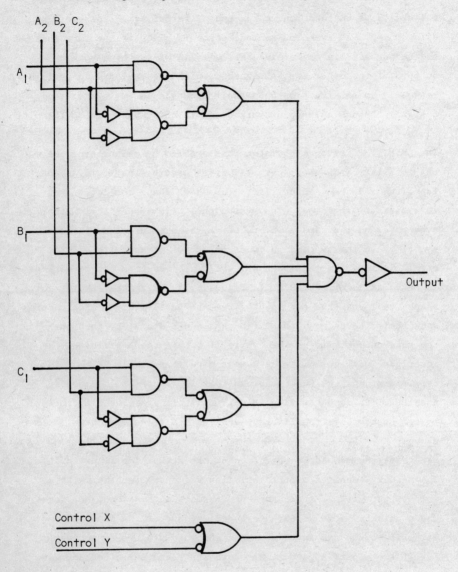

Figure 8.56

go HIGH if the 3 bit words $A_1B_1C_1$ and $A_2B_2C_2$ are identical, and an active LOW, Control X or Control Y, is received.

8.9 Sequential networks

The logic networks so far considered are examples of combinational logic networks. In a combinational network, the output is always determined by the combination of input variables present at that time. This is not exactly true, in that real gates take a finite time to change state, and invalid transient outputs may briefly occur. However, the statement is true once all transients have died away. It is possible, however, to create a gate network whose output is not decided only by the particular input logic values at any time.

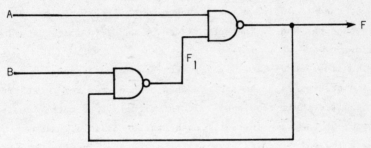

Figure 8.57

Consider the network of Figure 8.57. This is again only a two NAND gate network. In order to examine this network further, let us assume that the output F is at logical 0, and that both the inputs, A and B are at logical 1.
The output function from each NAND gate is given by

$$F_1 = \overline{BF} = \overline{B} + \overline{F}$$
and
$$F = \overline{AF_1} = \overline{A} + \overline{F_1}$$

Thus, with the values we have assumed, $\overline{F} = 1$, and hence $F_1 = 1$, giving $\overline{F_1} = 0$, and since $\overline{A} = 0$, then $F = 0$. The assumed values are, therefore, perfectly consistent.

Alternatively, let us assume, initially, that the output F is at logical 1, and that both the inputs A and B are at logical 1. Thus both $\overline{B}$ and $\overline{F}$ are at 0, giving $F_1 = 0$, and hence $F = 1$. These assumed initial values are also perfectly consistent.

We have, therefore, shown that, with $A = B = 1$, the output F may be either 0 or 1, i.e. the output value is no longer specified solely by the values applied at any time to the inputs.

This property of the network arises from the feedback connection from the output F back into an input of one of the gates. The connection allows the state of the output to modify the input variables of the network. It can be shown that the output value is now a function of the values of the input variables, but is also a function of the sequence in which the input variables are applied. In effect, the extra dimension of time has been added to the operation of the network, because, as the output value is now a function of the sequence of application of the inputs, it can be used to indicate the sequence which occurred, long after the sequence has ended. The network can be used, in fact, as a binary memory.

Let us consider an application for such a memory. Assume that a certain process is to be started when each of two remote stations have signalled their readiness. It is unlikely that both signals will occur at the same instant. Each station could signal by placing a logical 1 on its signal line. If both lines are then connected to the inputs of a 2 input AND gate, the gate output will go to logical 1, when both stations are ready, and this can start the process. This circuit is a combinational logic solution to the problem.

If, however, the operator at each station signifies his readiness by pushing momentarily a push button, only a short duration transient logical 1 will occur on the line. Some memory device must be included to remember that a transient 1 has occurred, and to provide a steady logical 1 output to drive the AND gate. This solution to the problem would be a sequential logic solution. The problem is, of course, trivial. However, the great majority of problems involving digital solutions are solved by sequential methods. The basic memory element in sequential systems is provided by one of several bistable circuits, which are so called because the circuit output will take one or two different, but quite stable

values, and can be arranged to switch between these values in
order to memorise the occurrence of some event.

A more commonly used name for this type of circuit is the 'flip
flop'. The operation of a sequential circuit is often more clearly
understood with the aid of a timing diagram. Figure 8.58 shows
the timing diagram for the process control problem.

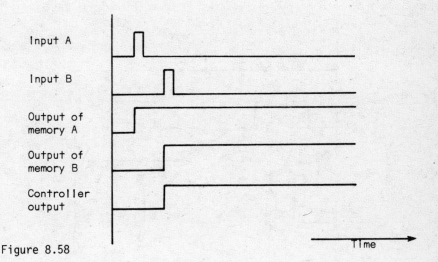

Figure 8.58

In drawing this timing diagram, we have again assumed perfect
devices. In practice, of course, propagation delays would occur
before each element assumes its new state. In a simple circuit,
the propagation delays would have little effect, and would result
only in a delay of a few nanoseconds before the output $F = 1$ is
obtained. In certain circumstances, however, in more complex
systems, such delays may cause malfunctions, or 'HAZARDS' as they
are termed, and the possible effects of delays upon any circuit
should be examined, using a timing diagram, before any design
is finalised.

8.9.1 The Set Reset bistable (flip flop)

A Set Reset flip flop is a bistable circuit which has two outputs,

322 Basic Electrical Engineering and Instrumentation for Engineers

one being the logical inversion of the other, usually labelled Q and $\bar{Q}$. It also has two inputs, the set input S being used to set the Q output to 1 ($\bar{Q}$ to 0), and the reset input R being used to reset the flip flop to the Q = 0 state.

The circuit diagram of an S R bistable using transistors is shown in Figure 8.59(a). The principle of operation of the circuit may, perhaps, be more clearly understood if it is redrawn as in Figure 8.59(b).

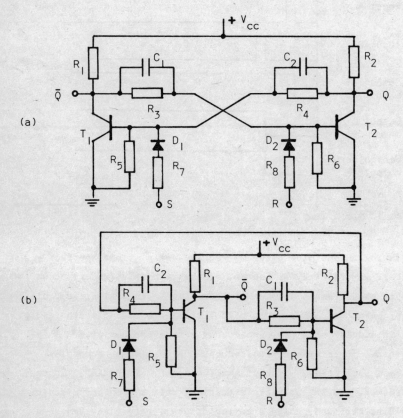

Figure 8.59 Transistor bistable memory circuit

With the exception of the two input connections, S and R, and the input components R_7, R_8, D_1 and D_2, the circuit is seen to consist of two simple transistor switching circuits; i.e. transistor T_1, with its collector load resistor R_1, and input resistor R_4, whose output is fed into a second similar circuit made up from T_2, R_2 and R_3. The two capacitors C_1 and C_2 are included to improve the switching speed of the circuit, and are known as 'speed up' capacitors. The output of the second circuit is, in turn, connected directly back into the input of the first. The transistors normally operate in either the cut off or saturated 'bottomed' states, the resistor values being chosen so that if T_1 is in the cut off state, the high voltage resulting at the collector of T_1 holds the transistor T_2 in the ON or saturated state; in turn the very low voltage existing at T_2 collector or in the saturated state holds transistor T_1 in the cut off state. The conditions in the circuit are, therefore, quite stable. They are also quite stable in the reverse condition, i.e. with T_1 the saturated transistor holding T_2 in the cut off state. These two stable possibilities form the bistable states of the circuit, either of the two states being obtainable by means of suitable triggering input signals. Transition between the two stable states occurs very rapidly since, as each transistor starts to change from the cut off state to the conducting state, or vice versa, it enters the high gain active region of operation. In the transitional state, therefore, the circuit consists of two high gain amplifiers in cascade, with output connected back into the input. This is a highly unstable condition, and the opposite stable state is very rapidly reached. The input circuits shown will allow the application of a positive control signal. Assuming positive logic, the application of a logical 1 voltage to the SET input will ensure that T_1 is switched ON and hence also that T_2 is switched OFF. Similarly, the application of a logical 1 voltage to the RESET input will switch T_2 ON and also T_1 OFF. A SET input will, therefore, result in a low voltage at T_1 collector and a high voltage at T_2 collector. The Q output connection is, therefore, taken from the collector of transistor T_2 and the $\overline{Q}$ output from the collector of T_1. The logical operation of the circuit is described by the truth table of Figure 8.60, which also shows the logic circuit symbol used to represent the network. Again, the circuit symbol makes no attempt to represent the actual circuit components used to

Inputs		Output change	
S	R	$Q \to Q_+$	$\bar{Q} \to \bar{Q}_+$
0	0	$0 \to 0$	$1 \to 1$
0	0	$1 \to 1$	$0 \to 0$
0	1	$0 \to 0$	$1 \to 1$
0	1	$1 \to 0$	$0 \to 1$
1	0	$0 \to 1$	$1 \to 0$
1	0	$1 \to 1$	$0 \to 0$
1	1	NOT ALLOWED	
1	1	NOT ALLOWED	

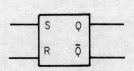

State table and logic symbol for S.R. flipflop.

Figure 8.60

create the flip flop. The box symbol, however, shows the two input connections S and R, and the two output connections, Q and $\bar{Q}$. The truth table for a sequential network is more commonly termed the state table. The state table, as in the case of the truth table, for combinational circuits, lists the output changes which occur for all possible combinations of the input variables. In fact, because, as we have just considered, the specification of the inputs to a sequential circuit does not specify uniquely the value of the output, each set of input variable values must be considered for both possible states of the output. Taking an example from Figure 8.59, the state table shows that if the input values $S = 0$ and $R = 0$ occur while the output $Q = 0$, then it will stay at 0.

Also, if the same input values occur while the output $Q = 1$, then it will stay at 1. The state table for the two input flip flop thus has eight entries.

Notice, in the state table, the use of the + subscript to denote the passage of time. Thus, Q_+ is the value of the variable

Q after the change, caused by the inputs, has finished. Similarly, $\bar{Q}_+$ is the new value of $\bar{Q}$.

It is, normally, not necessary to list, in the output column, the value of the output $\bar{Q}$, as well as the value of the output Q, because of course, $\bar{Q}$ is always the logical negation of Q (except, perhaps, during the actual switching transient).

The operation for the S.R. flip flop may be summarised as follows:

With S = 0 and R = 0, the output is not affected, and remains at its previous value.

With S = 1 and R = 0, the output Q will change to Q = 1, if it was previously at 0, and will remain at 1 if it was previously at 1. The output value Q = 1 is termed the SET condition. Notice also that the term HIGH is again often used to describe the 1 condition at any terminal.

With S = 0 and R = 1, the output Q will go LOW (to Q = 0), if it was previously HIGH (at Q = 1), and will remain at LOW, if it was previously LOW.

The input condition S = 1, R = 1 is not allowed with this flip flop. With both S and R inputs at 1, both transistors are made fully conducting, resulting in 0 values at both collectors, which form the outputs Q and $\bar{Q}$. Obviously, both Q and $\bar{Q}$ can not take the logic value 0 at the same time; this combination of the input variables must, therefore, be prevented from occurring. An S.R. flip flop may also be conveniently constructed from two two-input NOR gates, or, alternatively, two two-input NAND gates, with input inverters as shown in Figure 8.61.

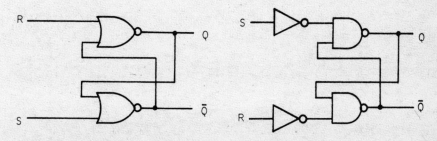

Figure 8.61 S.R. flipflops using cross coupled gates

326 Basic Electrical Engineering and Instrumentation for Engineers

As an exercise, show that the inputs S = 1 and R = 1 would, in both cases, result in the Q and $\bar{Q}$ ouputs having the same value. This condition, obviously, must not be allowed.

8.9.2 The T trigger flip flop

The trigger flip flop is another form of bistable circuit. Like other flip flops, it has two outputs, Q and $\bar{Q}$, but only one input T. The logic symbol, and the state table for the T flip flop are shown in Figure 8.62.

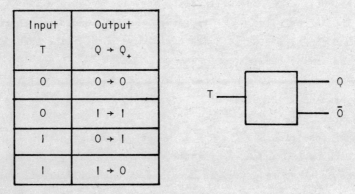

Input	Output
T	Q → Q_+
0	0 → 0
0	1 → 1
1	0 → 1
1	1 → 0

Figure 8.62 State table and logic symbol for T flipflop

The T flip flop changes state on each $\hat{T}$ input signal, and remains in that state as long as the T input remains LOW.

8.9.3 The J.K. flip flop

The S R flip flop and the T flip flop were developments of circuits commonly constructed using discrete transistors. The advent of integrated circuit techniques led to the development of the J.K. flip flop, which, since it can realise the truth tables of both the S R flip flop and the T flip flop, is the circuit in most common use in current practice.

Figure 8.63 shows the state table and the logic symbol for the J.K. flip flop. The state table is identical to that of the S R flip flop, with the exception that the input condition $J = 1$, $K = 1$ is allowed, the flip flop acting as a trigger T flip flop for these inputs. The J input can thus be considered as the SET input, while the K input acts as a RESET input.

The J K flip flop is intended to be used in a clocked (synchronous) mode of operation. The J and K inputs do not themselves initiate a change of logic state in the device, but are used as control inputs to determine the change of state which is to occur at a time decided by a pulse input to the CLOCK terminal. This mode of operation allows the precise timing of the state changes in a sequential circuit. The signal applied to the clock terminal is of pulse form, and the change of state of the flip flop will occur normally coincident in time with the rear edge of the pulse. Different designs of J K flip flop are available, however, which are clocked by a clock pulse edge of either polarity.

The J K flip flops of the TTL logic family are available in several forms, in 14 and 16 pin packages. The forms are dictated by the number of pins required to make the connections to the circuit.

Inputs		Output
J	K	$Q \to Q_+$
0	0	$0 \to 0$
0	0	$1 \to 1$
0	1	$0 \to 0$
0	1	$1 \to 0$
1	0	$0 \to 1$
1	0	$1 \to 1$
1	1	$0 \to 1$
1	1	$1 \to 0$

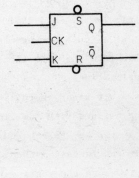

Figure 8.63 State table and logic symbol for J.K. flipflop

One or two flip flops are enclosed in a package; these may have single J and K inputs, or multiple J and K inputs to allow the logical ANDing together of input signals. They may also be provided with one or both of SET and CLEAR inputs which can be used to set output Q to 1, or clear output Q to 0 at any time. In the logic symbol of Figure 8.63, the SET and CLEAR inputs, marked S and C, are shown with a small circle against the input. This is to indicate that these inputs are activated by a LOW input, i.e. in a positive logic system by a 0 level input, and in fact, during the normal operation of the flip flop, these inputs must be held at the logic 1 level.

8.10 Registers and counters

We have seen that the bistable circuits we have considered may be used to remember that an event has happened, i.e. to remember that a certain signal has occurred at the input terminals. It can be arranged, relatively easily, that the occurrence of a logic 1 level on a certain line will set the flip flop so that its Q output is also at logic 1. The flip flop, in this way, will function as a single bit memory, memorising the fact that the logic 1 level has occured. For a binary signal of length n bits, i.e. an n bit binary word, n bistable circuits would be required to construct a word memory. When used in this manner, the bistables together are commonly called a REGISTER.

Different circuit arrangements are used to enter the value of each bit into the flip flops forming the register, depending upon the application.

If the n bit binary word is available in parallel form, i.e. available on n lines with one line for each bit, each flip flop in the register may be entered separately, usually at the same time. This is the fastest method, but since there are as many data input connections to the register as there are bits in the binary word, this is also a relatively expensive method. This form is known as a PARALLEL ENTRY register.

Alternatively, the binary word may be in serial form, i.e. available on one data line in time sequence. The values are entered serially, via the one data input connection, and would be timed into the register by a system clock. In this form the register is known as a SERIAL ENTRY or SHIFT register. The term SHIFT register arises from the action of entering one data bit into the first flip flop, which we can call A, on the first clock timing pulse; on the second clock pulse, the data contents of the first flip flop are shifted into the second flip flop (B), and the second data bit is entered into the first flip flop (A). On the third clock pulse the value in B is shifted into flip flop C, the value in A is shifted into B, and the value of the next bit, bit 3, is entered into flip flop A. The serial entry register thus requires as many operations (shift and store) as there are bits in the binary word to be stored.

Registers of the types described are available with either serial or parallel entry, and with either serial or parallel read out. Yet another type of register is the counting register, which is a number of flip flops arranged to store a binary word which represents the number of pulses applied to the input. Such counting registers can be arranged to count in any code, although, of course, counting according to the decimal code or the simple binary code is very common.

Using n flip flops, a total count of 2^n may be made. A four bit counter could, therefore, count up to 16 counts before any stored word would occur twice. The number of input pulses or counts needed to cycle a counter through its full range, and which would return the counter state to its starting value is called the MODULUS of the counter, often abbreviated to MOD. Thus a MOD 8 counter would have a count cycle of 8 counts.

Sequential networks are of two basic types; these are
(a) synchronous networks in which the state changes in the various parts of the network occur at the same instant, timed by the system clock,
(b) asynchronous networks, in which the state changes in the various

parts of the network do not occur at the same instant, and in fact may occur at a time decided by a state change in another part of the same system.

Figure 8.64 shows the state table and the timing diagram for a 3 bit counting register (counter), which counts according to the simple binary code. The 3 flip flops used to realise the counter are labelled, for convenience, A, B and C, with flip flop A representing the least significant bit of the binary coded signal. Examination of the state table shows that each flip flop is required to change state when the next less significant flip flop changes from the 1 state to the 0 state. Using J K flip flops, 1 to 0 voltage steps at each flip flop Q output may, therefore, be used

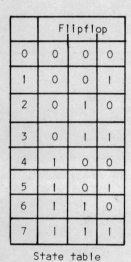

State table

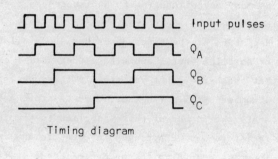

Timing diagram

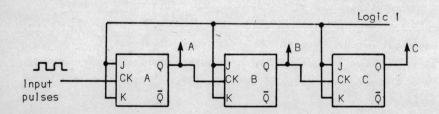

Figure 8.64 Asynchronous 3 bit binary counter

to clock the next more significant flip flop, if the control inputs
J and K are both held continually in the logic 1 state. The J K
flip flops are thus being used as T flip flops in this application.
The resulting circuit is also shown in Figure 8.64.

Notice carefully that in this network, the flip flop A has to
respond to the input pulse before it provides an output level
change to trigger the next flip flop. In particular, when in
state 7 with $Q_A = Q_B = Q_C = 1$, each flip flop will change state
in turn on the receipt of the next input pulse. Thus flip flop
A will change from 1 to 0, triggering flip flop B to change from
1 to 0, triggering flip flop C to change from 1 to 0. The network
is, therefore, asynchronous in nature and is often termed a 'ripple
counter', because of the way in which the change of state 'ripples'
along the flip flop chain. Care must be taken in reading out the
counter state because, of course, between the state 7 (111), and
the next state (000), a series of very short duration transient
states will occur. The sequence will, in fact, be

$$111 — 011 — 001 — 000$$

An alternative synchronous realisation of this state table is
shown in Figure 8.65.

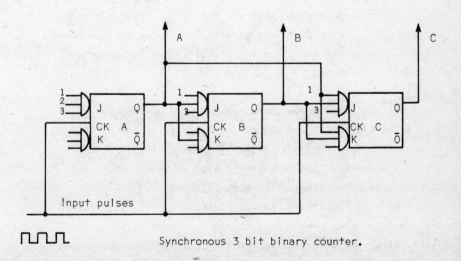

Figure 8.65 Synchronous 3 bit binary counter

Examination of the state table shows that flip flop A is required to change state at every input pulse. If the input pulses are fed to the CLOCK input, a J K flip flop A will change state at every clock pulse, if the control inputs J_A and K_A are connected to logic 1. Similarly, flip flop B is required to change state on every input pulse which occurs while flip flop A is at logic 1. The control inputs J_B and K_B are, therefore, connected to the output of flip flop A, Q_A. Flip flop C is required to change state on every input pulse when BOTH A AND B are at logic 1. The control inputs J_C and K_C are, therefore, driven by a logic function obtained by ANDing together the outputs Q_A and Q_B. This may be done by using AND gates, or, alternatively, using J K flip flops with multiple J and K inputs, as shown in the figure. Each flip flop in the figure has 3 J and 3 K inputs. Unused inputs may be left open as they automatically assume a logic 1 state in the open-circuit condition.

The basic principle of this type of synchronous circuit is that the logic levels available at the outputs Q and $\bar{Q}$ of each flip flop are used as inputs to logic gate networks to produce the required control signals for the J and K inputs of each flip flop, so as to ensure that all flip flops change state correctly, according to the desired state table when the input clock pulses occur. Because all the flip flops change state at the same time, the difficulties due to the transient states in the asynchronous 'ripple' counter do not occur.

Example

Realise a synchronous MOD 10 counter using J K flip flops. Because of the common use of the decimal system, the MODULUS 10 counter is often required. Figure 8.66 shows the state table for such a counter, counting in the simple binary code. The flip flops may be directly programmed as follows:

Flip flop A is required to change state on every input pulse. Its J and K inputs may, therefore, be left open, i.e. in the 1 state. Flip flop B is required to change state on every second input pulse, i.e. every time A is at logic 1, except that the change from 0 to 1, which would occur after state 9, must be inhib-

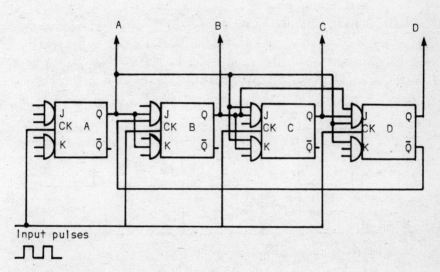

Figure 8.66 Synchronous MOD 10 counter

ited. The inputs J_B and K_B may be connected directly to the output of flip flop A, with the extra J_B input from $\bar{Q}_D$. The input functions are thus

$$J_B = A\bar{D} \quad K_B = A$$

Flip flop B will thus set to 1 only when A is at 1 and D is at 0. Flip flop C is required to change state every time both A and B are at logic 1. Its input functions are

$$J_C = K_C = AB$$

Flip flop D is required to set to 1 when all other flip flops are at logic 1. It is required to reset once in the count cycle, when changing from state 7 to state 8. However, if $K_D = A$ it will attempt to reset on every clock pulse which occurs while A is at logic 1. However, since it is only set to 1 itself once in the cycle, at state 8, it will, therefore, only be able to reset in state 9.

The input functions for flip flop D are thus

$$J_D = ABC \quad K_D = A$$

The resulting network is also shown in figure 8.66.

Example

Realise a shift register using J K flip flops.

The shift register was described earlier as a serial entry n bit word memory in which the data stored in one flip flop is shifted to the next flip flop on the command of the clock pulse.

A shift register may be constructed using J K flip flops as shown in Figure 8.67.

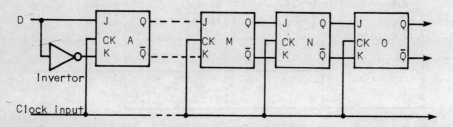

Figure 8.67 Shift register using J.K. flipflops

Consider flip flop N as representative of any flip flop in the register.

If the previous flip flop M is in state 1, then the inputs to flip flop N will be

$$J_N = 0 \quad K_N = 1$$

These will result in flip flop N resetting to state 0 on the next clock pulse (or staying at state 1 if it is already reset). Similarly, if the flip flop M is in state 1, then the inputs to flip flop N will be

$$J_N = 1 \quad K_N = 0$$

On the next clock pulse, flip flop N will set to state 1, or will stay set to 1 if it is already in that state.

For the first flip flop in the register, these same input conditions can be arranged by providing the control signal K_A from the data input D via an inverter.

Thus

$$J_A = \text{DATA VALUE}$$
$$K_A = \overline{\text{DATA VALUE}}$$

8.11 The multivibrator

The astable multivibrator is a basic pulse generator which is often used as a clock pulse generator for sequential networks. It is similar in form to the bistable memory circuit of Figure 8.59, and consists of two transistor switching circuits coupled in cascade. The coupling is, however, made via capacitors, rather than via resistors, with the result that each stable state of the circuit is only stable for a finite time, at the end of which the circuit returns to the active region of operation, and is rapidly switched to its other temporarily stable state. The circuit of a two transistor multivibrator is shown in Figure 8.68.

In order to understand the mode of operation of the circuit, consider initially that transistor T_2 is in the cut off state, with its collector potential almost equal to V_{CC}, being held thus by the charge on capacitor C_1. Transistor T_1 is heavily conducting, and is in the saturated state, with its collector potential very low. Transistor T_1 base will be at a potential of about 0.6 V; the capacitor C_2 will be charged therefore, in the polarity shown, to a voltage equal to the collector supply voltage V_{CC} less 0.6 volt.

The potential at transistor T_2 base is rising positively from a negative value, as the capacitor C_1 charges via resistor R_1 from the supply voltage V_{CC}. Eventually the base potential goes positive with respect to the 0 line. Transistor T_2 commences

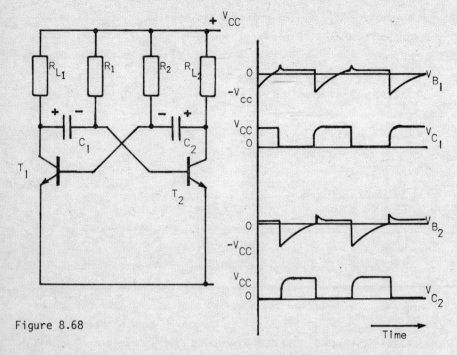

Figure 8.68

to conduct, its collector potential starting to fall. There can be no instantaneous change of voltage across capacitor C_2, because to change a capacitor voltage requires the movement of a finite quantity of charge. As T_2 collector potential falls, therefore, the fall in voltage is transferred also to the base of T_1. It is driven rapidly negative, so cutting off T_1. The potential of T_1 collector rises rapidly to the $+V_{CC}$ value, this rise being transferred also to the base of T_2, so reinforcing the rapid switching ON of T_2. It is this positive feedback, when the circuit is in the active region of its characteristic, which gives the rapid switching action between the two extreme states. The timing cycle then repeats. Figure 8.68 also shows, on a timing diagram, the waveforms of the collector and base voltages. It can be seen that the collector waveforms are approximately square in nature, and have a 1:1 mark to space ratio if the two halves of the circuit are identical.

The duration of each half period of the waveform may be estimated as follows.

When one of the transistors is rapidly switched ON, for example T_1, its collector potential falls from $+V_{CC}$ to almost zero. This voltage change is transferred via the coupling capacitor to the other transistor (T_2) base, which is, therefore, driven negative to a voltage of very nearly $-V_{CC}$ volts. This holds the transistor cut off until the capacitor C_1 charges from the collector supply voltage $+V_{CC}$ via the resistor R_1 to a voltage at T_2 base of zero volts.

Effectively, therefore, the capacitor is charging from a voltage of $-V_{CC}$ towards a voltage of $+V_{CC}$ (i.e. through $2V_{CC}$ volts) and the half period ends when it has charged through V_{CC} volts.
Thus

$$v = V_{CC} = 2V_{CC}(1-e^{-\frac{t}{C_1 R_1}})$$

or $t = 0.69\, C_1 R_1$ 8.16

The total period is, therfore,

$$T = 0.69(C_1 R_1 + C_2 R_2)$$ 8.17

Figure 8.69 shows a further modification of the same basic circuit,

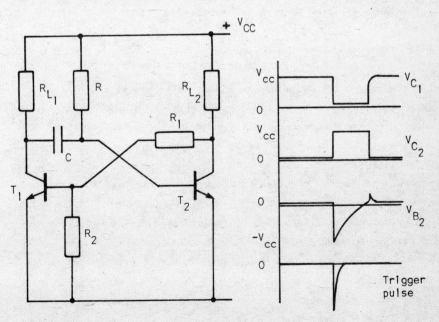

Figure 8.69

in which one coupling between the switching transistors is via a resistor, while the other coupling is via a capacitor. This results in a network with one completely stable state, and one temporarily stable state. The circuit thus waits in its stable state until triggered into its temporarily stable state; after a fixed timing period it then automatically reverts into its original stable state.

It is of interest to note that the operation of this circuit is the origin of the name 'flip flop', i.e. the circuit is 'flipped' into its temporary state, and eventually 'flops' back into its stable state. The circuit is called the monostable multivibrator, one use being to produce output pulses of predetermined duration from input trigger pulses of variable duration.

The duration of the temporary state is estimated as in the case of the astable multivibrator, and is given by

$$T = 0.69\ CR \qquad 8.18$$

The circuit may be triggered by a negative going trigger pulse, applied to the base of transistor T_2 or a positive going trigger pulse applied to the base of transistor T_1.

8.12 Integrated circuits

The logic circuits described so far in this chapter can all be constructed using standard electronic components i.e. resistors, capacitors and discrete transistors. However, circuits constructed in this way, especially in relatively large systems, suffer from major disadvantages.

    a) The labour required to produce such circuits is prohibitively expensive.

    b) The reliability of circuit connections made manually is relatively poor.

    c) The physical size and weight of the resulting circuit is excessive.

The first major reduction in circuit size and increase in its reliability occurred with the introduction of the transistor in the late 1940s, replacing the thermionic valve as the active component in electronic circuits. A typical thermionic valve operates with 200 to 300 volt power supplies and consumes power of the order of 2 watts to heat its cathode. A major drawback of large valve systems therefore was the size of the power supplies required and the large amount of waste heat which had to be removed from the equipment.

The next major circuit improvement came with the introduction in the early 1960s of the first commercial integrated circuits. An integrated circuit is manufactured by a process which allows, not only the active circuit components, such as transistors, but also the passive components, the resistors, capacitors and their interconnections to be formed upon a tiny silicon slice. As a result, an integrated circuit is a complete circuit in itself, requiring normally only the addition of the correct power supplies to produce a functioning system. An integrated circuit can be defined therefore as 'the physical realisation of a number of electrical elements, inseparably associated on or within a continuous body of semiconductor material to perform the functions of a circuit'. The actual production process is quite complex and requires many individual manufacturing steps. Basically, thin slices of pure or intrinsic semiconductor material, usually silicon, and prepared with a size of about 0.5 to 1 inch in diameter, are treated with a succession of dopant diffusion and chemical etching processes to form hundreds of identical circuits upon the single water. The silicon slice is then cut to produce individual dice, each containing one circuit.

The circuits are produced in large quantities and at this stage, the cost of each individual circuit on a single die, is very low. However, a relatively large proportion of the circuits produced will prove to be unsatisfactory for various reasons. A large part of the total cost of the final integrated circuit therefore comes in the testing of the separate circuits, the scrapping of the faulty dice and then in the mounting of the good circuits upon a carrier.

The dice, or chips, which are now of the order of 50 thousandths of an inch square, are mounted upon a carrier, connected using gold wires to the pins of the carrier and, after further extensive testing, are sealed to produce the complete packaged integrated circuit.

Largely as a result of the American aerospace projects, immense sums of money have been spent on research into production methods which allow greater and greater packing densities for components in an integrated circuit, together with increased reliability. The first integrated circuits, with from 1 to 10 transistors per chip are now designated as of S.S.I. (small scale integration) technology. Improved manufacturing methods eventually produced M.S.I. (medium scale integration) and then L.S.I. (large scale integration). V.L.S.I. (very large scale integration) technology devices are now commercially available and newer devices will soon enter the S.L.S.I. (super large scale integration) era. The different technologies may be roughly classified as follows:-

S.S.I.	1 to 10 transistors per chip
M.S.I.	10 to 100 transistors per chip
L.S.I.	100 to 10 000 transistors per chip
V.L.S.I.	10 000 to 50 000 transistors per chip
S.L.S.I.	Over 50 000 transistors per chip

Several different technologies have been used in the production of integrated circuits.

A large proportion of the small and medium scale integration devices, including the ever popular 7400 TTL logic family is based upon the bipolar technology described in section 4.3. This is one of the fastest technologies available today but suffers from the disadvantages of relatively large power consumption and relatively low packing density. It is not, therefore, in its basic form, particularly suitable for L.S.I. devices.

Current L.S.I. systems are based mainly upon the M.O.S. (metal oxide semiconductor) technology described in section 4.8.2. The first L.S.I. devices were based upon the P.M.O.S. technology in which a p-channel transistor is formed in an n-type substrate.

This is a well understood and reliable process but has the disadvantage that, being based upon the mobility of positive charge carriers to provide conduction, P.M.O.S. devices are relatively slow in operation.

Later devices are based upon the N.M.O.S. technology described in section 4.8.2. and, using the more mobile negatively charged electrons rather than the positive holes as the charge carriers, are capable of about a twofold increase in speed. However, even the N.M.O.S. devices are approximately 10 times slower in operation than are the bipolar devices.

The C.M.O.S. (complementary M.O.S.) technology, described in section 8.4.3 uses a combination of a p-channel transistor and an n-channel transistor, and thus achieves speeds between those of the P.M.O.S. and N.M.O.S. devices. The C.M.O.S. device has excellent noise immunity and very low power consumption making it suitable for use in electrically-noisy environments or where low power consumption is very important. The C.M.O.S. process was indeed developed directly for the aerospace industry, where extreme operating conditions are common.

Examples

1. Convert the following decimal numbers into their simple binary equivalents:

$$4596_{10} \quad 322_{10} \quad 49.875_{10}$$
$$(100111110100_2, \ 11000010_2, \ 110001.111_2)$$

2. Convert the following into their decimal equivalents:

$$2710_8 \quad 11010110_2 \quad 1101_3$$
$$(3226_8, \ 214_{10}, \ 37_{10})$$

3. Complete the following additions

$$11001_2 + 01111_2$$
$$0111_2 + 1100_2 + 1001_2$$
$$457_8 + 112_8$$
$$(101000_2, \ 11100_2, \ 571_8)$$

4. Complete the following subtractions

$11000_2 - 00111_2$

$651_8 - 377_8$

$(10001_2, 252_8)$

5. Express 349_{10} in its equivalent form in base 16 using A B C D E F as the base 16 characters corresponding respectively to the decimal characters 10 to 15.

(15D)

6. By constructing the truth tables for the functions, show that

$B + D = \bar{A}B + B\bar{D} + AD + \bar{A}\bar{B}D$

7. By the method of De Morgan, negate and simplify the following functions:

$A + \bar{B}, \bar{C}D, AB + \bar{A}\bar{B}, A(B + \bar{C})$

$(\bar{A}B, C + \bar{D}, (\bar{A} + \bar{B})(A + B), \bar{A} + \bar{B}C)$

8. Three members of a panel game are asked to record their opinion of a record using ON/OFF switches. Instead of recording individual preferences, however, data processing is required, such that a HIT lamp is lit if a majority of the panel like the record, while a MISS lamp is lit if a majority dislikes the record. Construct a truth table for the problem and develop and minimise the logical equations to the lamp signals.

$HIT = BC + AC + AB, MISS = \overline{HIT}$

9. Produce a state table for a count of 7 (Modulo 7) counter using the simple binary code. Develop the logical equations to the input signals for the J K flip flops required to realise the synchronous counter to perform the state table.

$J_A = \bar{B} + \bar{C}, \quad J_B = A, \quad J_C = AB$

$K_A = 1 \quad\quad K_B = A + C, \quad K_C = B$

N.B. A is the least significant bit.

CHAPTER NINE

Microprocessors

9.1 <u>Digital computers</u>

Without doubt, the digital computer has had a profound effect
upon modern life and continues to hold out exciting prospects
for the future. The basic idea of a calculating machine interested
scientists as early as the 17th Century and several attempts
were made to produce working machines using mechanical devices.
The modern digital computer, although a very complex digital system,
consists only of combinations of the same basic logical circuits
which we have considered in Chapter 8; i.e. AND, OR, NAND, NOR
gates and SHIFTING and COUNTING registers.
The production of commercially viable machines awaited the development of the manufacturing techniques which allowed the production
of large numbers of reliable gate and register circuits. Modern
digital computers fall into two categories:- general purpose and
special purpose machines. General purpose computers are designed
to be programmed to perform many different and quite unrelated
tasks, while a special purpose computer is designed to operate
on one specific problem and its circuits are minimised to provide
only the facilities required for the particular task. As a con-

sequence, a special purpose machine can be smaller and cheaper than a general purpose machine.

A digital computer operates by performing upon its input data, a sequence of relatively simple arithmetical or logical operations. The available operations are, in general, quite limited in scope and consist of operations such as ADD, SUBTRACT, COMPLEMENT, SHIFT LEFT, SHIFT RIGHT etc.

In order to perform a task a method must be devised whereby the task is broken down into a sequence of the operations which the particular computer is able to perform; i.e. an ALGORITHM is devised which will allow the task to be performed by the computer. The required sequence of operations for the task is the computer PROGRAMME and this is stored in the computer MEMORY.

The structure of a basic computer is illustrated in Figure 9.1. It consists of five fundamental units. The central unit is the CENTRAL PROCESSING UNIT, usually referred to as the C.P.U. This unit itself has two parts; the CONTROL UNIT (the C.U.) and the ARITHMETIC AND LOGIC UNIT (the A.L.U.). The function of the A.L.U. is to perform the basic computer arithmetic and logic operations, i.e. the ADD, COMPLEMENT, SHIFT operations etc., upon the data presented to it.

The function of the control unit is to FETCH from memory, in turn, the individual sequence instructions, to DECODE the instructions and to generate the resulting sequence of control signals which will arrange for the correct data flow to and from the A.L.U. and which will synchronise the EXECUTION of the instruction by the A.L.U. The computer operation thus consists of a sequence of FETCH, DECODE, EXECUTE cycles under the control of the control unit. The memory is shown as a separate unit of the computer. Its function is to store until required any data which may be used by the computer. It is used to store the programme which is to be executed and will, in general, also be used to store data which is required for the programme. The size of the memory provided for a computer will depend upon the task or tasks for which the machine is to be programmed; a special purpose machine may require only a short programme with little input/output data, while a general purpose machine may store several very long programmes at the same time, together with extensive data tables. The remaining two units shown in Figure 9.1 are the INPUT and OUTPUT units which provide the input and output interfaces to the external world, by means of which new data and new programmes can be entered into the machine and the results of the computation obtained.

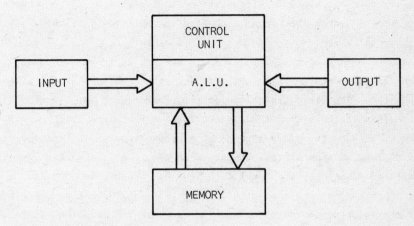

Figure 9.1 Basic Computer Structure

Data signal flow between the basic units of the computer takes place upon sets of signal lines which are called buses. The three standard buses in a computer are the data bus, the address bus and the control bus.

The DATA BUS provides the highway upon which data is pased between the various units of the computer, the data signal flow being controlled by signals produced by the control unit and distributed upon the CONTROL BUS. The ADDRESS BUS provides the highway for signals, produced by the control unit, which select the correct store location within the MEMORY UNIT. Each memory store location within the memory unit is given a unique address, by means of which the control unit can select any individual location, within the memory, as the source or destination of data or as the source of the next programme instruction.

The application of modern L.S.I. manufacturing techniques to computer circuits has now brought about a revolution in the use of computing techniques. The L.S.I. microprocessor, in which the functions of the C.P.U. of a digital computer have been realised upon a single silicon chip has enabled the production of complete digital processing facilities, with costs often well below £100. This has enabled the use of digital computing methods to be applied to a wide range of applications, from industrial process controls to musical door chimes and has changed the philosophy of the design of digital systems.

The development of the microprocessor device, as it exists today, grew out of research into L.S.I. devices intended for the calculator industry. The first general purpose microprocessor chip, the 8008, was introduced in 1972 by the Intel Corporation. Within the succeeding two years a range of microprocessors was introduced by various manufacturers, but the largest part of the market was captured jointly by the Intel 8080 and the Motorola 6800. These microprocessors, together with the later Zilog Z80, can be described as the industry standards. Since 1974 some rationalisation has taken place and more advanced microprocessors have been introduced by all of the three major manufacturers mentioned above.

A microprocessor is an L.S.I. device in which the functions of an A.L.U. and its associated control unit are produced on a single chip. It is possible to include, on the same chip, a clock generator circuit and also sufficient memory to allow the device to be described correctly as a single chip computer, and which would be adequate in power to perform small programmes. However most microprocessor systems require two or perhaps three L.S.I. devices; in addition to the L.S.I. central processing unit itself, clock generator circuits, memory devices and serial and parallel input/output devices are all available in L.S.I. form.

There are great similarities in many respects between the second generation microprocessors, the Intel 8080 and the Motorola 6800, and perhaps the best way to study the microprocessor is to look at one particular device in some detail. Consequently the following sections deal generally with microprocessors and specifically with the Motorola 6800 system.

Details of the Motorola devices discussed are reproduced by kind permission of Motorola Semiconductors Ltd. to which Company we express our gratitude.

9.2 The Power of a C.P.U.

The power of a central processor unit, i.e. its ability to process binary data, is dependent upon three main factors. These are:

(a) The order code of the C.P.U.

The order code or instruction code of a central processor unit is the list of arithmetic and logic operations which it can perform. To perform any given task, a programme must be devised which will achieve the required result, the programme being a sequence of the operations provided by the C.P.U. Thus a C.P.U. which has an extensive and well-designed order code may be able to perform the work with a shorter programme.

(b) The speed of operation of the C.P.U.

The speed of operation of the C.P.U. is determined by the technology by which it is manufactured. Thus the Motorola 6800 C.P.U., manufactured in N.M.O.S. technology achieves instruction times of the order of from 2 to 5 microseconds.

(c) The word length of the C.P.U.

The word length of the C.P.U. is the number of binary digits (bits) which it can process at any one time. Typically, large main frame computers have a word length of 32 or 64 bits while the microprocessors are commonly of 8 or 16 bits in length. The Motorola 6800 C.P.U. has an 8-bit word length. Figure 9.2 shows the data accuracy which can be obtained with various word lengths.

In section 8.2 the accuracy of encoded data was discussed. It can be seen that an 8-bit C.P.U. would be able to process data with an accuracy of 1 part in 256 i.e. with a data accuracy of better than $\frac{1}{2}$%.

If an accuracy better than this value is required, using an 8-bit C.P.U., this can be arranged, but only by using more than

Number of Bits	2^n	Data Accuracy $\frac{1}{2^n} \times 100\%$	Known As
4	16	6.250	NIBBLE
8	256	0.391	BYTE
10	1024	0.098	
16	65536	0.002	

Figure 9.2   Data Accuracy of Different Word Lengths

one C.P.U. word to represent the data. The programme could be written to express the data in, for example, two 8-bit words which can be considered to be placed side by side, to give effectively 16-bit accuracy. However the use of 'double precision' methods, in this manner, requires longer programmes and consequently the rate of data processing will be reduced.

It is common practice to refer to a binary word of 4-bits, as a 'nibble' and a word of 8-bits as a 'byte'. No other specific word is in common use for other lengths.

9.3   Memory organisation

The memory of a computer system is used to store both the programmes which are to be executed by the processor and the data which is to be operated on by the processor.

The memory can be thought of as a number of memory locations, each of the same word length as the C.P.U. itself. Each location will store one binary word and, in order that the location can be accessed to store or retrieve data at will, each store location is given a unique address by which it may be referenced. The memory addresses are, indeed, binary words themselves. An 8-bit C.P.U. therefore, if using one 8-bit word as the memory address can reference only 2^8 (256) locations. A memory of only 256 locations

would severely restrict the processor to small programmes. It is therefore common practice for 8-bit microprocessors to use two data words, i.e. 16-bits for the memory location addresses. The size of memory which can be uniquely addressed using a 16-bit address is thus 2^{16} (65536) locations in length
But

$$2^{16} = 2^6 \times 2^{10}$$
$$= 64 \times 1024$$

The memory is said to be 64K in size, where

1K (kilobyte) = 1024 locations.

A memory address is therefore a 16-bit binary word. Such a 16-bit string of 0's and 1's is inconvenient and clumsy to deal with, being difficult to remember and lengthy to write. It is common practice however to express the 16-bit binary string in its equivalent hexadecimal form, using the conversion techniques described in section 8.2.1. The 16-bit word is thus contracted to 4 hexadecimal characters which, with a little practice, can easily be converted back into binary whenever required.
For example, the 16-bit binary word

0110   1010   0110   0011
 6   A   6   3

converts to the hexadecimal form 6A63. Bear in mind that the use of the hexadecimal form in this way is only as a shorthand method of writing down long binary words i.e. the microprocessor itself operates purely in binary. Figure 9.3 shows a representation of the computer memory in which all of the memory locations are together shown as a vertical column, containing, as we have seen, 2^{16} = 65536 separate locations.

The binary addresses are shown at the Left-hand side, with the equivalent hexadecimal versions of the addresses shown at the Right-hand side.

Notice that each hexadecimal address requires two bytes, i.e. each hexadecimal character requires one nibble. The memory addresses extend therefore from address 0000_{16} to address $FFFF_{16}$.

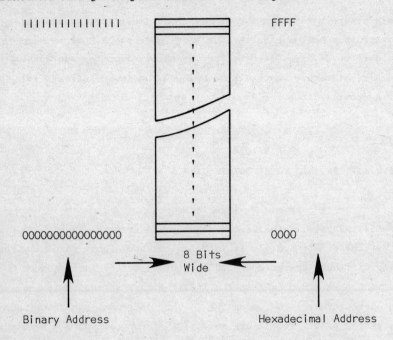

Figure 9.3 Representation of 16-bit Addresses on a Memory Map

A representation of the system memory addresses in this way is termed a 'memory map'.
In practice the main memory of a microprocessor system is normally implemented by one or more L.S.I. M.O.S. memory chips with, in a special purpose design, only sufficient memory connected to perform the required task. Thus the main memory may range in size from, say, only 128 bytes to the maximum of 64 kilobytes.

9.3.1 <u>Microprocessor memory devices</u>

Two main types of memory devices are used in microprocessor systems.
RAM - Random Access Memory.
The RAM is essentially a read/write memory, i.e. the word stored at any location may be read or alternatively a new word may be stored or written into the location.
The main disadvantage of a RAM is that it is 'volatile', which

means that whenever the power supply is removed from the RAM device, the stored contents of the chip are lost.

When power is applied to a RAM, all memory locations will assume completely random stored values.

Because of their volatility, complete programmes are seldom stored in RAM, for if the power were to be removed from the chip, by accident or due to supply failure, the programmes would be destroyed and would have to be re-entered into the memory before execution could begin again.

RAM is therefore used essentially to store data, which is of a temporary nature, or alternatively is used for the temporary storage of programmes which are under development and which therefore are being edited or changed. It is possible to purchase RAM devices which are provided with battery backup to provide a retention of stored data under power fail conditions. In order to reduce the battery drain, memory devices used with battery back-up are usually of the C.M.O.S. technology. For example, 4K memory cards are available which specify a retention time of 30 days starting with a fully charged battery. RAM devices are available in various configurations. The Motorola 6810 RAM is a 1K bit chip, i.e. contains 1024 single bit cells, which is organised for byte oriented systems and has 128 memory locations each 8 bits wide. One such chip will therefore be all that is required for very small systems. Alternatively 1K bit devices may be obtained, organised to have say 256 locations, each 4 bits in width. Thus two such chips will be required to provide a 256 location 8-bit memory. The latest memory devices available commercially are 64K bit devices, holding out the prospect of large memories with very few chips.

ROM - Read Only Memory.

A read-only-memory is a memory which, once programmed, can only have its contents read: i.e. the stored contents are permanent. Thus a read-only-memory is non-volatile and is used mainly for the storage of programmes.

There are several types of ROM in common use. The mask-programmed ROM is programmed permanently during manufacture. The required pattern is supplied by the purchaser to the manufacturer who prepares a mask which will establish the required bit pattern. Because of the cost of making the programming mask, the process is only economically feasible where quantities of the order of 1000 or more similar devices are to be purchased. These devices are thus intended for large scale manufacturing industry.

The programmable read-only memory, the PROM, is supplied by the manufacturer with all its memory calls in the same state. It can be user-programmed by use of a special equipment, a PROM-programmer. Each memory cell in the chip is provided with a fusible link which can be 'fused' or 'blown' using the PROM-programmer. However the programming of such a device is a once-only operation; a fusible link once blown cannot be replaced. A further, very commonly used, user programmable PROM is known as an EPROM, or Erasable PROM. The most common form of EPROM is a device which is manufactured with a quartz window on the top of the chip. The memory cells may be set all to the same logic value by exposing the chip to ultra-violet (U.V.) light for a period of 5 to 20 minutes. Once cleared in this way the EPROM is reprogrammable using the PROM programmer. This type of memory chip, known as a U.V. Erasable PROM, is well suited for use in the development of programmes, since, once the programme has been developed and edited in RAM, it may be programmed or 'burnt' into EPROM and checked for operation in its final form. The EPROM thus provides a quick, easily programmed form of ROM for development purposes or for use in manufactured systems where the number of ROM's required does not warrant the cost of a mask-programmed ROM.

The clearing and reprogramming process may be repeated a large number of times.

The memory in a typical microprocessor system will in general be made up of both ROM devices and RAM devices. The time of access of data from the memory into the C.P.U. is in the range 200 to 600 nanoseconds.

Example
In an 8-bit microprocessor system memory, three 1 k-bit RAM memory devices are addressed so that they occupy the lowest addresses on the memory map. What are the addresses located in each device?

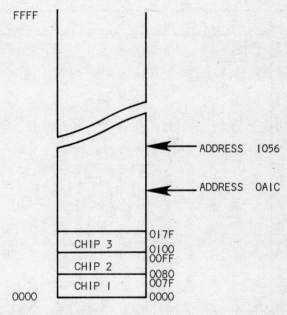

Figure 9.4 Memory Map for Examples

Figure 9.4 shows the three devices configured in the lowest addresses of the memory map.
Each memory chip contains 1024 memory bit cells. When used on an 8-bit system, each chip will provide 1024/8 = 128 memory locations. In hexadecimal notation, each device provides

$$128_{10} = 80_{16} \text{ locations}$$

The addresses, therefore, in chip 1 will be the first 80_{16} locations, i.e. addresses 0000 to 007F.
In chip 2, the addresses will be 0080 to 00FF.
In chip 3, the addresses will be 0100 to 017F.

HEXADECIMAL SUBTRACTION TABLE

$$X - Y = ?$$

X =

Y =

	0	1	2	3	4	5	6	7	8	9	A	B	C	D	E	F
1	F	0	1	2	3	4	5	6	7	8	9	A	B	C	D	E
2	E	F	0	1	2	3	4	5	6	7	8	9	A	B	C	D
3	D	E	F	0	1	2	3	4	5	6	7	8	9	A	B	C
4	C	D	E	F	0	1	2	3	4	5	6	7	8	9	A	B
5	B	C	D	E	F	0	1	2	3	4	5	6	7	8	9	A
6	A	B	C	D	E	F	0	1	2	3	4	5	6	7	8	9
7	9	A	B	C	D	E	F	0	1	2	3	4	5	6	7	8
8	8	9	A	B	C	D	E	F	0	1	2	3	4	5	6	7
9	7	8	9	A	B	C	D	E	F	0	1	2	3	4	5	6
A	6	7	8	9	A	B	C	D	E	F	0	1	2	3	4	5
B	5	6	7	8	9	A	B	C	D	E	F	0	1	2	3	4
C	4	5	6	7	8	9	A	B	C	D	E	F	0	1	2	3
D	3	4	5	6	7	8	9	A	B	C	D	E	F	0	1	2
E	2	3	4	5	6	7	8	9	A	B	C	D	E	F	0	1
F	1	2	3	4	5	6	7	8	9	A	B	C	D	E	F	0

FOR RESULTS BELOW THE STEPPED LINE BORROW A ONE FROM THE NEXT HIGHER ORDER DIGIT.

Figure 9.5 Hexadecimal Subtraction Table

Example
On the memory map of Figure 9.4, two memory addresses are shown.
These are locations 0A1C and 1056. How many memory locations are
there between these two addresses?

This result may be obtained by subtracting the lower address from
the higher, but the subtraction must be performed in hexadecimal
mathematics. To assist in this, Figure 9.5 shows a hexadecimal
subtraction table.

$$\begin{array}{r} 1056 \\ - \ 0A1C \\ \hline 063A \end{array}$$

The difference is $063A_{16}$ locations.

9.4 The Motorola 6800 microprocessor

The M6800 microprocessor is an 8-bit C.P.U. packaged in a 40-pin
D.I.L. chip. In addition to the Arithmetic and Logic Unit (A.L.U.)
and the Control Unit (C.U.), several internal registers are provided.
Registers internal to the C.P.U. provide the fastest data memory
which is available to the system, having a data access time of
less than 100 nanoseconds. Internal registers are thus very suitable
for the very fast storage of temporary data and an adequate number
of such registers can increase programme speed by reducing the
number of relatively slow transfers of data to and from the main
memory.

Figure 9.6 shows a representation of the internal registers of
the M6800 C.P.U.

Accumulators A and B: Accumulators are general purpose registers
used as temporary storage registers and are used in most arithmetic
and logic operations to hold one of the operands.

Thus if a number X, stored in memory location L_1 is to be added
to a number Y, stored in memory location L_2, with the result of
the addition to be stored in location L_3, then the operation sequence

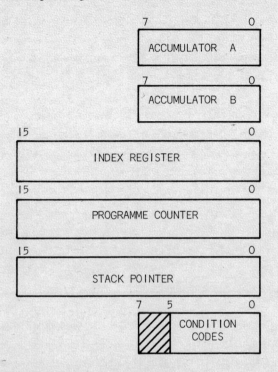

Figure 9.6   The 6800 C.P.U. Registers

would be performed, using one of the accumulators, as follows:-
- STEP 1. Load the number X from location L_1 into Accumulator A.
- STEP 2. Add to the contents of Accumulator A, the contents of location L_2. The result of this addition will be the new contents of Accumulator A.
- STEP 3. Store the contents of Accumulator A in the memory location L_3.

In this example, the accumulator is used in a manner analogous to the use of a scribbling pad when arithmetic is performed manually, i.e. it is used to note the intermediate results of calculations. For this reason, general purpose registers used in this way, and which in different microprocessors are not always called accumulators, are often referred to as 'scratchpad' memory.

Index Register X: This 16-bit register is used to hold the address of an operand in the main memory. Its use will be considered later.
Programme Counter P.C: This again is a 16-bit register which always contains the address in memory of the next programme step to be carried out. Remember that the C.P.U. performs a sequence of FETCH-DECODE-EXECUTE operations. Each operation consists of a FETCH of the 8-bit binary contents of the address in memory which is currently held in the Programme Counter. The Programme Counter contents are then incremented (i.e. increased by 1) to point to the next memory address. The binary word obtained in the FETCH is then decoded by the C.P.U. and it is thus determined whether the operation can be executed immediately or whether the next memory location, to which the Programme Counter is now pointing, contains necessary data which must also be read. If this is so, the data is read into the C.P.U. and the Programme Counter contents again increased by one. The function of the Programme Counter is therefore to keep track of the address in memory of the next byte of the programme. A Programme Counter is thus an indispensable part of any processor.

Stack Pointer: It is often convenient to make use of a section of read/write memory, in the main memory, for use as further temporary storage for data. Such a section of memory is usually termed a 'stack'. In order to keep track of the position of the top of the stack, within the memory, a 16-bit register, the Stack Pointer is provided which points to the top, unused memory location in the stack area.

The Condition Code Register: This register is also an 8-bit register with only the six lower bits, i.e. bits 0 to 5, having any significance. The two upper bits, i.e. bit 6 and 7 set to 1 when power is applied to the processor.

(Note that the 8 bits of any register or memory are always numbered b_0 to b_7 with b_0 being the least significant bit. See section 8.2.)

The six active bits, bit 0 to bit 5, are used as 'flags', i.e. they are used in the same manner as the Royal Standard on Buckingham Palace. Whenever the Monarch is in residence at the Palace the

Royal Standard is raised to indicate that fact. Each of the six
flag bits is similarly set to logic 1 to indicate the occurrence
of a certain event within the microprocessor system. The flags
are then used as the basis for decisions taken by the C.P.U.
The use of each flag bit will now be considered separately.

Bit 0 - the CARRY (C) flag:

The carry flag, C, is used in arithmetical and logical operations
effectively as the ninth bit of the system to do with spillover
from the 8-bit operations performed in the accumulators. Thus,
if, for example, an addition is to be performed as follows:-

	b_7	b_6	b_5	b_4	b_3	b_2	b_1	b_0	
Carry	1	1	0	1	1	0	0	1	
Flag	1	0	0	0	0	0	0	0	+
1 ←	0	1	0	1	1	0	0	1	

it is necessary to carry a 1 from b_7 into b_8 which, of course,
does not exist. The carry flag is set to 1 to flag the occurrence
of this event. The Accumulator, in which the addition has been
performed, will then hold the lower 8 bits of the result. The
logic state of the carry flag may be reset under programme control.
The C flag is, similarly, used as the ninth bit extension of the
8-bit system in logical operations such as Rotate Left. In this
operation, the contents of the Accumulator or Memory Location
being used, are shifted by one bit to the Left, with the contents
of the most significant bit, b_7, transferring into the C flag
and with the previous contents of the C flag becoming the new
contents of the least significant bit, b_0. This, and other logical
operations, are illustrated in Figure 9.7.

Bit 1 - the Overflow (V) flag: In section 8. it was shown that
the number range covered by an 8-bit word using the 2's complement
method extended from -128 to +127 decimal. Obviously it is quite
likely that the result of an arithmetic operation upon numbers
which are valid will result in an answer which lies outside the
allowable number range. For example the decimal addition 126 + 126
= 252 cannot be accommodated in an 8-bit 2's complement range.

In binary the addition becomes

$$0\ 1\ 1\ 1\ 1\ 1\ 1\ 0 = 126_{10}$$
$$0\ 1\ 1\ 1\ 1\ 1\ 1\ 0 = 126_{10}$$
$$\overline{1\ 1\ 1\ 1\ 1\ 1\ 0\ 0} = -4_{10}$$

The result is obviously incorrect since the sum of the two positive numbers appears to give a negative answer, i.e. the result has a 1 in the sign bit. This effect is due to a 'carry' propagating within the word into the most significant bit, so changing the sign of the result.

Whenever this overflow occurs, the V flag is set to 1.

Bit 2 - the Zero (Z) flag. The Z flag is set to 1 whenever the results of an operation, arithmetic or logical, in either an Accumulator or a Memory Location is zero.

Bit 3 - the Negative (N) flag. Whenever an operation results in a 1 in the most significant bit (bit 7) of an Accumulator or a Memory Location, the N flag will also be set to 1. The flag is called the N flag because, in arithmetic operations, the bit 7 is taken to be the sign bit and a 1 in this bit would indicate a Negative result.

Bit 4 - the Interrupt Mask (I). This is a special purpose flag which is used by the processor in dealing with Interrupt Programmes. The use of this flag will not be considered further in this section.

Bit 5 - the Half-carry (H) flag. This flag is used when dealing with Binary Coded Decimal characters. The B.C.D. code was described in Section 8.2.3. Each B.C.D. character requires 4 bits; an 8-bit word will therefore conveniently represent two B.C.D. characters. When performing B.C.D. arithmetic in the binary processor, an addition, for example, might generate a carry from bit 3 of the result into bit 4 i.e. from the first B.C.D. digit into the second B.C.D. digit. This would result in an incorrect answer and must be corrected. The H flag therefore sets to 1 to indicate a carry from bit 3 into bit 4 in the result of a binary addition.

ACCUMULATOR AND MEMORY		IMMED			DIRECT			INDEX			EXTND			INHER			BOOLEAN/ARITHMETIC OPERATION (All register labels refer to contents)	COND. CODE REG.					
OPERATIONS	MNEMONIC	OP	~	#	OP	~	#	OP	~	#	OP	~	#	OP	~	#		H	I	N	Z	V	C
Add	ADDA	8B	2	2	9B	3	2	AB	5	2	BB	4	3				A + M → A	↕	•	↕	↕	↕	↕
	ADDB	CB	2	2	DB	3	2	EB	5	2	FB	4	3				B + M → B	↕	•	↕	↕	↕	↕
Add Acmltrs	ABA													1B	2	1	A + B → A	↕	•	↕	↕	↕	↕
Add with Carry	ADCA	89	2	2	99	3	2	A9	5	2	B9	4	3				A + M + C → A	↕	•	↕	↕	↕	↕
	ADCB	C9	2	2	D9	3	2	E9	5	2	F9	4	3				B + M + C → B	↕	•	↕	↕	↕	↕
And	ANDA	84	2	2	94	3	2	A4	5	2	B4	4	3				A • M → A	•	•	↕	↕	R	•
	ANDB	C4	2	2	D4	3	2	E4	5	2	F4	4	3				B • M → B	•	•	↕	↕	R	•
Bit Test	BITA	85	2	2	95	3	2	A5	5	2	B5	4	3				A • M	•	•	↕	↕	R	•
	BITB	C5	2	2	D5	3	2	E5	5	2	F5	4	3				B • M	•	•	↕	↕	R	•
Clear	CLR							6F	7	2	7F	6	3				00 → M	•	•	R	S	R	R
	CLRA													4F	2	1	00 → A	•	•	R	S	R	R
	CLRB													5F	2	1	00 → B	•	•	R	S	R	R
Compare	CMPA	81	2	2	91	3	2	A1	5	2	B1	4	3				A − M	•	•	↕	↕	↕	↕
	CMPB	C1	2	2	D1	3	2	E1	5	2	F1	4	3				B − M	•	•	↕	↕	↕	↕
Compare Acmltrs	CBA													11	2	1	A − B	•	•	↕	↕	↕	↕
Complement, 1's	COM							63	7	2	73	6	3				M̄ → M	•	•	↕	↕	R	S
	COMA													43	2	1	Ā → A	•	•	↕	↕	R	S
	COMB													53	2	1	B̄ → B	•	•	↕	↕	R	S
Complement, 2's (Negate)	NEG							60	7	2	70	6	3				00 − M → M	•	•	↕	↕	①	②
	NEGA													40	2	1	00 − A → A	•	•	↕	↕	①	②
	NEGB													50	2	1	00 − B → B	•	•	↕	↕	①	②
Decimal Adjust, A	DAA													19	2	1	Converts Binary Add. of BCD Characters into BCD Format	•	•	↕	↕	↕	③
Decrement	DEC							6A	7	2	7A	6	3				M − 1 → M	•	•	↕	↕	④	•
	DECA													4A	2	1	A − 1 → A	•	•	↕	↕	④	•
	DECB													5A	2	1	B − 1 → B	•	•	↕	↕	④	•
Exclusive OR	EORA	88	2	2	98	3	2	A8	5	2	B8	4	3				A ⊕ M → A	•	•	↕	↕	R	•
	EORB	C8	2	2	D8	3	2	E8	5	2	F8	4	3				B ⊕ M → B	•	•	↕	↕	R	•
Increment	INC							6C	7	2	7C	6	3				M + 1 → M	•	•	↕	↕	⑤	•
	INCA													4C	2	1	A + 1 → A	•	•	↕	↕	⑤	•
	INCB													5C	2	1	B + 1 → B	•	•	↕	↕	⑤	•

Figure 9.7 M6800 Order Code

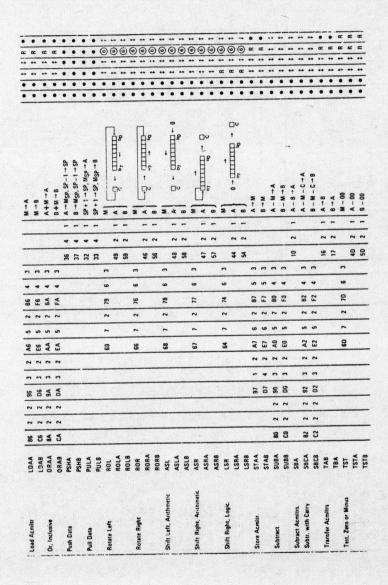

Figure 9.7 (cont.)

INDEX REGISTER AND STACK POINTER OPERATIONS

POINTER OPERATIONS	MNEMONIC	IMMED OP	~	#	DIRECT OP	~	#	INDEX OP	~	#	EXTND OP	~	#	INHER OP	~	#	BOOLEAN/ARITHMETIC OPERATION	5 H	4 I	3 N	2 Z	1 V	0 C
Compare Index Reg	CPX	8C	3	3	9C	4	2	AC	6	2	BC	5	3				$(X_H/X_L) - (M/M+1)$	•	•	⑦	↕	⑧	•
Decrement Index Reg	DEX													09	4	1	$X - 1 \to X$	•	•	•	↕	•	•
Decrement Stack Pntr	DES													34	4	1	$SP - 1 \to SP$	•	•	•	•	•	•
Increment Index Reg	INX													08	4	1	$X + 1 \to X$	•	•	•	↕	•	•
Increment Stack Pntr	INS													31	4	1	$SP + 1 \to SP$	•	•	•	•	•	•
Load Index Reg	LDX	CE	3	3	DE	4	2	EE	6	2	FE	5	3				$M \to X_H, (M+1) \to X_L$	•	•	⑦	↕	R	•
Load Stack Pntr	LDS	8E	3	3	9E	4	2	AE	6	2	BE	5	3				$M \to SP_H, (M+1) \to SP_L$	•	•	⑦	↕	R	•
Store Index Reg	STX				DF	5	2	EF	7	2	FF	6	3				$X_H \to M, X_L \to (M+1)$	•	•	⑦	↕	R	•
Store Stack Pntr	STS				9F	5	2	AF	7	2	BF	6	3				$SP_H \to M, SP_L \to (M+1)$	•	•	⑦	↕	R	•
Indx Reg → Stack Pntr	TXS													35	4	1	$X - 1 \to SP$	•	•	•	•	•	•
Stack Pntr → Indx Reg	TSX													30	4	1	$SP + 1 \to X$	•	•	•	•	•	•

JUMP AND BRANCH OPERATIONS

OPERATIONS	MNEMONIC	RELATIVE OP	~	#	INDEX OP	~	#	EXTND OP	~	#	INHER OP	~	#	BRANCH TEST	5 H	4 I	3 N	2 Z	1 V	0 C
Branch Always	BRA	20	4	2										None	•	•	•	•	•	•
Branch If Carry Clear	BCC	24	4	2										$C = 0$	•	•	•	•	•	•
Branch If Carry Set	BCS	25	4	2										$C = 1$	•	•	•	•	•	•
Branch If = Zero	BEQ	27	4	2										$Z = 1$	•	•	•	•	•	•
Branch If ≥ Zero	BGE	2C	4	2										$N \oplus V = 0$	•	•	•	•	•	•
Branch If > Zero	BGT	2E	4	2										$Z + (N \oplus V) = 0$	•	•	•	•	•	•
Branch If Higher	BHI	22	4	2										$C + Z = 0$	•	•	•	•	•	•
Branch If ≤ Zero	BLE	2F	4	2										$Z + (N \oplus V) = 1$	•	•	•	•	•	•
Branch If Lower Or Same	BLS	23	4	2										$C + Z = 1$	•	•	•	•	•	•
Branch If < Zero	BLT	2D	4	2										$N \oplus V = 1$	•	•	•	•	•	•
Branch If Minus	BMI	2B	4	2										$N = 1$	•	•	•	•	•	•
Branch If Not Equal Zero	BNE	26	4	2										$Z = 0$	•	•	•	•	•	•
Branch If Overflow Clear	BVC	28	4	2										$V = 0$	•	•	•	•	•	•
Branch If Overflow Set	BVS	29	4	2										$V = 1$	•	•	•	•	•	•
Branch If Plus	BPL	2A	4	2										$N = 0$	•	•	•	•	•	•
Branch To Subroutine	BSR	8D	8	2											•	•	•	•	•	•
Jump	JMP				6E	4	2	7E	3	3				} See Special Operations	•	•	•	•	•	•
Jump To Subroutine	JSR				AD	8	2	BD	9	3					•	•	•	•	•	•
No Operation	NOP										01	2	1	Advances Prog. Cntr. Only	•	•	•	•	•	•
Return From Interrupt	RTI										3B	10	1		⑩					
Return From Subroutine	RTS										39	5	1		•	•	•	•	•	•
Software Interrupt	SWI										3F	12	1	} See special Operations	•	S	•	•	•	•
Wait for Interrupt	WAI										3E	9	1		•	⑪	•	•	•	•

Figure 9.7 (cont.)

CONDITIONS CODE REGISTER		INHER			BOOLEAN	5	4	3	2	1	0
OPERATIONS	MNEMONIC	OP	~	=	OPERATION	H	I	N	Z	V	C
Clear Carry	CLC	0C	2	1	0 → C	•	•	•	•	•	R
Clear Interrupt Mask	CLI	0E	2	1	0 → I	•	R	•	•	•	•
Clear Overflow	CLV	0A	2	1	0 → V	•	•	•	•	R	•
Set Carry	SEC	0D	2	1	1 → C	•	•	•	•	•	S
Set Interrupt Mask	SEI	0F	2	1	1 → I	•	S	•	•	•	•
Set Overflow	SEV	0B	2	1	1 → V	•	•	•	•	S	•
Acmltr A → CCR	TAP	06	2	1	A → CCR	⑫					
CCR → Acmltr A	TPA	07	2	1	CCR → A	•	•	•	•	•	•

LEGEND:
OP Operation Code (Hexadecimal);
~   Number of MFU Cycles;
=   Number of Program Bytes;
+   Arithmetic Plus;
-   Arithmetic Minus;
•   Boolean AND;
M_SP Contents of memory location pointed to be Stack Pointer;
+   Boolean Inclusive OR;
⊕   Boolean Exclusive OR;
M̄   Complement of M;
↑   Transfer Into;
0   Bit = Zero;

00 Byte = Zero;
H   Half-carry from bit 3;
I   Interrupt mask
N   Negative (sign bit)
Z   Zero (byte)
V   Overflow, 2's complement
C   Carry from bit 7
R   Reset Always
S   Set Always
‡   Test and set if true, cleared otherwise
•   Not Affected
CCR Condition Code Register
LS Least Significant
MS Most Significant

CONDITION CODE REGISTER NOTES:
(Bit set if test is true and cleared otherwise)

① (Bit V) Test: Result = 10000000?
② (Bit C) Test: Result = 00000000?
③ (Bit C) Test: Decimal value of most significant BCD Character greater than nine? (Not cleared if previously set.)
④ (Bit V) Test: Operand = 10000000 prior to execution?
⑤ (Bit V) Test: Operand = 01111111 prior to execution?
⑥ (Bit V) Test: Set equal to result of N ⊕ C after shift has occurred.
⑦ (Bit V) Test: Sign bit of most significant (MS) byte of result = 1?
⑧ (Bit V) Test: 2's complement overflow from subtraction of LS bytes?
⑨ (Bit N) Test: Result less than zero? (Bit 15 = 1)
⑩ (All) Load Condition Code Register from Stack. (See Special Operations)
⑪ (Bit I) Set when interrupt occurs. If previously set, a Non-Maskable Interrupt is required to exit the wait state.
⑫ (ALL) Set according to the contents of Accumulator A.

Figure 9.7 (cont.)

9.5 Machine Code Programming

The M6800 instruction set, or order code, contains 72 unique instructions. Many of the instructions are provided with more than one addressing mode, which alters the manner in which the C.P.U. obtains the data upon which to perform the operation. The C.P.U. recognises each instruction as an 8-bit binary code and a programme can be written and presented to the processor as a sequence of such 8-bit binary words. A programme written, in this manner is said to be a machine code programme, because it is written in the binary language which the C.P.U. itself understands. This is the lowest level of programming language and although it is necessary to understand the principles of 'machine code programming' the use of this level of language for other than small programmes would be very tedious. A higher level of programming language uses MNEMONICS to represent the individual instructions and this, together with other important facilities, makes ASSEMBLER language programming, as it is called, a much more convenient process.
In fact ASSEMBLER language is used for the great majority of programming for industrial designs.
Figure 9.7 lists the order code of the M6800 processor and shows, for each instruction, the various addressing modes, the machine codes in hexadecimal notation, the instruction mnemonics and other details. At the right hand side of the figure, each condition code register flag is shown with an indication of the effect of the instruction upon the flag.

9.5.1 The addressing modes

In studying the addressing modes of the M6800 processor, reference should be made to the order code list given in Figure 9.7. Each mode is illustrated by a short programme segment, which is shown as it would be entered into the microprocessor memory. The memory locations shown are chosen purely arbitrarily; no significance should be attached to the actual numerical addresses indicated.

Inherent addressing mode

The inherent mode includes those operation codes which require no additional data other than the operation code. Inherent mode operation codes are therefore only one byte in length.

Example

Enter all zero's into Accumulator B

Memory	Machine Code	Mnemonic
0100	8F	CLRB
0101	Next instruction	

Example

Reduce the contents of Acc A by 1

Memory	Machine Code	Mnemonic
0050	4A	DECA

Immediate addressing mode

The Immediate mode is used where the operand, i.e. the data upon which the operation is to be performed, is to be given directly. Immediate mode instructions require two or three bytes of machine code; these bytes are the one-byte instruction together with one or two bytes of data.

Example

Load the data byte D3 (hexadecimal) into ACCA

Memory	Machine Code	Mnemonic
0030	86 D3	LDAA
0032	Next instruction	

Because the Accumulator A is an 8-bit register, the LDAA Immediate mode instruction is a two byte instruction i.e. one byte contains the operation code itself while the next byte must contain the 8-bit binary data. The complete instruction, when entered into memory, must therefore require two memory locations.

Example

Add to the contents of ACCA the data 56 and then load the Index Register with the data ABCD

Memory	Machine Code	Mnemonic
0020	8B 56	ADDA
0022	CE AB CD	LDX
0025	Next instruction	

The first instruction is again a two-byte instruction since it refers to the 8-bit Accumulator A. However the second instruction is a three-byte instruction, since it refers to the Index Register which is 16 bits in length. The LDX instruction therefore must specify the data to be loaded into all the 16-bits.

Note that, although it is conventional to write the instruction and the associated data on the same line as shown, the memory is actually loaded as follows:-

Memory	Contents
0020	8B
0021	56
0022	CE
0023	AB
0024	CD
0025	next instruction

Direct and Extended addressing modes

Whereas the Immediate mode instruction provides the processor with the actual numbers which are to be operated on by the instruction, the Direct and Extended modes provide the processor with a memory address which contains the data to be operated on or where the data is to be stored.

The Extended addressing mode instruction is a three-byte instruction, one byte being the instruction code itself, while the second and third bytes contain the full 16-bit (4 hexadecimal character) address to be used.

Example

Load ACCA with the data contained in memory address 89 AB

Memory	Machine Code	Mnemonic
0075	B6 89 AB	LDAA
0078	Next instruction	

Example

Store the data contained in ACCB into the memory location 1234.

Memory	Machine Code	Mnemonic
0060	F7 12 34	STAB

Example
Rotate left the data contained in memory location 5000.

Memory	Machine Code	Mnemonic
0000	79 50 00	ROL

These Extended Mode three-byte instructions each occupy three memory locations and take 5 or 6 clock cycles to complete. The Direct addressing mode provides a faster operation and, because the instructions are only two bytes in length, occupies less memory. The Direct mode assumes that the first byte of the relevant address is 00; only the second byte of the address must be specified. The Direct mode is thus a shorter and quicker form of the Extended mode, but which is applicable, only when the relevant address is in the 'first page' of the memory i.e. in the range 0000 to 00FF.

Example
Load ACCA with the data contained in memory address 00AB

Memory	Machine Code	Mnemonic
2000	96 AB	LDAA
2002	Next instruction	

Example
Store the data contained in ACCB into the memory location 0034

Memory	Machine Code	Mnemonic
1020	D7 34	STAB
1022	Next instruction	

NOTE: The Direct mode of operation is not provided for all the operations. For example it is not applicable to ROL, ROR, ASL etc.

The Relative addressing mode

In normal operation, each step in the programme is performed in sequence. At certain points in the programme however, decisions may be taken, usually as a result of certain tests, to alter the course of the programme, i.e. to branch into a different programming sequence.

In order to facilitate branching decisions, a range of Branch operations is included in the Order Code of the M6800.

Each Branch instruction tests the state of one or more status flags in the Condition Code Register.
Examples are:-

Operation	Mnemonic	Branch Test
Branch if carry clear	BCC	C = 0
Branch if minus	BMI	N = 1
Branch if zero	BEQ	Z = 1

Note that a Branch instruction tests the state of the relevant status flags as set by the previous programme instructions. The Relative addressing mode is used by the M6800 processor, solely for the destination of the Branch type of instruction.

A Relative mode instruction requires two bytes of machine language code. The first byte is the instruction code itself; the second byte gives the number of memory locations to be moved by the Programme Counter RELATIVE to its present position. Now remember that before obeying an instruction, the Programme Counter moves on to the position of the next instruction, i.e. the Programme Counter moves on after an instruction is FETCHED but before the instruction is EXECUTED.

The second byte of the Relative mode instruction is therefore an OFFSET which is the number of memory locations to move from the present position of the Programme Counter to the desired new position of the Programme Counter. The number is an 8-bit 2's complement binary number.

The available branch range is therefore from -128 to +127 memory locations from the position of the instruction step after the Branch instructions.

Example

Branch forwards five places if the result of the previous instruction is zero.

	Memory	Machine Code	Mnemonic
	1850	27 05	BEQ
	1852	Next instruction	
	1853		
Branch	1854		

```
5 places  1855
          1856
          1857              Next instruction
                            after branching
```
N.B. Perhaps the easiest way to determine the correct value for the OFFSET is to subtract the address of the instruction following the Branch instruction from the address to which it is desired to branch. Thus:

```
                  1857
                - 1852
                  0005
```

The correct OFFSET is the two least significant hexadecimal characters of the result.

Note also that the maximum forward offset is given by 7F (+ 127) and the maximum backward offset is given by 80 (- 128).

See section 8.2.3.

Example

From a Branch instruction in address 2000, branch to address 1FD2.

Memory	Machine Code	Mnemonic
1FD0		
1FD1		
1FD2		
'		
'		
'		
'		
1FFF		
2000	20 D0	BRA
2002		

```
Note:       1FD2
          - 2002
            FFD0
```

The required offset byte is the two least significant hexadecimal characters of the result, i.e. D0.

The Indexed addressing mode

This mode is similar to the Direct and Extended modes in that the machine code specifies an address from which the data required by the operation can be obtained. In the Direct and Extended modes however, the specified address is given as a constant written into the programme and which cannot be varied during the running of the programme.

The Indexed mode of addressing specifies that the address of the operand is given by the contents of the Index register, which of course is a 16-bit register. As the contents of the index register can be altered, under programme control, then the instruction operand can be varied while the programme is running. The indexed mode instruction is a two-byte instruction. The first byte is the operation code itself; the second byte is an unsigned 8-bit binary number, termed the OFFSET, whose value is added to the contents of the Index register to give the effective address of the operand.

Note that the OFFSET may have a minimum size of 0O hexadecimal (00 decimal) and a maximum size of FF hexadecimal (255 decimal).

Example

Load ACCA with the contents of the memory location 09 plus the Index register contents.

Memory	Machine Code	Mnemonic
0100	CE 01 00	LDX
0103	A6 09	LDAA
0105	Next instruction	
0106		
0107		
0108		
0109	55	
010A		

The Index register contents 0100 are added to the offset 09 to give the address 0109. The contents of address 0109 (in this

example, an arbitrary 55) are loaded into Accumulator A.
The Indexed mode is especially suitable for accessing data tables where successive entries in the table can be obtained by incrementing the Index register.

9.5.2 Machine code programming examples

All programmes intended for use in dedicated microprocessor systems, e.g. traffic light controller, automatic washing machine controller etc., would, when completely developed and debugged, be 'burned' into ROM so that the programme is permanent and available as soon as the system is switched on. In order that such a programme will start automatically, as soon as the power is applied, the address of the first programme instruction is entered into memory in two memory locations, FFFE and FFFF, known as the RESET vector. The M6800 processor, upon switch-on, will immediately examine the two locations to obtain the start address of the programme, and will commence operating at that address. Thus if the programme start address is 0500, then the byte 05 would be entered into location FFFE and the byte 00 would be entered into location FFFF. Programmes under development however will normally reside in the system RAM and microprocessor development systems, which are intended for use in developing new programmes, are often provided with extensive sections of RAM memory. Consequently the addresses in RAM where a new programme under development may be placed, is largely a matter of convenience and is rather arbitrary. However it is convenient to use the 'first page' of memory when possible, i.e. addresses 0000 to 00FF, in order to make use of the efficient Direct mode of addressing. In the following examples, the chosen memory addresses are in the 'first page' but other than that are arbitrary.

Example
Add together the contents of memory locations 0050 and 0051 and place the result in location 2000.

Memory	Machine Code	Mnemonic	Mode
0000	96 50	LDAA	Direct
0002	9B 51	ADDA	Direct
0004	A7 20 00	STAA	Extended
0007	3E	WAI	Inherent

N.B. When the programme segment is run by the processor, some instruction must be included at the end of the segment to effectively stop the processor. If this is not done, the processor will attempt to decode the data randomly held in the next memory locations and the desired programme operation will not in general be obtained. As the M6800 microprocessor cannot be stopped merely by stopping the system clock, some programming device must be used to place the processor in a short closed programming loop which effectively inhibits the processor operation. One such method is to use the Machine code instruction 3E (mnemonic-WAI-wait for interrupt). In practice this difficulty will not arise since most practical programmes are non-ending. For example on completing a washing programme, the washing machine controller will automatically perform a programming loop, examining the system inputs and waiting to be told the next washing sequence.

In all the examples in this section, the programme is ended with the code 3E to perform the 'stop execution' function.

Example

Add together the first ten hexadecimal numbers and place the result in location 0000.

Memory	Machine Code	Mnemonic	Mode
0000			
0001	4F	CLRA	Inherent
0002	5F	CLRB	Inherent
0003	1B	ABA	Inherent
0004	5C	INCB	Immediate
0005	C1 11	CMPB	Immediate
0007	26 FA	BNE	Relative
0009	97 00	STAA	Direct
000B	3E		Inherent

The steps of this programme will perhaps be more easily followed by referring to Figure 9.8 which is a FLOW MAP for the example.

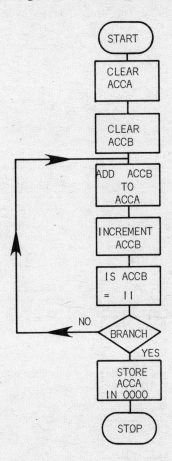

Figure 9.8   Flow Map

In this example, the ADD and INCREMENT loop must continue until the ACCB has been incremented to 11; i.e. when the contents of ACCB are equal to 11, this is NOT added to the ACCA.
To test for this condition, the COMPARE instruction is used. When the two quantities being compared are equal (i.e. ACCB=11) then the Zero flag Z is set to 1. This is so because the COMPARE instruction effectively performs the Exclusive OR function between the quantities being compared.

374 Basic Electrical Engineering and Instrumentation for Engineers

The Z flag can therefore be tested using the BNE branch instruction. The branch offset may be calculated as previously explained by subtracting the source address of the Branch from the desired address.

Thus

$$\begin{array}{r} 0003 \\ -0009 \\ \hline FFFA \end{array}$$

The branch offset is the lowest two hexadecimal characters of the result, i.e. FA.

For other than trivial problems, it is usually much easier if the programme is designed in the form of a flow map before the machine code programme is attempted.

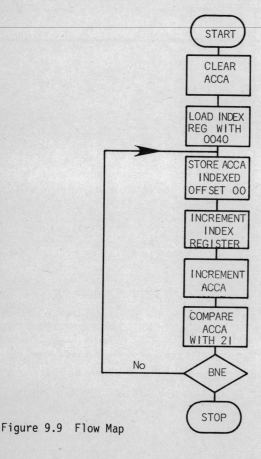

Figure 9.9 Flow Map

Example

Store the hexadecimal numbers 0 to 20 in the memory locations 0040 to 0060

Figure 9.9 shows a flow map which represents a possible solution to this exercise.

The solution uses the Index Register as a pointer to the memory location in which the data is to be stored. By incrementing the Index Register once per loop, the data is stored in successive locations in the memory.

Memory	Machine Code	Mnemonic	Mode
0000	4F	CLRA	Inherent
0001	CE 00 40	LDX	Immediate
0004	A7 00	STAA	Indexed
0006	08	INX	Inherent
0007	4C	INCA	Inherent
0008	81 21	CMPA	Immediate
000A	26 F8	BNE	Relative
000C	3E	WAI	Inherent

Branch Offset   0004
              -   000C
                  FFF8

N.B. The same store location would have been obtained if the index register had been loaded with the base address 0000 and the offset used in the indexed store instruction had been 40.

9.6 Entering the programme

The microcomputer system will consist of the microprocessor and memory unit, together with peripheral equipment to allow programmes to be entered into the memory. The peripheral equipment will consist of either a teletypewriter (T.T.Y.) or a keyboard and visual display unit (V.D.U.). The teletypewriter consists of:

(a) a keyboard, similar to a typewriter keyboard, which produces individual A.S.C.I.I. coded alphanumerical characters at a rate of 10 per second for transmission serially to the C.P.U.

(b) a printer which accepts similarly coded alphanumerical characters from the C.P.U.

(c) optionally, a 10 character per second paper tape reader.

(d) optionally, a 10 character per second paper tape punch.

The microprocessor, in its unprogrammed state cannot understand even the input signals from the teletypewriter. In order to use the teletypewriter as an input/output peripheral, a programme must be written and entered into the memory, usually in the form of preprogrammed ROM devices, which will include the peripheral handler programmes. Such a programme is referred to as the system MONITOR, and it will also usually include routines which will allow the user to perform, via the teletypewriter peripheral, several important functions:

(a) examine the contents of any memory locations,

(b) change the contents of any RAM memory locations,

(c) examine a register contents,

(d) load a programme via the paper tape reader,

(e) dump a programme to the paper tape punch,

(f) run a programme starting in any address.

Using the Memory Examine and Memory Change routines, the machine code programme, once developed can be entered and run in the microprocessor system. It is unusual for any but trivial programmes to run correctly first time; in general the development of a working programme is a process of repeated trial and correction.

The process of fault finding an incorrect programme is helped if it can be determined at which point in the programme the operation departs from the expected. For this reason, facilities are often incorporated in the system monitor, to insert 'breakpoint' i.e. to temporarily insert stops in the programme to allow small portions of the programme to be run individually. Alternatively, 'trace' facilities which can allow the programme to be run normally, but with each C.P.U. register contents to be printed upon the system output consol after the completion of every programme step. Using a teletypewriter for this purpose can be tedious, due to the slow rate of printing and data interchange. The use of a separate keyboard and V.D.U. however will allow much faster operation, but in turn, suffers the disadvantage of not providing a 'hard copy' print-out, i.e. a print-out on paper. All fault finding i.e. programme debug, must therefore involve the use of the V.D.U. screen.

9.7 Assembly Language programmes

The use of machine code for programming is tedious, if the programme is not short, for several reasons;

(a) for a programme with several-loops, the calculation of branch offsets becomes repetitive,

(b) any error which, when corrected, requires the insertion of extra programme bytes, will require changes to the branch offsets,

(c) the machine code instructions are not easily recognisable; the machine code programme is not easily readable.

For the majority of programming applications, therefore, a higher level language is chosen, and in most cases the language used is the ASSEMBLY language.

This is a language based upon the MNEMONICS already introduced. It is thus more akin to an English language programme and includes facilities for introducing explanatory comments etc. to make the purpose of each programme segment clear.

A programme written in Assembly language is termed a SOURCE programme. The process does, however, require the use of another computer programme, which may be included in the system MONITOR, which will allow the microcomputer to accept the SOURCE programme via the system input keyboard or tape reader and from it produce the machine code or OBJECT programme which always is the programme which finally runs in the machine.

The SOURCE programme

The source programme consists of a sequence of STATEMENTS, successive statements being separated by the Carriage Return (C.R.) character.

A source statement contains from one to four fields. These are

           LABEL   OPERATOR   OPERAND   COMMENT

The successive fields in a statement are separated by either one or more space (SP) characters or a horizontal (TAB).

The LABEL field

An entry in the LABEL field is used as a name by which to refer to a particular address in the memory, or occasionally refer to a particular value of data. The name or label can, for example, be used as the branch address of a branch instruction and the actual branch offset byte will be calculated automatically during the assembly process, i.e. when the SOURCE Programme is being 'assembled' into machine code.

The following rules apply to labels:

A label consists of from 1 to 6 alphanumeric characters, the first character must be alphabetic. The label must begin in the first character position of a statement.

A label must not consist of a single character A, B or X as these reserved names refer to the accumulators or the index register.

All labels within a programme must be unique.

Two special rules should be noted:

A space (SP) in the first character position indicates that no label is present in the label field.

An asterisk (*) in the first character position indicates that the entire statement is a 'comment' i.e. is a comment included to make the purpose of the programme more clear. A comment will be ignored by the ASSEMBLER during the assembly process.

The OPERATOR field

This consists of the MNEMONIC found in the Order Code (Figure 9.7) and which has already been introduced, e.g. LDAA, CMPA, STAB etc.

The OPERAND field

The kind of information placed in the operand field depends on the particular mnemonic operator and upon the addressing mode required.

The assembler will recognise numbers, symbols and expressions in the operand field.

Numbers are accepted by the assembler in the following formats:

Operand	Assembler Assumes	Example
Number	Decimal	584
$ Number	Hexadecimal	$8E
Number H	Hexadecimal	$4F
@ Number	Octal	@37
Number O	Octal	37O
Number Q	Octal	37Q
% Number	Binary	%1101
Number B	Binary	1101B

where NUMBER is a positive integer.

Symbols are accepted by the assembler with the same rules as for labels.

Expressions

An expression is a combination of symbols and/or numbers separated by one of the arithmetic operators +, -, *, or /.

The assembler evaluates the expressions algebraically from left to right. The rules of precedence, for example, of multiply over addition which apply in normal arithmetic, do not apply.

The addressing mode chosen by the assembler for any instruction depends upon the entry in the OPERAND field, according to the following table.

Mode	Operand Field
Immediate	# Number
	# Symbol
	# Expression
Relative	Label
Indexed	X
	Number, X
	Symbol, X
	Expression, X
Direct or Extended	Number
	Symbol
	Expression

In choosing between the DIRECT and EXTENDED modes, the assembler will select the DIRECT mode if the address falls within the memory 'first page' i.e. between 0000 and 00FF.

The COMMENT field

The fourth field, the COMMENT field, is optional and allows the programmer to include any comments which may make the programme more understandable to a reader. The comment will be ignored by the assembler during the production of the machine code OBJECT programme, but would be listed as part of the SOURCE programme.

Assembler Directives

In addition to the statements which will be assembled to produce machine code, facilities exist for including lines which give information to the assembler as to how the machine code OBJECT programme is to be produced; these are called Assembler Directives. Assembler directives, which are placed in the OPERATOR FIELD are used in the organisation of the format of the assembler output, for equating numbers to labels, and for reserving blocks of memory, for example, for use in data tables.

Some of the allowable directives are:

ORG — Origin. Defines the numerical memory address for the programme steps following the ORG directive.

EQU — Equates a symbol to a numerical value, another symbol, or to an expression.

FCB — Form Constant Byte. Locates constants in particular memory locations.

FCC — Form Constant Characters. Translates strings of characters into A.S.C.I.I. code (see Appendix 4).

FDB — Form Double Constant Byte. Stores the 16-bit binary number from the operand in two successive memory locations.

RMB — Reserve Memory Bytes. Reserves a block of memory for future use.

END — Indicates the end of the source programme.

NAM — Name. Gives the programme name. This must be the FIRST statement in the SOURCE programme.

Example

Repeat, using assembler language, the programme to store the hexadecimal numbers 0 to 20 in the memory locations 0040 to 0060. The source programme and the assembled programme are shown in Figure 9.10.

SOURCE PROGRAMME

```
  NAM STORE HEX NUMBERS
  ORG 0
BEGIN CLRA
  LDX #0 INITIALISE INDEX REGISTER
MORE STAA $40,X
  INX INCREMENT INDEX REGISTER
  INCA INCREMENT ACCA
  CMPA #$21 ARE ALL 20 NUMBERS STORED?
  BNE MORE STORE NEXT NUMBER
  WAI STOP PROCESSING
  END
```

ASSEMBLER OUTPUT

Line No.	Address	Machine Code	Label	Operator.	Operand.	Comment
00001				NAM	STORE	HEX NUMBERS
00002	0000			ORG	0	
00003	0000	4F	BEGIN	CLR A		
00004	0001	CE 0000		LDX	#0	INITIALISE INDEX REGISTER
00005	0004	A7 40	MORE	STA A	$40,X	
00006	0006	08		INX		INCREMENT INDEX REGISTER
00007	0007	4C		INC A		INCREMENT ACCA
00008	0008	81 21		CMP A	#$21	ARE ALL 20 NUMBERS STORED?
00009	000A	26 F8		BNE	MORE	STORE NEXT NUMBER
00010	0000	3E		WAI		STOP PROCESSING
00011	0000			END		

TOTAL ERRORS 00000

Figure 9.10

9.8 Input/output programming

The programming exercises used so far involve the manipulation
of data completely within the microprocessor system, i.e. no input
signals are required (other than the programme itself) during
the running of the programme and no output signals are produced.
The microprocessor system is being used in a manner analogous
to a desk top calculator in the sense that no interaction with
the outside world, other than with the operator, is involved.
The main application of microprocessors is in the real time control
of external devices or processes and this requires the ability
at various states of the programme to output control signals to
external equipment or tc read into the microprocessor control
signals generated by external equipment.

Such input/output data transfers are arranged via an additional
L.S.I. device which can be added to the microprocessor system
and which shares the same data, address and control buses with
the microprocessor. The device, marketed by Motorola Semiconductors
Ltd., for use with the 6800 and other microprocessors is known
as a Peripheral Interface Adaptor (MC6820 or MC6821). The P.I.A.
is a general purpose input/output device and provides 16 input/
output connections, configured as two 8-bit input/output ports.
Each individual connection of the two ports may be used separately
as an input or as an output.

Together with the two ports, four other connections are provided
which are intended for use in controlling the input/output data
transfers.

The P.I.A. is packaged as a 40 pin D.I.L. chip. All inputs and
outputs are designed to be TTL device compatible which makes interfacing the microprocessor to external logic relatively easy.

The P.I.A. is represented in Figure 9.11.

From the point of view of the external connections, the P.I.A.
divides into two almost identical parts, each containing one input/
output port and two control connections. The two sections can
be differentiated by the letters A and B.

Microprocessors 383

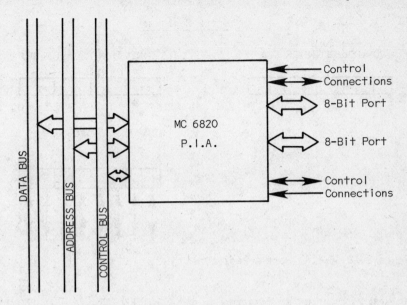

Figure 9.11 Peripheral Interface Adaptor

Thus the A and B input/output ports are basically similar in operation; the slight differences which do exist are in their electrical characteristics and in the operation of the control connections. For a first understanding therefore, the two sections A and B may be considered identical, and the following discussion of the A section applies equally well to the B section.

9.8.1 Data transfers via the Ports A and B

Associated with each Port there are three 8-bit registers internal to the P.I.A. These are illustrated for Port A in Figure 9.12. The equivalent Port B registers are similar.
The P.I.A. is intended to be used as a MEMORY-MAPPED device, i.e. the chip appears to the microprocessor as though it was four memory locations. The microprocessor, in fact, has no way of knowing that anything other than memory locations exists.
Two of the memory locations are assigned to the input/output registers to which the external connections are made. These

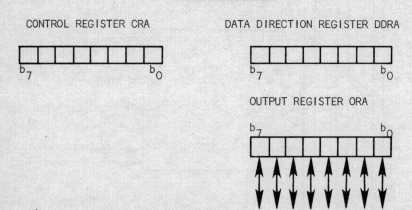

Figure 9.12 P.I.A. Port A Registers

registers are normally referred to as OUTPUT REGISTER A - (MNEMONIC (ORA)) and OUTPUT REGISTER B - (MNEMONIC (ORB)) even though they can be used for data input as well as data output. The names and mnemonics normally used for the remaining registers are CONTROL REGISTER A (CRA), CONTROL REGISTER B (CRB), DATA DIRECTION REGISTER A (DDRA) and DATA DIRECTION REGISTER B (DDRB).

By suitable arrangement of the connection of the Address Bus to the P.I.A., any of the 64K allowable addresses may be assigned to the P.I.A. registers.

A typical memory assignment is as follows:

Register		Assigned Address
Output Register A	ORA	$8008
Data Direction Register A	DDRA	$8008
Control Register A	CRA	$8009
Output Register	ORB	$800A
Data Direction Register B	DDRB	$800A
Control Register B	CRB	$800B

The output registers, i.e. the data input/output connections, can be used directly for data transfers, as follows:-
To input data to the microprocessor, use the programme step

$$\text{LDAA} \quad \text{ORA}$$

Either accumulator, of course, may be used and the data transfer is made effectively by the microprocessor reading the contents of the ORA memory location.

To output data from the microprocessor, use the programme step

               LDAA   #DATA
               STAA   ORA

Again either accumulator can be used. The data is loaded first into the accumulator and the accumulator contents are stored in the ORA memory location.

The data transfers are thus very easily arranged.

One small complication however is introduced by the need to INITIALISE the PIA prior to using it in this way. This is necessary because the individual register bits must be set up first as either inputs or outputs. In other words, the system must be told which Output Register bits are to be inputs and which outputs, prior to being used. Any bit in either output register may be used as an input or as an output.

The specification of an output register bit as an input or output is arranged by placing a pattern of bits in the Data Direction Register. Thus if an 0 is placed in a bit in the DDRA, then the equivalent bit of the ORA is configured as an input. Conversely, if a 1 is placed in a bit in the DDRA, the equivalent bit of the ORA is configured as an output.

As an example, the pattern hexadecimal 0F ($0F) placed in the DDRA would configure bits 0 to 3 of the ORA as outputs and bits 4 to 7 as inputs.

Refer again to the table of register assigned addresses. You will note that the Data Direction Register and the Output Register in each section of the PIA share the same address.

How can two registers be used separately at the same address? Obviously they cannot, without some extra switch to connect the desired register to the address connections. The switch used is bit 2 of the Control Register.

With bit 2 of the ORA set to 0, the address $8008 connects to the Data Direction Register, DDRA. With bit 2 of the ORA set to 1, the address $8008 connects to the Output Register, ORA. The

remaining bits of the Control Register are used in interrupt programmes, which will not be dealt with here.
These bits should therefore be set always to 0.
An initialisation segment, to configure ORA for 4 inputs (bits 4 to 7) and 4 outputs (bits 0 to 3) is as follows:-

* INITIALISATION SEGMENT

	CLR	CRA	Switches address to DDRA
	LDAA	#$0F	
	STAA	DDRA	Configures 4 inputs, 4 outputs
	LDAA	#$04	
	STAA	CRA	Switches address to ORA

* END OF INITIALISATION SEGMENT

Example

Write a program to generate a square waveform at each bit of ORA. Estimate the period of the square waveform.

A square wave voltage can be generated by alternately outputting from the ORA logic 0 levels and logic 1 levels. A delay would be required between the changes of output level in order to determine the square wave periods.

The solution can initially be represented as a flow map as in Figure 9.13.

One circuit of the main programme loop produces one half period of the waveform, with the data output from ACCA being changed in successive half periods by the Complement ACCA instruction. The box marked DELAY is itself a small segment of programme which is shown separately in the Figure. The delay is obtained by decrementing ACCB down to zero, each decrementing loop taking 6 clock cycles.

Note that the time of operation of each code is given in the Order Code table in Figure 9.7 in the columns headed $\sim$. As most M6800 systems operate with a 1 MHz system clock, then one clock cycle is of 1 μsecond duration.

The delay counter, ACCB, is set to an initial value by loading the Accumulator from the RAM buffer location assigned the label 'PERIOD'. The square wave period can thus be changed by changing

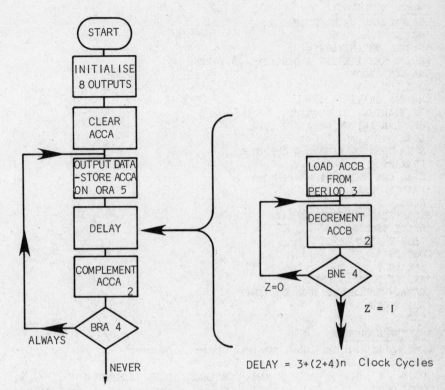

Figure 9.13 Flow Map for Square Wave Generator

the contents of the buffer location. Note also that the maximum value of delay will be obtained when the value inserted in 'PERIOD' is 00. You should verify that the maximum half-period is then equal to 1.55 milliseconds.

The Source programme and the Assembly listing for this example is shown in Figure 9.14.

Figure 9.14

```
SOURCE PROGRAMME
 NAM SQUARE WAVE GENERATOR
 ORG 0
*SYSTEM DEFINITIONS
 PERIOD RMB 1 RESERVE ONE BYTE AS BUFFER
 ORA EQU $8008
 DDRA EQU ORA
 CRA EQU ORA+1
*INITIALISATION SEGMENT
 CLR CRA SWITCH $8008 TO DDRA
 LDAA #$FF
 STAA DDRA CONFIGURE 8 OUTPUTS
 LDAA #$04
 STAA CRA SWITCH $8008 TO ORA
*MAIN PROGRAM
 CLRA
 STORE STAA ORA OUTPUT DATA
*DELAY SEGMENT
 LDAB PERIOD
LOOP DECB
 BNE LOOP
*END OF DELAY SEGMENT
 COMA   COMPLEMENT ACCA CONTENTS
 BRA STORE
 END
```

ASSEMBLER OUTPUT

```
Line  Addr-  Machine  Label   Oper-   Oper-    Comment
No.   ess.   Code             ator.   and.
00001                         NAM     SQUARE   WAVE GENERATOR
00002 0000                    ORG     0
00003                        *SYSTEM DEFINITIONS
00004 0000   0001    PERIOD   RMB     1        RESERVE ONE BYTE AS BUFFER
00005        8008    ORA      EQU     $8008
00006        8008    DDRA     EQU     ORA
00007        8009    CRA      EQU     ORA+1
00008                        *INITIALISATION SEGMENT
00009 0001   7F 8009          CLR     CRA      SWITCH $8008 TO DDRA
00010 0004   86 FF            LDA A   #$FF
00011 0006   B7 8008          STA A   DDRA     CONFIGURE 8 OUTPUTS
00012 0009   86 04            LDA A   #$04
00013 000B   B7 8009          STA A   CRA      SWITCH $8008 TO ORA
00014                        *MAIN PROGRAM
00015 000E   4F               CLR A
00016 000F   B7 8008  STORE   STA A   ORA      OUTPUT DATA
00017                        *DELAY SEGMENT
00018 0012   D6 00            LDA B   PERIOD
00019 0014   5A       LOOP    DEC B
00020 0015   26 FD            BNE     LOOP
00021                        *END OF DELAY SEGMENT
00022 0017   43               COM A            COMPLEMENT ACCA CONTENTS
00023 0018   20 75            BRA     STORE
00024        0000             END

TOTAL ERRORS 00000
```

9.9 The use of microprocessors in digital system design

As was stated at the beginning of this chapter, the advent of microprocessors has changed the philosophy of digital system design. Before the introduction of microprocessors, digital hardware was mainly TTL device based and the production of a working system followed a well-defined path.

(a) The logic problem is defined.
(b) The required logic hardware is designed.
(c) The logic devices are obtained - hopefully from stock but special devices from suppliers.
(d) A printed circuit layout is drawn.
(e) The printed circuit is manufactured.
(f) The circuit is assembled.
(g) The circuit is tested.
(h) The complete system is assembled and tested.

Any mistakes which occur in any step necessitate very difficult changes in expensive hardware, and unfortunately the easiest place to make a mistake is in step (a) in the problem definition. The equivalent steps for a microprocessor-based design are,

(a) The problem is defined.
(b) An algorithm to solve the problem is worked out.
(c) The software (programme) is written.
(d) The software is checked and run in a microprocessor development system.
(e) The complete system is assembled and tested with the programme running in the development system.
(f) The complete system is assembled with the development system replaced by standard microprocessor circuit cards obtained from the manufacturer, and with the tested programme 'burned' into ROM.

The important points are these:

1. The microprocessor 'hardware', i.e. the circuit cards containing the microprocessor, the memory and the input/output modules, are identical for each and every problem, apart from

interface electronics needed to interface the input/output modules to the external components, e.g. transducers, stepping motors, etc. The microprocessor hardware, being standard can be bought ready and tested from suppliers and in fact is not needed until the complete design has been tested.

2. Errors in any step in the design will necessitate changes only in the system software, i.e. the programme, and will not affect the microprocessor hardware. Such changes are usually easy to make and are relatively cheap.

Digital system design has thus become very much a software oriented subject and electronic design engineers have made a remarkable change in professional expertise over a very short period.

CHAPTER TEN
Electrical Transducers

In any electrical measurement or control system, the first link in the chain must be a device which converts the physical quantity to be measured, and which forms the input quantity of the system, into an electrical signal which may subsequently be processed. In performing the conversion, the device abstracts energy from the input and converts it into a new form; the converting device is called a transducer. Because the transducer always abstracts some energy from the input quantity, the use of a transducer always interferes with the input system and, theoretically, a perfectly accurate measurement is impossible. Well designed transducers will, however, cause minimum disturbance.

Transducers may be divided into two basic types. A passive transducer abstracts energy from the input and converts it directly into the output signal. An example of a passive transducer is the mercury in glass thermometer which abstracts heat energy from the input medium; the energy is used to provide a visual indication of the medium temperature by expanding the mercury column, causing a change in its length. In doing this the thermometer will cause a disturbance to the input medium, i.e. heat will be absorbed from the medium when the colder thermometer is inserted.

An active transducer uses the energy it abstracts from the input as a control signal to control the transfer of energy from a power supply to produce the transducer output signal. An active form of thermometer could consist of a small piece of resistance wire, made perhaps of platinum, which is inserted into the input medium. The resistance of the transducer will change as its temperature changes. If a constant current is passed through the resistor, the voltage produced across it is dependent upon the temperature of the input medium. The output signal energy is provided, therefore, by the power supply. The thermal mass of the resistance wire sensor can be small in comparison with that of the mercury in glass thermometer, and will allow a measurement of the temperature to be made with a smaller disturbance of the input medium.

In Section 8.1, two other transducers were described which can be used to produce an electrical signal containing information regarding the depth of water in a tank. Both the transducers are active transducers, requiring an external power supply, but differ in that one produces an output signal which is in analogue form, i.e. a voltage whose amplitude is proportional to water depth, while the other produces a digital output signal requiring more than one data output line. These are good examples of a division of transducers in two further classes, those providing an analogue output signal, and those providing a digital output signal. Inside these sub-divisions there are many different types of transducer, and in fact, each physical variable may be capable of measurement, using transducers working on many different principles. In trying to describe current practice in this subject, therefore, it is, perhaps, best not to attempt to describe all the transducers which may be used for a particular input variable, but rather to describe the principles upon which a type of transducer is based, and which can be adapted to suit a wide range of physical variables.

The work in this chapter will be restricted to those transducers which produce an output signal in electrical form. Success in adapting a basic principle, to give a working transducer system, often relies heavily upon the ingenuity of the engineer.

10.1 Static properties of transducers

Before considering the properties of individual types of transducer, certain properties of transducers in general may be defined.

Range. The range of a transducer may be defined in terms of the range of the input variables over which the transducer will operate with acceptable characteristics. Thus a linear displacement transducer may give acceptable readings over an input range of, say, 0.01" to 4.0".

Linearity. The input/output relationship for a transducer system may be in the general form

$$y = a_0 + a_1 x + a_2 x^2 + a_3 x^3 + \ldots\ldots + a_n x^n \qquad 10.1$$

wehre y is the output variable and x is the input variable. The coefficients $a_0 \ldots\ldots a_n$ are the calibration factors, whose values determine the form of the relationship between y and x. IF a_2 and all higher order coefficients are zero, then the relationship reduces to

$$y = a_0 + a_1 x \qquad 10.2$$

which represents a linear relationship between the variables. Further, when a_0 is also zero, then the relationship

$$y = a_1 x \qquad 10.3$$

represents a linear relationship which passes through the origin, giving a zero output signal when the input quantity is zero. This ideal linear relationship is rarely achieved in practice, except over a very limited range of input variation.

Figure 10.1 illustrates the form of a linear input-output relationship, and also of two non-linear relationships, one containing only even order powers and one only odd powers of the variable x beyond the linear term. An arrangement of two identical transducers, differentially connected (i.e. with one deflected in a positive sense and the other deflected in a negative sense) can sometimes be used, and can be shown to give cancellation of the even order powers to give the odd-power only characteristic shown in the figure.

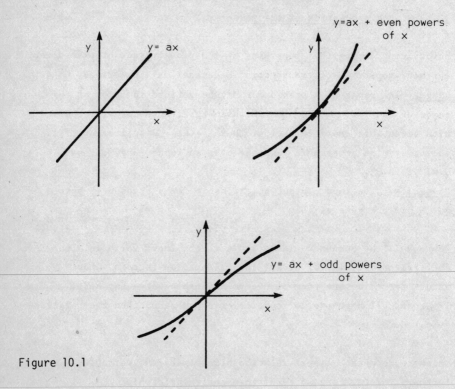

Figure 10.1

Sensitivity. The sensitivity of a system specifies the magnitude of the output variable change caused by a unit change of the input variable.

In the equation $\quad y = a_1 x \quad$ 10.3
the sensitivity is specified by the magnitude of the constant a_1. For example, if the output quantity y is measured in volts, and the input quantity x is a temperature, measured in °C, then unit of the constant a_1 is

$$a_1 = \frac{y}{x} \text{ volts per °C}$$

Resolution. The resolution of a measurement system is the smallest change in the input variable to which the measuring system will respond.

Accuracy. The accuracy of a measurement system is a measure of the difference between a measured value and the so called true value of the quantity. Bear in mind, however, that our knowledge

of the true value of any quantity is only known as a result of
one or more mreasurements of the value, even though none may,
in fact, be stated to be the true value. The true value can only
be inferred by a statistical examination of a series of measurements.
The accuracy of a system will also, in general, vary throughout
the range of the system, and may be specified over the range by
means of a calibration curve.

Drift. The drift of a system describes the gradual change in
the characteristics of the system, with age, temperature, or other
external property. Such an external change may affect the system
in either or both of two ways.

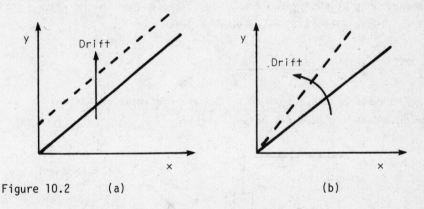

Figure 10.2   (a)   (b)

Zero drift - as shown in Figure 10.2(a), a drift in the zero level
of a system results in a shift in the whole characteristic.
Sensitivity drift - illustrated in Figure 10.2(b) results in a
change in the value of the constant a_1 and hence in the slope
of the characteristic.

Repeatability. This property describes the degree to which a
measurement of a given input value may be repeated.

The characteristics briefly mentioned in this section can be measured
by the application of steady or perhaps slowly varying signals.
Where more than one external quantity may affect the input-output
relationship, only one variable is allowed to change while a characteristic is measured. In this way a family of static characteristics covering the operation of a measuring system may be obtained.

10.2 Dynamic properties of transducers

Under dynamic conditions, when input variables may change rapidly, the system characteristics may be quite different from the static characteristics. In practice, the variation of an input quantity will not, in general, follow a simple mathematical relationship with time; a study of the dynamic behaviour of systems can, therefore, be complex. However, a knowledge of the response of a system to simple time varying inputs can allow a reasonable prediction of the response of the system to more complex time variations. In particular, a knowledge of the step response of a system, i.e. the response of the system to a sudden step change in the value of the input variable, can be very useful.

10.2.1 A zero order transducer

The simplest possible form of transducer system is one in which no time delaying elements of any sort occur.

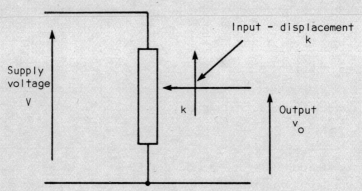

Figure 10.3

Figure 10.3 shows a simple potentiometer which is to be used to give an output voltage proportional to the displacement of the slider from the zero position. The equation for a zero order system is the ideal equation

$$y = a_1 x \qquad\qquad 10.3$$

The potentiometer output voltage may be written

$$V_0 = Vk \qquad 10.4$$

where k is the fractional displacement of the slider.
Thus when k = 0, the output voltage is zero, and when k = 1, the output voltage is V. The equation is a zero order equation, but is only true so long as there are no elements in the construction of the potentiometer which will prevent the output voltage following instantly the movement of the slider. We must ensure that the potentiometer winding has no inductance; that there is no stray capacitance between the winding or the output terminal and any other part of the circuit. These requirements are obviously impossible to fulfil. Further, the system must be considered as a whole, from both the electrical and the mechanical point of view. It must be ensured, therefore, that the moving parts must have no mass, so that no inertia time delays occur, and also that the moving parts are perfectly rigid, so that no elastic bending effects can occur. The zero order instrument is obviously an unrealisable objective, but may be approximated in some conditions for very slowly varying inputs. A consideration of this basic instrument, however, leads to a very important conclusion. In considering a transducer, the complete system must be examined, i.e. the mechanical and the electrical properties of the transducer must be examined together before its operation can be fully understood.

If a truly zero order instrument could be achieved, it would respond to any input change without delay, and produce an output signal of the same form as the input signal.

10.2.2 A first order transducer

A first order system is one which contains only one time delay element. Thus a mechanical system could contain either mass or stiffness, in addition to friction, while an electrical system could contain either inductance or capacitance in addition to resistance.

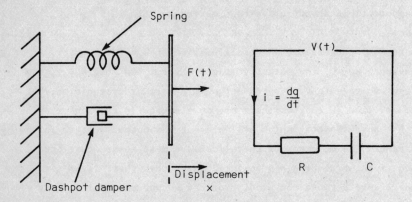

Figure 10.4 Mechanical and electrical first order systems

Figure 10.4 shows two simple systems which are of first order. The mechanical system consists of a moving plate, which has negligible mass, connected to a frame by a spring. The spring has a stiffness of k, where stiffness is defined as the force required to displace the spring by unit distance. The system has also mechanical friction represented by the dashpot damper, producing a resistive force equal to $r\frac{dx}{dt}$, where $\frac{dx}{dt}$ is the velocity of the plate.

The electrical first order system consists of a capacitor C in series with an electrical resistor R across a voltage source. The inputs of the two systems are a time varying force F(t) in the mechanical case, and a time varying voltage V(t) in the electrical case.

Each system may be analysed as follows.

<u>Mechanical system</u>

Equating the forces on the plate

$$F(t) = r\frac{dx}{dt} + kx$$

or
$$\frac{1}{k}F(t) = \frac{r}{k}\frac{dx}{dt} + x \qquad 10.5$$

The term $\frac{r}{k}$ is known as the 'time constant' τ of the system.

The term $\frac{1}{k}$ is known as the 'static sensitivity'.

The general form of the equation is

$$\frac{1}{k}F(t) = \tau\frac{dx}{dt} + x \qquad 10.6$$

Electrical system

Summing the voltages round the circuit

$$V(t) = R\frac{dq}{dt} + \frac{1}{C}q$$

where q is the charge and $\frac{dq}{dt}$ is the circuit current

$$CV(t) = CR\frac{dq}{dt} + q \qquad 10.7$$

The product CR is again known as the 'time constant'
The term C also becomes the 'static sensitivity'.
Again, the general form of the equation is

$$CV(t) = \tau\frac{dq}{dt} + q \qquad 10.8$$

It can be seen that equations 10.6 and 10.8 are of the same form.
If corresponding terms are taken from the equations for the two systems, it can be seen that a correspondence may be drawn between

electrical resistance	and	mechanical resistance
capacitance	and	$\frac{1}{\text{stiffness}}$
voltage	and	force
electrical charge	and	mechanical displacement
current	and	velocity

Solution to the equations may be obtained for various inputs.

Steady state solution.

The steady state solution is obtained when the force or voltage has been applied for a sufficiently long time for all changes in the system to have been completed. All derivative terms are then zero. The equations reduce to

$$\frac{1}{k}F(t) = x \qquad 10.9$$

and
$$CV(t) = q \qquad 10.10$$

The equations have reduced to the ideal static equations of section 10.1. If the voltage or the force is taken as the input quantity of equation 10.3 then

$$x = \frac{1}{k}F(t) \qquad 10.9$$

and
$$q = CV(t) \qquad 10.10$$

Solution when the input is a step function.

A step function input is produced when the input quantity makes an abrupt change of magnitude.
Consider a force input which is zero up to time t = 0 when it

becomes instantly a force of F Newtons.

The eventual steady state value of the displacement, x_s, is given by the steady state equation

$$x_s = \frac{1}{k}F \qquad 10.9$$

The response of the first order mechanical system to such a step input is

$$x = x_s(1 - e^{\frac{-t}{\tau}}) \qquad 10.11$$

Similarly the response of the first order electrical system to a step voltage input which rises instantly from zero to a value V at time t - 0 is given by

$$q = q_s(1 - e^{\frac{-t}{\tau}}) \qquad 10.12$$

where, again, q_s is the steady state solution given by the steady state equation

$$q_s = CV \qquad 10.10$$

These responses are illustrated in Figure 10.5. The responses show that the first order mechanical and electrical systems are completely similar in operation, and can be considered as analogous systems. A useful dynamic characteristic can be derived from the first order responses. The 'settling time' of a system is defined as the time which elapses after the application of a step input for the output of the system to reach and stay within a given tolerance

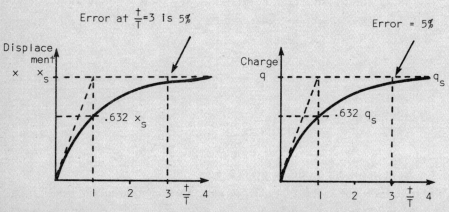

Figure 10.5 Step responses of first order mechanical and electrical systems

band about its final value. A system or instrument with a small settling time has, therefore, a fast response. For a first order instrument, a 5% tolerance settling time is equivalent to approximately 3 time constants.

10.2.3 A second order transducer

If the mass of the moving system is not negligible, or if the electrical system has inductance included, the two systems become second order. These systems are illustrated in Figure 10.6.

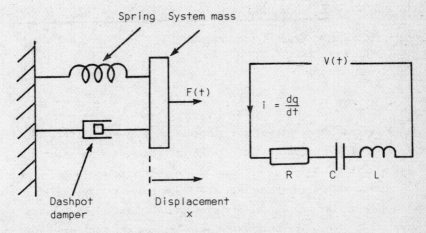

Figure 10.6 Mechanical and electrical second order systems

The equations to the two systems are:
for the mechanical system

$$F(t) = m\frac{d^2x}{dt^2} + r\frac{dx}{dt} + Kx \qquad 10.13$$

where m is the mass of the moving parts,
and for the electrical system

$$V(t) = L\frac{d^2q}{dt^2} + R\frac{dq}{dt} + \frac{1}{C}q \qquad 10.14$$

where L is the added series inductance.

The response of the second order system to a step input is shown in Figure 10.7. The steady state response, when all the transients have ended, is the same as in the previous examples, and is

402 Basic Electrical Engineering and Instrumentation for Engineers

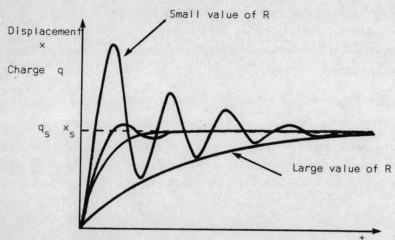

Figure 10.7 Second order system response

$$x_s = \frac{1}{k}F \qquad 10.9$$

or

$$q_s = CV \qquad 10.10$$

With a second order system, the response becomes oscillatory if the value of the mechanical or electrical resistance is reduced below a critical value. The choice of the correct constants, i.e. the correct values for the mass, stiffness and the frictional resistance of the mechanical circuit, is important if the system output is to reach its true value quickly, and yet with minimum overshoot.

In general, a transducer system will consist of both mechanical and electrical components, and its dynamic response is determined by the combined effects of all of its physical constants. The analysis of such systems, to determine their dynamic properties, is a difficult process, and is outside the scope of this book. It must be remembered, however, that a simple analysis of a transducer, which may give quite accurate results for operation of the system at low frequencies, will usually become useless as the operating frequency is increased.

10.3 Variable resistance transducers

In this section, transducers which depend for their operation upon a changing resistance value will be considered.

10.3.1 Variable potentiometer transducers

Variable potentiometers are in wide use in measurement systems, mainly because of the ease with which they can be adapted to give an indication of linear or angular displacement, but also because of their relatively high electrical output. Potentiometers for measurement purposes are usually made by winding a resistance wire upon a rigid former, the variable sliding contact being a pressure contact upon the wire surface. Alternatively, the resistive element may be a carbon film, deposited upon a former. The output voltage per unit deflection of the sliding contact is dependent upon the supply voltage to the potentiometer; within limits imposed by the heating of the element, the sensitivity may be increased by increasing the supply voltage.

The resolution of the potentiometer is limited by electrical noise generated in the potentiometer. Electrical noise is defined as small random voltages generated within the system, which mask smaller voltages generated correctly by the transducer. Such noise voltages are generated in several ways.

(a) Contact noise. A major source of noise is the sliding contact. If the potentiometer is wire wound, the sliding contact will, in general, contact several adjacent wound turns at the same time. If this were not so, then the slide would necessarily be narrow and pointed, and would, therefore, penetrate into the valleys between adjacent turns, causing rapid wear. As the contact slides over the turns, it alternately short circuits several adjacent turns, causing small but definite steps to occur in the output voltae characteristic, which cannot be distinguished from similar small steps in the output voltage due to small displacements of the slider. Another source of contact noise is wear or dirt upon the track.

404 Basic Electrical Engineering and Instrumentation for Engineers

(b) Thermal or Johnson noise. Thermal noise is characterised by randomness of amplitude and frequency distribution. The noise signal is developed because of the random motion of the current carriers in the electrical conductor. Noise of this type is spoken of as 'white noise', its frequency distribution being constant over the whole spectrum. Consequently, the total noise power developed is proportional to the frequency bandwidth of the signal processing system.

The root mean square noise voltage generated in a resistor is given by

$$E_{noise} = \sqrt{4kTR\delta f} \text{ volts} \qquad 10.15$$

where

T is the absolute temperature, °K

R is the resistor value, ohms

δf is the frequency bandwidth of the signal processing system, Hz

k is the Boltzmann constant, of value equal to 1.38×10^{-23} joule/°K

The linearity of a potentiometer device is affected by the loading applied to the transducer output.

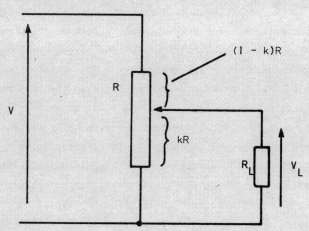

Figure 10.8

Figure 10.8 shows a potentiometer of total resistance R ohms, loaded at its output by a resistor R_L.

The linearity is determined by the ratio

$$m = \frac{\text{total potentiometer resistance}}{\text{load resistance}}$$

$$= \frac{R}{R_L} \qquad 10.16$$

If the load resistor R_L is disconnected, the output voltage V_{L_0} is given by

$$\frac{V_{L_0}}{V} = k \qquad 10.17$$

With the load resistor reconnected, the output voltage is

$$\frac{V_L}{V} = \frac{\frac{R_L kR}{R_L + kR}}{(1-k)R + \frac{R_L kR}{R_L + kR}}$$

or

$$\frac{V_L}{V} = \frac{1}{1 + mk(1-k)} \qquad 10.18$$

The difference between the unloaded output voltage and the loaded voltage, expressed as a fraction of the unloaded voltage is

$$\frac{V_{L_0} - V_L}{V_{L_0}} = \frac{mk(1-k)}{1 + mk(1-k)} \qquad 10.19$$

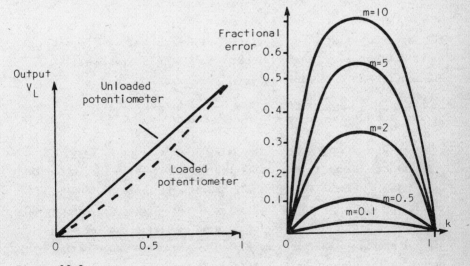

Figure 10.9

Figure 10.9 shows typical unloaded and loaded output voltage characteristics, and also a plot of the fractional error for different values of the ratio m.

The following conclusions may be drawn:

(a) The maximum error occurs at mid-deflection, i.e. when $k = 0.5$,

(b) For a maximum error of 1%, the load resistor value must not be lower than about 25 times that of the complete potentiometer. Variable potentiometers of the types described are most useful for measuring relatively large displacements. Resolutions of the order of 10^{-3} inch are obtainable. A linearity of the order of 0.01% of full scale is typical.

10.3.2 Strain gauges

For small deflections, the use of strain gauges is very common. Strain is defined as the change in length per unit length of the body. A wire resistance strain gauge consists of a length of fine wire, usually folded lengthwise, so that although small in overall dimensions, its effective length in the direction it is to be strained is large.

The resistance of a length of uniform wire is

$$R = \rho \frac{L}{A} \text{ ohms} \qquad 10.20$$

where L and A are its length and cross sectional area, and ρ is the material resistivity. When the wire is strained, the resistance of the wire changes, due to a change in all of the parameters. The sensitivity of a gauge is usually expressed in the form

$$\text{Strain sensitivity} = \frac{\delta R/R_0}{\delta L/L_0} = S, \text{ the Gauge Factor} \qquad 10.21$$

For most wire strain gauges, the value of S is about 2, so that

Fractional change in resistance = 2 x fractional change in length

Commercially available wire strain gauges are formed by bonding the folded wire grid to an insulating former. The gauges are commonly constructed from copper-nickel foil upon a plastic backing strip. Special adhesives have been developed by which the gauge may be bonded to the element being strained. For maximum sensitivity, the length of the gauge must be mounted in the direction of the strain. Gauges are available in single or multiple element rosettes, the multiple elements allowing the resolving of strains of unknown direction. Gauges are available with a wide range of resistance values, with 120 Ω, 350 Ω, 600 Ω and 1000 Ω as standard preferred values.

A typical gauge specification is

$R = 120 \ \Omega \pm 2\%$ GAUGE FACTOR $S = 2.1 \pm 1\%$

Sensitivity is again limited by heating, normal current levels being of the order of 10 to 30 mA.

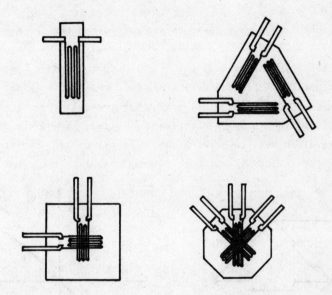

Figure 10.10

Figure 10.10 shows several different styles of construction which are commercially available.

In the early 1960s, semiconductor strain gauges became available. These strain gauges are normally produced from single crystal silicon. A distortion of the crystal lattice, due to externally applied stress, produces large changes in the electrical resistivity of the material. This effect is known as the piezoresistance effect. In bulk form, silicon is a brittle material, but in very thin strip, is sufficiently flexible to be bent round a radius of as little as 0.25 inch.

Silicon strain gauges are very small, typical dimensions being

0.3 inch overall length, .01 inch width, and .001 inch thick. The great advantage of the semiconductor strain gauge is in its sensitivity; the gauge factor S may exceed 200. Gauges are also available with both positive and negative gauge factors. A maximum strain of the order of 3000 μ strain (i.e. fractional change in length of 3000×10^{-6}) is normally quoted.

Strain gauge techniques are applicable down to strains of the order of 10^{-6} or 10^{-7}.

The Wheatstone bridge provides a convenient method for the measurement of gauge resistance. The bridge will normally be balanced while the gauge is unstrained; the bridge output as the gauge is then strained can provide a sensitive and reasonably linear indication over a wide range of strain. Figure 10.11 shows a Wheatstone bridge witih one arm consisting of a strain gauge. The gauge has an unstrained resistance of R ohms; the bridge can, therefore, be balanced by including resistors of equal value in the other arms.

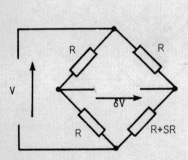
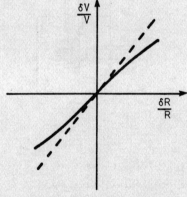

Figure 10.11

When strained, the gauge resistance changes by δR ohms. If the bridge detector is of high impedance, so that the detector current may be neglected, then the bridge output voltage is

$$\frac{\delta V}{V} = \frac{\delta R}{R} \times \frac{R}{2(2R + \delta R)} \qquad 10.22$$

The voltage/defection characteristic is, therefore, non-linear, but for small defelctions, i.e. if δR << 2R, the equation reduces

to
$$\frac{\delta V}{V} = \frac{\delta R}{R} \times \frac{1}{4} \qquad 10.23$$

The characteristic may then also be written as

$$\frac{\delta V}{V} = \frac{S}{4} \times \text{strain} \qquad 10.24$$

If the bridge is arranged to have two active arms, i.e. if gauges in adjacent arms are strained in the opposite sense, or if gauges in opposite arms are strained in the same sense, then the sensitivity is doubled, giving

$$\frac{\delta V}{V} = \frac{S}{2} \times \text{strain} \qquad 10.25$$

With four active arms, with gauges strained with the appropriate sense,

$$\frac{\delta V}{V} = S \times \text{strain} \qquad 10.26$$

The voltage supply to the bridge may be either direct or alternating current. If, however, a.c. supply is used, the effect of stray capacitance may make a null balance difficult to achieve. A small trimmer capacitance across one or other arms of the bridge may be adjusted to give a better balance. It must be remembered, however, that an a.c. bridge can only be balanced at one frequency, i.e. the balance is frequency sensitive. Any harmonic distortion components existing in the bridge supply will tend to mask the balance null, since the bridge will be unbalanced at these frequencies.

An increase in temperature will affect the gauges in the following ways:

(a) The resistance of the gauge increases.
(b) The gauge will change in length due to the effect of the temperature upon the test piece to which it is attached.
(c) The gauge factor changes.

Corrections for these effects may be applied through good measuring circuit design.

Gauges can also be made which have gauge factors of opposite sign, but with temperature coefficients of resistivity of the same sign. If such gauges are connected in adjacent arms of a measuring bridge, the bridge will tend to remain balanced over reasonable temperature changes.

Temperature compensation can also be achieved by circuit design in a variety of ways. An identical but unstrained gauge may be connected into an adjacent arm of the bridge. If this inactive gauge is mounted close to the active gauge on the test piece, but in such a direction as to be unstrained, it will compensate for temperature effects in the active gauge.

A method which is useful when equal positive and negative strains are available, for example on opposite surfaces of a bending beam, is to use four identical gauges, one in each arm of the measuring bridge. Adjacent arms of the bridge should include gauges which are subjected to strains of opposite sign.

10.3.3 Temperature sensitive resistors

A general equation giving the variation of resistance of a material with temperature is

$$R = R_o (1 + \alpha_o t + \beta_o t^2 + \gamma_o t^3 \ldots\ldots\ldots) \qquad 10.27$$

For many materials, the values of the constants α_o, β_o, etc. are very small with the result that the equation

$$R = R_o(1 + \alpha_o t)$$

is a fair approximation to the resistance characteristic over a restricted range of temperature variation. The constant α_o is the resistance temperature coefficient of the material, at the base temperature from which the temperature change t is calculated. R_o is the resistance of the piece of material at the same base temperature. The metals in particular show reasonably linear resistance/temperature characteristics over a restricted temperature range. For temperature sensing applications, however, the value of the temperature coefficient is rather small.

Material	Resistance temperature coefficient near room temperature/$^{\circ}$C
Aluminium	0.0045
Copper	0.0043
Gold	0.0039
Mercury	0.00099
Silver	0.00414
Platinum	0.00391
Lead	0.0041
Tungsten	0.00486
Manganese	± 0.00002
Carbon	− 0.00070

Figure 10.12

Figure 10.12 lists the values of α_o for the common metals, and also for carbon, which exhibits a negative temperature coefficient. The most commonly used material for resistance thermometers is pure platinum. Platinum is linear within ± 0.3% over the temperature range −20°C to + 120°C.

If linearity of the resistance-temperature characteristic is not of prime importance, much greater sensitivity may be obtained by using a semiconductor sensing element instead of a metal sensor. As explained in section 4.1, the resistivity of semiconductor materials decreases with an increase in temperature.

Many semiconductor temperature sensors are now commercially available in the form of beads, open or glass encapsulated, discs or rods, or washers. The materials are various mixtures of manganese, nickel, cobalt, copper and other metals, formed with binders into the various shapes and sintered at very high temperature (1100 to 1400°C). They become a hard ceramic material. Leads can be attached if required.

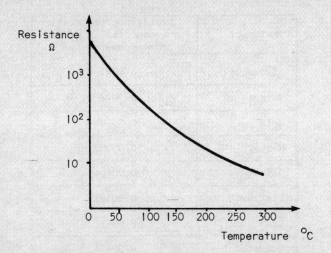

Figure 10.13 Thermistor resistance/temperature characteristic

Figure 10.13 shows the resistance-temperature characteristic for a typical thermistor. Because of the wide resistance variation, the resistance axis scale is plotted in logarithmic form. The general form of the thermistor resistance characteristic is

$$R_{Th} = R_o e^{(\frac{B}{T} - \frac{B}{T_o})} \qquad 10.28$$

where R_{Th} is the thermistor resistance in ohms.
R_o is the thermistor resistance at a base temperature of T_o degrees absolute.
B is the thermistor constant, which is a characteristic of the particular material. Typical values for the constant B lie in the range 2000 to 4500.

The effective temperature coefficient may be derived using the definition

$$\alpha = \frac{1}{R_{Th}} \frac{d R_{Th}}{dT} \qquad 10.29$$

giving

$$\alpha = \frac{1}{R_o e^{(\frac{B}{T} - \frac{B}{T_o})}} R_o e^{(\frac{B}{T} - \frac{B}{T_o})} (-\frac{B}{T^2})$$

or
$$\alpha = -\frac{B}{T^2} \text{ ohms/ohm/}°C \qquad 10.30$$

For a typical value of B of 4000
$$\alpha = -.047 \text{ ohms/ohm/}°C \text{ at } 20°C$$

In temperature sensing applications, thermistors are commonly used in a Wheatstone bridge circuit. The basic bridge techniques are similar to those described for use with strain gauges.

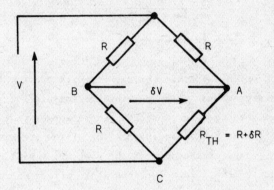

Figure 10.14

Figure 10.14 shows a simple Wheatstone bridge with the thermistor sensor forming one arm of the bridge.

With a high impedance detector, the bridge output is

$$\frac{\delta V}{V} = \frac{\delta R}{R} \frac{R}{4R + 2\delta R} \qquad 10.22$$

The rate of change of the bridge output voltage may be estimated as follows:

Variation is due to the R - R_{Th} side of the bridge only.

With respect to point C, the potential of point A is

$$V_A = V \frac{R_{Th}}{R + R_{Th}} \qquad 10.31$$

whence
$$\frac{dV_A}{dT} = V \frac{R}{(R + R_{Th})^2} \frac{dR_{Th}}{dT} \qquad 10.32$$

For use around a given temperature T, it is useful to be able to maximise the rate of change of output voltage.

At the temperature T, the thermistor resistance R_{Th} is fixed. The bridge sensitivity may, therefore, be maximised by choosing the best value for R; this is the value of R which makes the factor

$$\frac{R}{(R + R_{Th})^2}$$

a maximum.

Thus, putting

$$\frac{d}{dT} \frac{R}{(R + R_{Th})^2} = 0$$

we obtain the optimum relationship

$$R = R_{Th} \qquad 10.33$$

Errors may occur in thermistor bridge circuits due to self heating of the thermistor or due to temperature or noise effects in long leads which may be used to connect a remote sensor to the bridge. Self heating errors are caused by heating of the thermistor above its ambient temperature because of the power dissipation in the thermistor due to the bridge currents. This effect can obviously be reduced by decreasing the bridge supply voltage, but this also reduces the sensitivity. If the temperature being measured is relatively constant, or only changing slowly, several measurements may be made at different transducer currents. A graph of measured temperature against transducer current may be extrapolated backwards to give a deduced value at zero transducer current. Another possibility is the excitation of the bridge circuit by short duration, high amplitude voltage pulses. This method relies on the measurement being completed during the excitation pulse before the temperature of the sensor has time to change appreciably.

Errors due to long sensor leads may be minimised as shown in Figure 10.15, which shows the inclusion of dummy leads in an adjacent bridge arm, the dummy leads being physically adjacent to the sensor leads. Because similar effects may be expected to occur in both the active and dummy leads, their effect on the bridge is negligible.

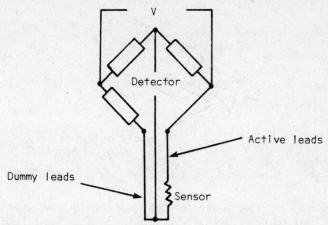

Figure 10.15

10.3.4 Hot wire measurements on fluids

Measurements of fluid velocity may be made by inverting a temperature sensitive resistor in the fluid and passing an energising current through it. The power dissipated in the resistor will cause its temperature to rise above that of the fluid. Analysis of the heat transfer between the resistor and the fluid is not simple, because in practice the heat transfer coefficient is complex in its dependence upon surface conditions. In the steady state, however, the resistor will attain a steady temperature in a time relatively long when compared to the thermal time constant of the system.
If the fluid is in motion the temperature attained by the resistor is a function of the velocity of the fluid. Measurements of the resistance of the transducer can, by calibration, be used to give velocity measurements.
A suitable sensor material will have a high temperature coefficient of resistance, a convenient resistivity to enable sensors of reasonable dimensions to be produced, robustness to withstand turbulent flow conditions, and high oxidation resistance. Tungsten, platinum, or platinum-iridium alloy wire sensors are commonly used in the form shown in Figure 10.16.

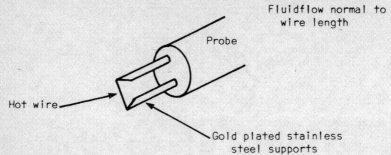

Figure 10.16

Such sensors can respond over a frequency range up to about 500 Hz. For applications which do not require such a rapid response, thermistors have found wide use, giving a frequency response up to about 2 Hz. Hot film sensors are available, consisting of a conducting film of perhaps platinum held on a ceramic substrate, and having a better frequency response than wires.

Effects due to a variation in the temperature of the fluid itself may be compensated by the inclusion of a further sensor in the fluid, shielded from flow effects, whose resistance is then a function of the fluid temperature only. The combination of the two signals, to give an output dependent upon the flow rate only, is, however, not a simple problem.

There are two basically different methods by which transducer systems such as the hot wire sensor system may be operated.

(a) The constant current method.

A constant known current is passed through the resistance sensor. The equilibrium temperature is reached when the power dissipated in the resistor is balanced by the heat loss from the sensor to the fluid. The resistance of the sensor at the balance temperature is measured by determining the voltage across it.

(b) The constant temperature method.

Figure 10.17 shows a bridge circuit which includes the hot wire sensor as one of its arms. The bridge can be balanced using the adjustable arm R_3. Once balanced, the bridge will unbalance again if the sensor resistance changes, i.e. if the sensor temperature

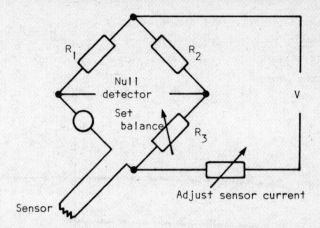

Figure 10.17

changes. At any other fluid flow rate, therefore, the temperature may be brought back to its balance value by adjusting the bridge supply voltage, so altering the current in the sensor. The sensor current, which may be measured by using an ammeter, is then a function of the fluid flow rate. Arrangements may conveniently be made for such a system to be self-balancing providing a continuous record of flow rate changes.

Notice carefully that flow rate measuring systems as described are not absolute measuring systems, but require calibration in fluids moving with known velocity. A calibration curve will be required to cover the full range of the device, because such systems are not linear in operation.

10.3.5 Other variable resistance transducers

Many other forms of variable resistance transducer have been adapted to suit different physical variables.
The silicon photodiode, operated with reverse voltage bias, functions in effect as a light sensitive resistor.
The diode consists basically of a light sensitive p-n junction. When irradiated with light, electron-hole pairs are generated on both sides of the junction. Extra minority carriers are then

418 Basic Electrical Engineering and Instrumentation for Engineers

available, and the effective resistance of the diode in the reverse biased sense is reduced. The cell resistance is, therefore, varied by the cell illumination, the resistance falling as the illumination intensity increases.

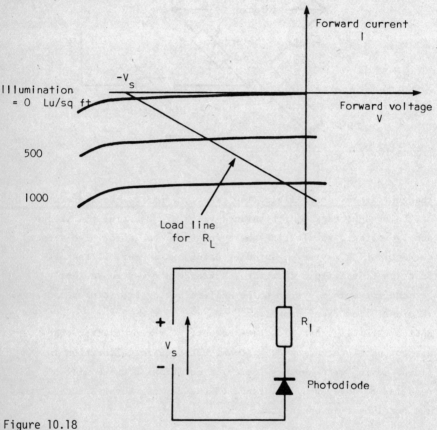

Figure 10.18

Figure 10.18 shows the characteristic of such a photodiode, together with a load line for the resistor load R_L. The reverse resistance of the diode is also temperature sensitive, reducing with increasing temperature.

Acoustic variable resistance transducers have been in common use for many years in the form of the carbon granule microphone, which is the standard microphone used in telephone systems. Microphones are of two basic types; pressure microphones, which operate because

of the acoustic pressure upon one side of the microphone diagram,
and velocity, or pressure gradient microphones, which depend for
their operation upon the acoustic pressure difference between
the two sides of the diaphragm. The carbon granule microphone
is a pressure device, and makes use of the fact that the resistance
of a mass of carbon granules varies with the pressure applied
to the granules. A direct current is passed through the granules,
which are mounted so that the diaphragm can exert pressure upon
them. Sound waves striking the diaphragm vary the pressure exerted
upon the granules, so varying the resistance of the microphone.
The current thus varies in accordance with the sound pressure
applied to the diaphragm.

10.4 Variable reactance transducers

Various transducers can be constructed which operate by means
of the production of a variable reactance, which may be measured
by means of alternating current techniques.

10.4.1 Variable capacitance transducers

The capacitance of a parallel plate capacitor, which consists
of a pair of parallel plates, each of area A square metres, spaced
a distance d metres, and separated by a material of relative permittivity ϵ_r is given by

$$C = \frac{\epsilon_o \epsilon_r A}{d} \text{ Farads} \qquad 10.34$$

where ϵ_o is the absolute permittivity of free space.
A variable transducer can be produced by any system by which the
input variable can alter the value of the capacitor variables
ϵ_r, A or d. The most common capacitor transducers are those in
which the spacing d is varied, giving a direct reading of distance.
The area A may be varied, perhaps by altering the overlap between
two plates. The dielectric constant may be varied by moving in
or out of the space between the two plates a wedge shaped dielectric.
These effects are illustrated in Figure 10.19, which shows three
different methods of producing a displacement transducer.

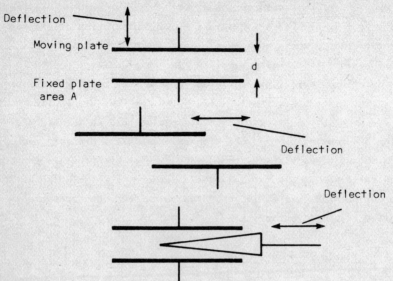

Figure 10.19

A major difference between variable resistor and variable capacitor transducers is the impedance level of the sensor. Resistive sensors can be produced, in general, with a resistance value of any order which is convenient to use. Capacitor devices are, on the other hand, normally of extremely high reactance.

Consider a parallel plate capacitor consisting of circular plates of diameter 1 cm, spaced by 1 mm. The capacitance of such a component is

$$C = \frac{0.7}{10^3} \text{ pF}$$

if the dielectric is air.

At a frequency of 1 kHz, the reactance of the capacitor is of the order of 230 MΩ.

The reactance of the capacitor may be reduced by using a higher measuring frequency. At higher frequencies, however, the capacitive sensor will cease to behave as a pure capacitor, and would be represented by the equivalent circuit of Figure 10.20.

In the figure, the resistor R_s and L_s represent the series resistance, and inductance of the capacitor leads respectively, while R_p repres-

Electrical Transducers 421

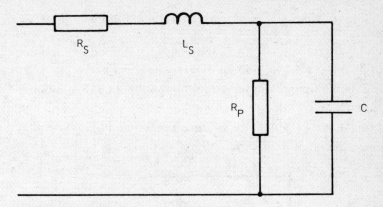

Figure 10.20 High frequency equivalent circuit of a capacitor

ents the effective leakage in the capacitor dielectric. The linearity of the variable spacing capacitor transducer is given, for very small variations in spacing, by differentiating equation 9.34, with respect to the spacing d.
Thus

$$\frac{dC}{dd} = \frac{-\epsilon_o \epsilon_r A}{d^2} = -\frac{C}{d}$$

This may be expressed as

$$\frac{\delta C}{C} = -\frac{\delta d}{d} \qquad 10.35$$

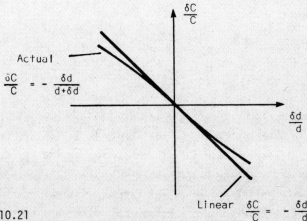

Figure 10.21

For larger changes in spacing, the relationship may be shown to be

$$\frac{\delta C}{C} = \frac{-\delta d}{d + \delta d}$$

and the relationship between displacement and variation in capacitance is no longer linear. The relationship is illustrated in Figure 10.21.

Alternating current bridge techniques can be applied satisfactorily to capacitive transducer measuring systems.

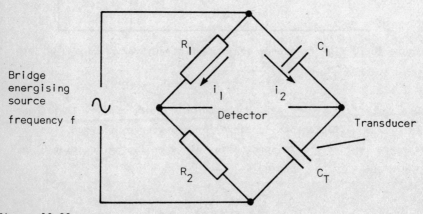

Figure 10.22

Figure 10.22 shows a relatively simple a.c. bridge. When the bridge is balanced

$$i_1 R_1 = i_2 \frac{1}{j\omega C_1}$$

and

$$i_1 R_2 = i_2 \frac{1}{j\omega C_T}$$

whence

$$C_T = C_1 \frac{R_1}{R_2} \qquad 10.36$$

The precautions outlined previously for a.c. bridges are equally applicable here; the energising source should be reasonably free from harmonic distortion. A problem commonly met when using a.c. bridges to measure small reactance changes is the effect of stray capacitances between various leads and earth, or between various leads. Readings taken with such a bridge can be checked by reversing connections, for example, to the detector, or from the bridge source, or to the ratio arms R_1 and R_2. If stray capacitance effects are negligibly small, the same measurement results will be obtained. For maximum bridge sensitivity, the two capacitors should be equal in value, and the resistor values should equal that of the capacitive reactance at the measuring frequency. A good balance will not be achieved unless both capacitors are equally loss-free.

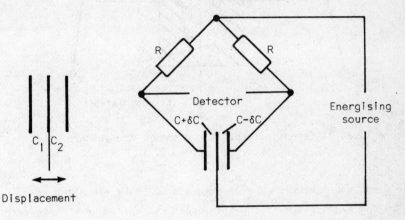

Figure 10.23

The linearity of the capacitor sensor may be improved by using a differentially connected transducer, as shown in Figure 10.23, which also shows how such a transducer could be connected in two adjacent arms of the a.c. bridge.

Displacement of the central plate will increase one capacitor and decrease the other. The change in capacitance of the two capacitors may be made additive by including them in adjacent arms of the measuring bridge.

Figure 10.24 shows schematically a differential capacitor pressure transducer.

An alternative measuring system is used with the capacitor microphone which, because of its stable characteristics and linearity, forms a standard piece of equipment in acoustic laboratories. The microphone, which is a pressure actuated device, consists of a very thin diaphragm, stretched to raise its resonant frequency.

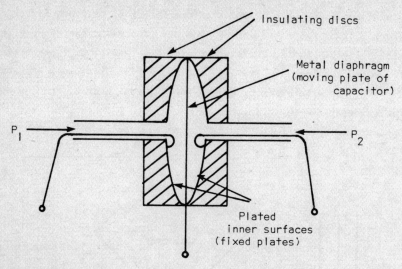

Figure 10.24 Differential capacitor pressure transducer

The capacitance between the diaphragm and a back plate, which forms the fixed plate of the transducer, is varied by the pressure variations exerted upon the diaphragm by the sound pressure waves. The capacitor microphone is polarised by a direct voltage of several hundred volts via a resistor.

The capacitor charges to the potential of the polarising voltage. Rapid variations in the capacitor value produce voltage changes across the capacitor, because the time constant of the capacitor-resistor circuit is too long for the charge stored in the capacitor to change. A slow change in capacitor value, however, produces only a very small capacitor voltage, and this method produces zero output voltage for static deflections of the transducer.

Such a frequency response is perfectly adequate for acoustic purposes. To prevent loading effects, the circuit connected across the capacitor transducer must be of very high impedance, usually of the order of 10 MΩ. For this purpose, many capacitor microphones are equipped with very high impedance buffering amplifiers at the microphone end of the microphone cable.

10.4.2 Variable inductance transducers

Many different types of transducer can be classified loosely under the heading of variable inductance transducer.
The inductance of a uniformly wound toroidal coil of length ℓ metres and cross-sectional area A square metres is given by

$$L = \mu_o \mu_r \frac{AN^2}{\ell} \text{ henrys} \qquad 10.37$$

where N is the total number of turns

μ_o is the absolute permeability of free space

and μ_r is the relative permeability of the magnetic core.

The inductance of the coil may be varied by changing any of the equation variables. However, changing the physical dimensions of an inductor is not normally a convenient process, especially as the coils are normally wound upon a ferromagnetic core in order to reduce the overall physical size.

Various methods are in use whereby the inductance may be varied. If the magnetic core path is completed, via a ferromagnetic armature

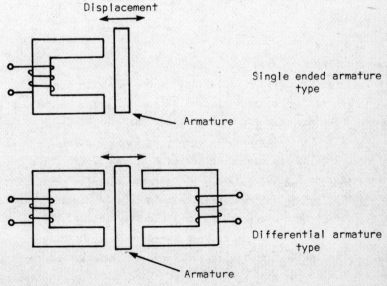

Figure 10.25

which may be moved relative to the remainder of the core, a change in the air gap in the magnetic circuit produces a change in the inductance. Figure 10.25 shows schematically both single-ended and differential forms of this type of transducer.

In the differential form, the inductance of the two coils will change in opposite senses; the coils may conveniently be included in adjacent arms of the measuring bridge.

A variation of this form of differential transducer, which is suitable for use in displacement measuring systems, is illustrated in Figure 10.26.

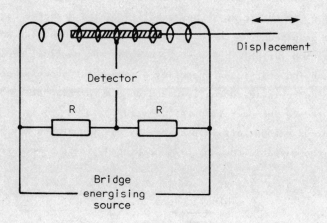

Figure 10.26

As in the previous case, the inductance, and hence the impedance, is varied by changing the properties of the magnetic circuit. This is achieved by making the core of the form of a ferromagnetic plunger which may be moved axially along the centre tapped inductor. The figure shows the transducer connected in an a.c. bridge with the two halves of the device in adjacent bridge arms. If the two halves are identical, and if the plunger is centrally placed in the inductor then the bridge will be balanced. A deviation in the position of the plunger will unbalance the bridge. The use of a differential system in this way produces a relatively linear output characteristic for a limited range of deviation of the plunger. One of the most commonly used inductive type transducers

is the linear variable differential transformer, or LVDT as it
is widely known. This device is also basically a displacement
measuring transducer. It is based upon a variable transformer
effect, and is made in differential form to improve the linearity.

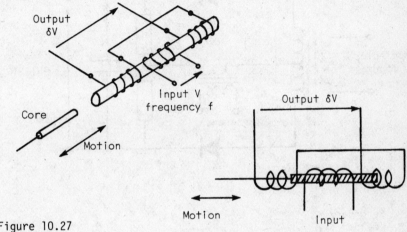

Figure 10.27

Figure 10.27 shows two identical secondary windings which are
connected in series but in opposing directions. The primary winding
is symmetrically placed with respect to the secondary windings.
With the sliding magnetic core in the central position, the whole
device is symmetrical; the e.m.f's generated in the secondary
windings are then equal, and the output voltage is, therefore,
zero. Deviation of the core from the central position causes one
secondary e.m.f. to increase and the other to decrease; the output
voltage, being the difference between the two, thus increases
for a deviation of the core on either side of the null position.
With correct design the magnitude of the output voltage is very
nearly proportional to the deviation of the core from the null.
With a sinusoidal input voltage the two secondary voltages are
also sinusoidal. Because of the differential connection, the phase
of the output voltage changes by 180° as the core moves through
the null position. If the detector used with the measuring system
is phase sensitive, it is possible to discriminate between deflections

on different sides of the null. Figure 10.28 shows a phase-sensitive detector which uses as its phase reference the phase of the a.c. source which energises the primary of the LDVT.

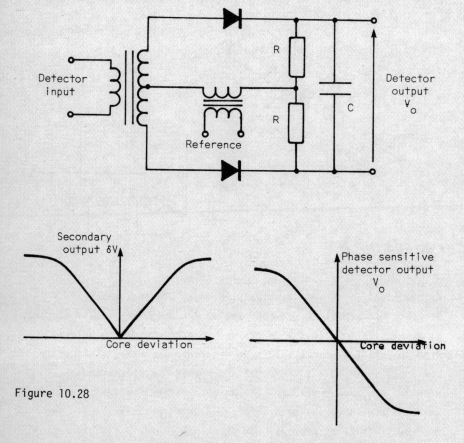

Figure 10.28

The detector circuit itself is basically a bridge circuit, which, from the point of view of the reference signal is balanced. With zero input to the detector, therefore, the output voltage V_o is also zero. The input to the detector should be either in phase with, or 180° out of phase with the reference signal. It then adds to the reference signal in one half of the bridge, and subtracts from the reference signal in the other half of the bridge, resulting in the output characteristic also shown in Figure 10.28. In order to achieve the correct phasing of the detector input signal, some phase correction will normally have to be applied to the output of the LDVT.

If the transducer is to be used only on one side of the null position i.e. accepting a smaller linear deflection range, then of course phase sensitive detection is not necessary.

The output voltage from the LDVT at the balance point will not, in general, be zero. The effect of harmonic distortion in the input voltage and also stray capacitance effects leads to a small minimum voltage which obscures the null point. LDVT's are available with ranges of up to several inches, with a linearity of about $\frac{1}{2}$% of full scale, and sensitivity of the order of 5 volts per volt per inch of deflection. With a normal energising voltage of, say, 10 V r.m.s. this will give an output voltage of 50 mV per thousandth of an inch deflection. It is possible, using these transducers, to resolve to about 10^{-6} inch.

10.5 Electrodynamic transducers

Direct velocity sensing transducers may be based upon the relationship

$$e = Blv \text{ volts} \qquad 10.38$$

where e is the e.m.f. induced in a conductor, length l metres, moving with velocity v metres per second in a magnetic flux density B teslas (Wb/m^2).

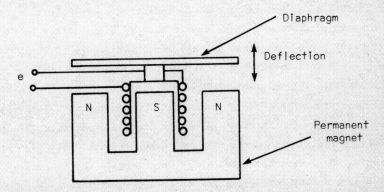

Figure 10.29

An example of such a transducer is illustrated in Figure 10.29 which shows a moving diaphragm or table which is supported by a spring mount so that a moving coil which is attached to it moves axially in the field produced by a permanent magnet.

For a sinusoidal velocity of the coil in the magnetic field the induced e.m.f. e will also be sinusoidal in form. The device is reversible in that the application of a sinusoidal e.m.f. to the coil will result in a sinusoidal velocity in the coil and diaphragm. Good examples of such transducers are the moving coil loudspeaker, which is almost universally used for audio applications (electrostatic-variable capacitor loudspeakers are also used) and the moving coil microphone. These two devices, based upon the same principle, are good examples of overall system design in that their construction must be carefully related to the conditions under which they are used. The moving coil loudspeaker is designed to accept relatively high power alternating current, and to convert this power into acoustic power in space in the form of a radiated sound wave. The design requires a large moving diaphragm or cone in order to couple efficiently into free space. Particular attention is paid to the flexible mount in order to produce a very free moving system with linear characteristics. In order to respond efficiently at very low frequencies (down to 10 Hz) the resonant frequency of the moving system must be very low. The cone must be light and yet rigid to prevent the 'break up' of the cone movement into complex oscillations at different driving frequencies. It is impossible to achieve these design aims in high quality units over a wide frequency range. Consequently, wide range systems will, in general, use 2 or 3 different transducers with the input audio signal split between them using filters.

In the microphone the design problem is to couple with relatively small sound pressure variations, and to produce a moving coil velocity proportional to the varying sound pressure. This is achieved by making the resonant frequency of the system in the range 500 to 2000 Hz, and by correcting the upper and lower frequency response by using resonant air chambers coupled acoustically to the diaphragm. It is possible to produce a relatively flat response over a frequency range of 40 – 13000 kHz.

10.6 Piezoelectric transducers

The operation of piezoelectric devices is based upon the fact that in certain crystalline materials, mechanical deformation of the crystal lattice results in a relative displacement of positive and negative charges within the material. This results in the generation of an e.m.f. between opposite faces of the crystal which is proportional to the degree of distortion of the material. The effect is reversible in that the application of an e.m.f. results in a physical deformation of the crystal.
Materials commonly used for such devices include quartz, Rochelle salt and barium titanate ceramics. The basic device is in effect a capacitor with the piezoelectric material as the dielectric between two electrodes. The charge Q produced by the deformation produces, across the device, a voltage V whose magnitude is dependent upon the capacitance of the system. A static deformation of the crystal produces a static charge, and hence a voltage across the device, but the voltage will only remain as long as the charge does not leak away via the measuring system connected across the transducer. The detector must, therefore, have a very high input impedance and specially developed circuits are used if very low frequency response is required, with an input impedance which may reach 10^{14} ohms. A true zero frequency response is, however, not possible.

10.7 Thermocouples

Another transducer wich produces directly an output e.m.f. is the thermocouple, which is widely used as a temperature measuring device, or more accurately a temperature difference measuring device.
If two wires of different materials are joined together, a small e.m.f. is developed across the junction which is dependent upon its temperature. Figure 10.30(a) shows a simple circuit loop consist-

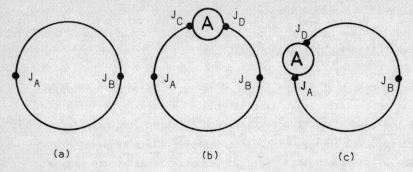

Figure 10.30

ing of two dissimilar metals. There are obviously two junctions, J_A and J_B, associated with the circuit. If the two junctions are at the same temperature the e.m.f's produced at the two junctions are equal and opposite, and hence no current flows in the circuit. Figure 10.30(b) shows an ammeter connected in one branch of the circuit.

This does not alter the current flowing in the circuit (other than by increasing the total resistance of the circuit loop). Although two extra junctions, J_C and J_D have been introduced into the loop, and in general the ammeter will be constructed of a third type of metal, the two new junction effects cancel out if the junctions are at the same temperature, which would normally be the case. This is also true if the ammeter is moved in the circuit so that junctions J_A and J_C are coincident, as shown in Figure 10.30(c). Although there are now only three junctions in the loop, and all three are between different materials, the ammeter material does not alter the loop current as long as the junctions J_A and J_D are at the same temperature. In use, one junction is used as the active probe, measuring the difference in temperature with reference to the temperature of the other junction, the reference junction. The accuracy of the measurement can be no better than the accuracy to which the temperature of the reference junction is known.

The thermocouple materials in common use are copper-constantin,

iron-constantin, chromel-alumel, and platinum-rhodium.
The sensitivity varies from a maximum of about 60 µV/°C for copper-constantin down to less than 10 µV/°C for platinum-rhodium. The usable temperature range varies for different couples, being from − 200°C to + 380°C for copper-constantin, and from 0 to 1500°C for platinum-rhodium.

Examples

1. A resistor bridge is initially balanced with all four arms equal in value. It is energised by a supply of voltage V. If one arm is deflected from its initial value of R ohms by a small amount δR, show that the out of balance voltage is given by

$$\frac{\delta V}{V} = \frac{\delta R}{R} \cdot \frac{R}{2(2R + \delta R)}$$

The detector impedance may be assumed to be infinite.

2. If two arms of the bridge in Question 1 are equally deflected an amount δR, show that the out of balance voltage is given by

$$\frac{\varepsilon V}{V} = \frac{\delta R}{R} \frac{R}{(2R + \delta R)}$$

3. If the bridge of Question 1 is used with a detector of zero resistance, show that the detector current is given by

$$\frac{\delta I}{V} = \frac{\delta R}{R} \frac{1}{(4R + 3\delta R)}$$

4. If the bridge of Question 1 is used with a detector of resistance R_D ohms, show that the out of balance voltage is given by

$$\frac{\delta V}{V} = \frac{\delta R}{R} \cdot \frac{1}{4 + 4\frac{R}{R_D} + R\left(\frac{3}{R_D} + \frac{2}{\delta R}\right)}$$

5. A first order low pass instrument is to respond to a frequency of 80 Hz with an accuracy of not less than 2%. What must be its minimum cut off frequency?

(2.475 kHz)

6. A copper resistor has a resistance of 90 Ω at 20°C and is connected to d.c. supply. By what percentage must the supply voltage be increased in order to maintain the resistor current constant if the temperature of the resistor rises to 60°C? The temperature coefficient of resistance of copper may be taken as 0.00428 at 0°C.

(15.8%)

7. A temperature measuring device consists of a thermistor and a 10 KΩ resistor connected in series across a 5 V direct supply. The voltage across the 10 KΩ resistor is measured by means of a very high resistance voltmeter. The thermistor has a resistance of 2000 Ω at 25°C and a B value of 3000 at 25°C. Determine the rate of change of output voltage with temperature (a) at 25°C, (b) at 50°C.

(23.5 mV/°C, 1.0 mV/°C)

8. A capacitor displacement transducer consists of two parallel plates of area A spaced a distance d in the air. Show that, if the spacing is altered by a small amount δd, then the fractional change in capacitance is given by

$$\frac{\delta C}{C} = -\frac{\delta d}{d + \delta d}$$

CHAPTER ELEVEN

Signal Processing

11.1 <u>Noise in systems</u>

We have already considered in section 10.3.1 the effect of noise generated internally in a transducer. Noise, defined as any unwanted signal, is, in fact, introduced at all parts of any signal system and, as previously discussed, the amplitude of the total noise signal determines the smallest detectable signal level in the system.

The noise level in a system is conveniently quoted as a signal to noise ratio, defined as the ratio of the magnitude of the wanted signal to that of the total noise signal, and is normally expressed in decibels, so that

$$\text{signal to noise ratio} = 20 \log_{10} \frac{v_s}{v_n} \text{ dB} \qquad 11.1$$

Although it is possible to recover signals which are hidden by noise, even with a signal to noise ratio which is less than unity, for normal instrumentation processes the minimum signal to noise level must be very much better; great care must, therefore, be taken in the design and physical layout of circuits where the signal is at its lowest level. For high quality audio systems, for example, signal to noise ratios of the order of 60 to 80 dB are expected. Bearing in mind that a decibel ratio of 60 dB corresponds to a ratio of 1000 to 1, the achievement of such noise levels requires extreme care to be taken with the low signal level amplifier stages, and in particular with the amplifier input stage. Wideband (white) noise will be generated, as shown in section 10.3.1, in the input components of the amplifier, and also in the transducer, which may be, for example, a tape recorder replay head, which is providing the signal for the amplifier input. In addition, white noise will also be introduced by the active devices in the amplifier circuit. Any semiconductor device or saturated vacuum device is a source of a further form of wideband (white) noise, known as 'shot noise', caused by the random generation of current carriers. Both shot noise and Johnson, or thermal, noise may be classed as internal noise, being generated by physical processes within the transducer and other input devices themselves. The effects of such internal noise may be minimised by choosing only good quality low noise components for use in low signal level circuits, and by operating the components under the optimum low noise level conditions specified by the component manufacturers. For some communication systems, a low internal noise level has been achieved by operating the low signal level amplifying devices at a low temperature.

In addition to internal noise, interfering signals may be introduced into the system via stray conduction, electric or magnetic fields which couple with the system. Such noise may be classified as external noise.

Perhaps the most common source of external noise is the local mains supply system. The large signal power level available from abundant sources which easily couple with any system causes interfering signals at power frequency (50 Hz) and its harmonics. Other low frequency or direct interfering signals are generated by, for example, thermoelectric effects, by electrolytic effects, or by ground leakage currents from adjacent systems. At higher frequencies, even low power noise sources may cause difficulties due to the easier coupling effects at higher frequencies via stray capacitance. External interference may itself be caused by either of two mechanisms.

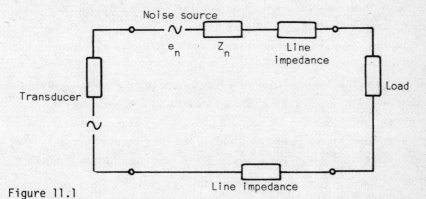

Figure 11.1

Differential mode interference is illustrated in Figure 11.1, and is caused when the noise source couples to one input line only. The figure shows a transducer coupled to its load, which, in most cases would be the input terminals of a high gain amplifier system. The noise generator, of voltage e_n and internal impedance z_n is shown in series with one input lead. The noise voltage produced across the load terminals is dependent upon the magnitude of e_n, and also upon the relative impedances of the transducer, the load and the noise source.

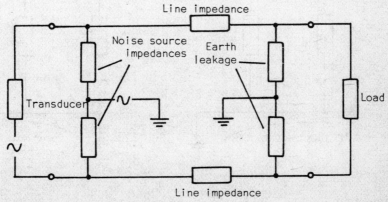

Figure 11.2

Common mode interference affects both input leads similarly. Currents flow from the noise source via the input lines, and via earth leakage paths; if the impedances of such paths are not identical a noise voltage may be produced across the load terminals. This mechanism is illustrated in Figure 11.2.

Interfering signals are coupled into signal circuits by conduction, electric or magnetic mechanisms.

Conductive effects may be minimised by arranging the circuit wiring so that no section of wiring which is carrying currents due to a possible noise source, is included in the low signal level circuits of the system. A good example of this effect is when a portion of the earth line of an audio amplifier is included in the input circuit.

If the earth line is also carrying earth loop currents, perhaps from the power transformer, then power frequency signals may be introduced into the amplifier input. Figure 11.3 illustrates this noise coupling mechanism, and also shows how, by good wiring layout, the fault may be eliminated. In Figure 11.3(b) the input circuit, although connected to the system earth, does not include any portion of the chassis or earth rail system which may be carrying noisy earth currents.

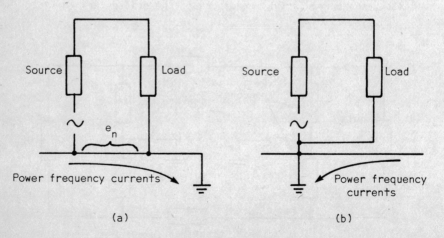

Figure 11.3

Interference due to magnetic coupling occurs usually due to normal
transformer action. A changing magnetic field is set up by an
alternating or a varying direct current. Any circuit loop which
links a portion of the varying field will have an e.m.f. induced
in it of magnitude equal to the rate of change of flux linkages
in the circuit loop. The coupling effect may be increased severely
by the presence nearby of ferromagnetic material, which can give
local increases in the magnetic flux density. Ferromagnetic material
may be used to screen a circuit if it surrounds the circuit completely and in a continuous manner (without airgaps). A useful
rule of thumb is that ferromagnetic material may encircle signal
circuits but signal circuits should never encircle ferromagnetic
material. Magnetic coupling is always in the differential mode,
the noise currents so produced being dependent upon the circuit
loop impedance. A test for electromagnetically coupled noise is
to short circuit the input to the section of network. If the coupled
noise increases, as indicated by an increase in the overall circuit
output, then the coupling mechanism is probably electromagnetic.
Electrostatic coupling takes place due to capacitance between
the signal system and the noise source, and may be either differential or common mode. If the coupling mechanism is electrostatic,
then short circuiting the input to the section of network will
probably result in a decrease in the noise signal output.
The use of screened cable is highly recommended for low level
signal circuits, in either single core form or in balanced double
core form. These are highly effective in reducing electrostatically induced noise. To reduce magnetically induced noise, the
area of the circuit loop must be kept to a minimum; this is
accomplished by using tightly twisted pairs. It is interesting
to note that coaxial cable in which a central conductor is entirely
encircled by a tubular outer conductor, and which is commonly
used for television aerial cables, is quite prone to electromagnetically induced noise effects.

11.2 Analogue/digital conversions

When studying analogue or digital techniques it must be borne in mind that the circuits considered are not necessarily complete in themselves and, in general, form part of larger systems. The overall system may not be entirely analogue or entirely digital in nature. Thus a digital system may be controlled by input signals which are the amplified analogue outputs, perhaps of some measuring transducers. Similarly, a digital system output may be required to control the measured analogue system via analogue control valves. Interfacing is thus required between the analogue and the digital sub-systems, and it is necessary to be able to convert an analogue signal into a digital equivalent signal, and vice versa. In section 8.1 the basic analogue to digital conversion process was considered, and it was shown that in a digital system, because of the time required to form each coded data value, it is only possible to produce digitally coded values of sample values of the analogue signal. An analogue signal, by which we mean a continuously variable signal, can not be represented exactly by a digital signal. The analogue signal must be sampled at intervals, the amplitude sample then being converted into a digital form. Samples must be taken at sufficiently frequent intervals for all the relevant information in the analogue signal variations to be retained. Sampling theory shows that in order to retain information which is contained in a signal, then at least two samples must be obtained per period of the highest frequency component. If the highest signal frequency component is f_s Hz, then the period of the sampling signal is given by

$$T \leqslant \frac{1}{2f_s} \qquad \qquad 11.2$$

Figure 11.4 shows a basic sample and hold circuit. The capacitor C is used as a store or memory to hold the value of the sample. It is connected to the analogue signal input via the resistor R, which in practice will include the output impedance of the analogue signal source. The time constant CR is chosen to be sufficiently short so that the capacitor voltage can follow the

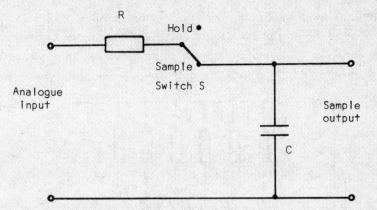

Figure 11.4 Basic sample and hold circuit

required analogue signal variations. At the instant that the sample is to be taken, switch S is changed into the HOLD position, and the sample voltage is available to the succeeding analogue to digital convertor. The disadvantage of this simple sample and hold circuit lies in the drift which occurs in the capacitor voltage during the HOLD period. This is mainly due to the load placed upon the capacitor by the following circuitry. The use of as large a capacitor as is allowed by the time constant design will minimise the voltage drift, but this simple circuit is still not satisfactory for accurate use. An improvement may be made by including a high impedance buffer amplifier between the memory capacitor and its applied load, but for high accuracy systems, sample and hold circuits include some error correction arrangements. Whatever circuit is used, however, a certain time is necessary in the sample position for the memory device to store an accurate sample value. This is known as the 'aperture time', and corresponds to the sample time, t_s, shown in Figure 8.8. An ideal sample and hold circuit would have a zero aperture time and an infinite hold time.

11.2.1 Digital to analogue convertors

The most common digital to analogue convertors are designed to operate with a digital input in a weighted code form, usually in simple binary coded form. For digital inputs which are coded in some other form, for example in GRAY binary code, it is necessary normally for the input signal to be converted into a weighted code form prior to the digital to analogue conversion.

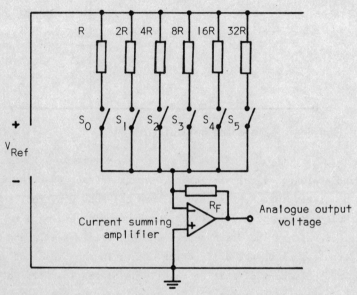

Figure 11.5 Weighted resistor digital to analogue convertor

Figure 11.5 shows a weighted resistor D to A convertor, which is arranged for digital input words of six bits.
The six switches S_0 to S_5 are driven by the digital input, being closed when the relevant input bit has the logical value 1, and open when the bit has the value 0. The digital input must, therefore, be available in parallel form. The resistors are weighted so that each allows a current to flow into the summing amplifier of value proportional to the weighting of the bit; for the simple binary coded form shown the resistor values are graded according to the powers of two. The analogue output voltage is then proportional to the value of the digital input, having its maximum value when all switches are closed.
A modification of the circuit, which will accept binary coded

Signal Processing 443

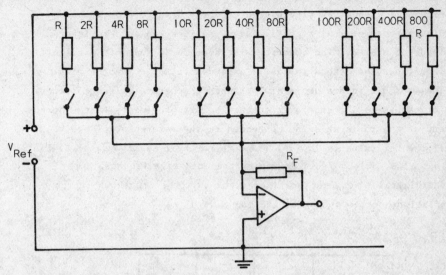

Figure 11.6 Binary coded decimal digital to analogue convertor

decimal inputs, is shown in Figure 11.6. Four graded resistors are used per decade, with a ratio of 10 between the values of adjacent decades. The disadvantage of convertors of this type is the need for a large number of accurately graded resistors.

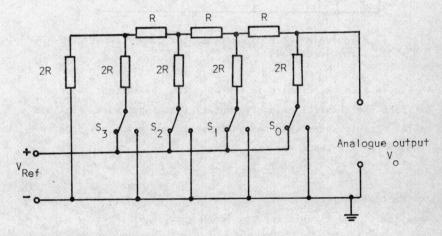

Figure 11.7 Ladder resistor digital to analogue convertor

444 Basic Electrical Engineering and Instrumentation for Engineers

Figure 11.7 shows an alternative type of convertor for simple binary coded digital signals, and which is known as a resistor ladder convertor. The convertor is shown for a four bit parallel input, but the method is easily extended to any number of bits. The four bit inputs are again arranged to operate switch contacts; for each bit, a change over contact set is provided with the bottom end of the resistor being connected to the reference voltage V_{REF} if the bit value is logical 1 and to earth if the bit value is logical 0. Switch S_o is driven by the most significant bit of the digital input. In the figure, the switches are shown in the positions corresponding to the input word

$$F = 1111$$

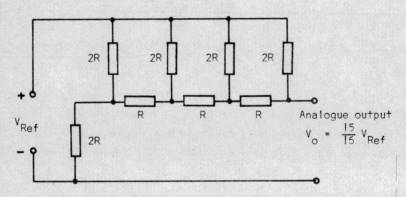

Figure 11.8 Equivalent circuit of resistor ladder convertor for input 1111

In Figure 11.8 the circuit is redrawn with the switches omitted, but with the connections equivalent to this input word. Analysis of the circuit shows that the analogue output voltage V_o corresponding to this input word is

$$V_o = \frac{15}{16} V_{REF}$$

The decimal equivalent value of the binary word 1111 is 15. Analysis of the circuit for any input word shows that the output voltage V_o is given by

$$V_o = \frac{V_{REF}}{16} \times \text{decimal equivalent value of input word} \qquad 11.3$$

For a ladder network of this type, the actual value of the resistors does not affect the value of the analogue output voltage as long as the ratio R:2R is correct, and as long as the output is not loaded significantly.

In the two types of convertor described, the switches have been shown as contact switches; in modern fast systems these switches would normally be electronic switches. Both the convertors require the digital input in parallel form - a serial digital input must therefore be converted into parallel form. This could be arranged by clocking the serial input into a parallel output shift register.

11.2.2 Analogue to digital convertors

Analogue to digital convertors are commonly based upon a comparison process, in which a locally produced digital word is converted into its analogue equivalent and is compared with the input analogue sample. The locally produced digital word is varied until agreement with the sample is reached.

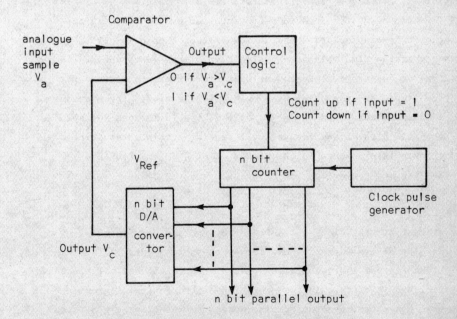

Figure 11.9 Continuous balance digital to analogue convertor

Figure 11.9 shows an analogue to digital convertor known as a 'continuous balance' convertor. Clock pulses from the generator are clocked into a counter, whose parallel digital output is converted into its analogue equivalent, the voltage V_C in the figure. Voltage V_C is compared with the analogue input sample voltage V_A in a comparator, whose output controls the gating of the clock pulses to the counter. The system forms a continuous control loop which maintains the word stored in the counter as the digital equivalent to the analogue sample input.

The full range of the counter would be equivalent to the full voltage range of the input sample. If, therefore, the allowed analogue range is from 0 to 10 volts, a 10 bit binary counter would have a resolution, equivalent to 1 least significant bit value of

$$v = \frac{10.0}{2^{10}} \text{ volts}$$

$$= \frac{10.0}{1024} = 10 \text{ millivolts approximately}$$

This is roughly a resolution of 1% of full scale.

The time taken for an A to D convertor to produce the digital output in response to a new analogue input sample is called the 'conversion time'. The conversion time of the continuous balance type of convertor is dependent upon the difference between the new sample and the previous sample. If, therefore, the previous sample was zero, and the new sample is the full range voltage, then all bits of the counter will require setting, requiring $2^n - 1$ clock pulses where n is the number of bits in the counter. The maximum conversion time is thus

$$t = \frac{2^n - 1}{f_c} \text{ seconds} \qquad 11.4$$

where f_c is the clock pulse frequency (pulses per second).

An alternative form of convertor is shown in Figure 11.10. The block diagram is identical to that of the continuous balance convertor, but it differs in the operation of the control logic. In this convertor, the counter is emptied by resetting all bits to zero before a conversion is started. When the new analogue sample

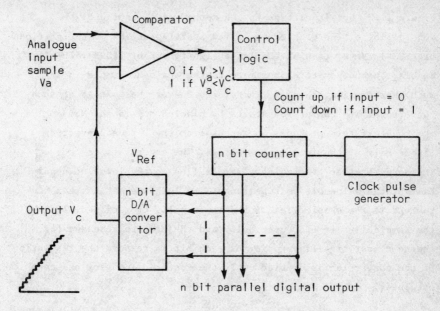

Figure 11.10 Staircase ramp analogue to digital convertor

is present, the control logic starts the count, i.e. clock pulses are fed into the counter. The counter digital output thus increases bit by bit, at the clock frequency. The output from the D to A convertor is a linear ramp, made up of equal incremental steps. The count continues until the generated staircase ramp exceeds the value of the analogue sample voltage, when the comparator output goes to logic 1 and stops the count. The counter output is, at this time, the digital equivalent of the analogue sample. The resolution and the conversion time are again decided by the number of bits in the counter, the range of the analogue input voltage, and the clock frequency; the comparator must, of course, be capable of resolving sufficiently small changes of input voltages. The conversion time of any convertor using a counter is relatively long, if the counter is filled in the normal manner, with each input pulse equivalent in value to the least significant bit. The 'successive approximation' type of convertor differs from the previously described convertors in the method of filling the counter. Its block diagram is again identical to that shown in

Figure 11.9, but it differs in the method of operation of the control logic. The counter is filled starting at the most significant bit (MSB) rather than at the least significant bit (LSB). This enables the output to approximate to the analogue sample in a much shorter time; this is, therefore, one of the fastest of such convertors. The conversion starts by placing a 1 in the MSB of the counter; this produces an output from the D to A convertor of one half of the full range voltage. The control logic thus indicates whether the analogue sample lies in the lower or upper half of the allowable voltage range. The 1 in the MSB is then removed if the sample lies in the lower half, or retained, if the sample lies in the upper half, and the process repeated for the next most significant bit. As each bit is tested, the resolution of the conversion is doubled until the required accuracy has been obtained.

11.3 Real signals

In the previous sections of this book, signals have been considered without detailed reference to their frequency spectrum. A continuous sinusoid contains no information other than the presence of the sinusoid; all real information carrying signals are, therefore, non sinusoidal in nature and consist of a voltage or current, some property of which is varied as an analogue of the information. The use of the Fourier series allows any repetitive signal waveform to be represented as a d.c. component together with an infinite series of sinusoidal components; the lowest sinusoidal term, the fundamental, being of the same frequency as the original signal waveform, and the remaining sinusoidal terms being multiples or harmonics of the fundamental.

Expressed mathematically, any complex signal may be written

$$v = V_o + V_1 \sin(\omega t + \phi_1) + V_2 \sin(2\omega t + \phi_2) + \ldots + V_n \sin(n\omega t + \phi_n)$$

$$11.5$$

where V_o is the d.c. component.

$V_1 \ldots V_n$ are the peak values of the sinusoidal terms

$\phi_1 \ldots \phi_n$ are the phases of these waves at time $t = 0$

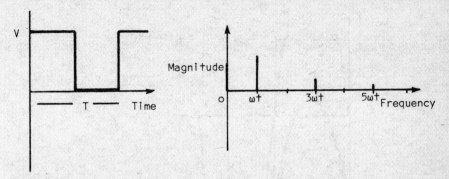

Figure 11.11

Of great interest, especially in dealing with digital systems, is the square pulse waveform shown in Figure 11.11 which shows a square wave of peak value V and period $T = 2\pi/\omega$ seconds. The d.c. component may be obtained by inspection, the average value of the wave being 0.5V. A Fourier expansion gives the wave as

$$v = \frac{V}{2} + \frac{2V}{\pi}(\sin \omega t + \frac{1}{3} \sin 3\omega t + \frac{1}{5} \sin 5\omega t \ldots) \qquad 11.6$$

The component frequences of a signal are conveniently displayed as a frequency spectrum, also shown in the figure, in which the length of the line is proportional to the magnitude of the frequency component. The wave may be reconstituted by adding together the d.c. component and the alternating components in the correct magnitude and phase; all components, with frequencies extending up to infinity must be included to obtain the original ideal square wave shape. Because, of course, all practical systems have a limited bandwidth, it follows that the realisation of an ideal square wave, in practice, is impossible, and that limits are imposed, by the system, on the maximum rate of change of signal level that can be achieved.

It is interesting to consider the synthesis of a square wave using only a very limited number of harmonics. Figure 11.12 shows the synthesis of a square wave by adding together a d.c. component, and fundamental and third harmonic terms, with amplitudes and phase according to equation 11.6.

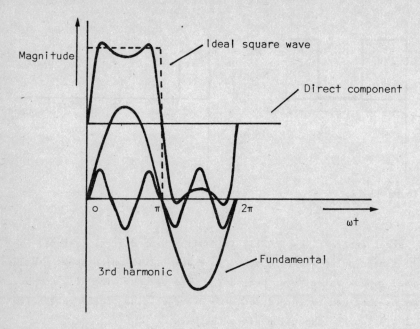

Figure 11.12

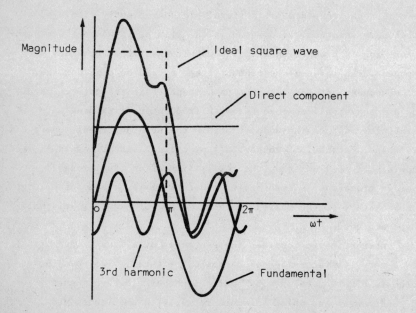

Figure 11.13

In Figure 11.13 the same synthesis is shown with the phase of
the third harmonic shifted by π radians. It can be seen that the
resultant wave is very different from that shown in Figure 11.12.
It is important that when signals are processed by a system, if the
waveform is not to be altered, then all the signal components
are processed in the same way, their relative magnitudes and phases
being maintained. By maintaining the same relative phase, we in
fact mean that each signal component is subject to the same time
delay in the system. Bearing in mind that one degree of phase
at the fundamental frequency is equivalent in time to three degrees
of phase at the third harmonic frequency, we can say that the
phase response of a system should be linearly proportional to
frequency.

In practice pulse waveforms are usually specified in terms of
their rise and fall times, which are defined as the times required
for the signal voltage or current to change between 10% and 90%
of its final value.

The frequency bandwidth required to process signals obtained from
transducers obviously depends greatly upon the original physical
variable. For a high quality audio signal, the required bandwidth
is approximately from 30Hz to 15kHz. For commercial telephony,
it is found that the normal speech bandwidth can be restricted
to the range 300 Hz to 3400 Hz without loss of intelligibility.
Signals originating from mechanical vibration measurements would
normally lie in the frequency range up to 3kHz, and may be required
to respond to static deflections, which would require the system
to have a response extending down to 0 Hz.

11.3.1 Data transmission

Data transmission between a signal source and its load may be
via connecting wires or via a radiated link.

If a d.c. voltage is applied to the source ends of a pair of conductors, the voltage received at the load end is equal to the source
voltage minus the voltage drop across the resistance of the con-

ductors. When an alternating signal voltage is applied to the
source end of the line, the reactive properties of the line also
affect the value of the voltage at the load. Capacitance between
the two conductors and the series inductance of each conductor
cause the line to take on different transmission properties. At
low frequencies the reactive effects will be small, and are usually
negligible. At high frequencies, however, the connecting wires
can have considerable effect upon the waveshape received at the
line termination. For systems in which the signal is in pulse
form, high frequency techniques are necessary even at relatively
low data rates, because the need to retain the pulse waveshape
requires, as discussed in the previous section, a bandwidth of
up to ten times the basic pulse repetition frequency.

As a rough guide, low frequency signals may be transmitted over
very long wires, a good example being the commercial telephone
system. For the TTL and other logic devices described in Chapter
8, frequency components in excess of 100 MHz are not uncommon,
and connecting wires of lengths of several centimetres must be
treated as transmission lines. In general, if the transmission
length forms a significant fraction of the wavelength of the signal,
then care must be taken. The velocity of propagation of the signal
on a line is dependent upon the reactive properties of the line.
The actual velocity is somewhat less than the velocity of light
(3×10^8 metres/sec), giving a propagation time along a cable
of the order of 5 nanoseconds per metre.

Figure 11.14 shows an eqivalent circuit which, when used to represent
successive lengths of a transmission line, gives similar transmission
properties. For good quality cables, the series resistance and
the shunt leakage components may be neglected in comparison with
the reactive components, giving, in effect, the loss less line
also shown in the figure.

An infinitely long length of such loss less cable presents, to
a signal generator at the source end, an input impedance which
is purely resistive and of value Z_o, where

$$Z_o = \sqrt{\frac{L}{C}} \text{ ohms} \qquad 11.7$$

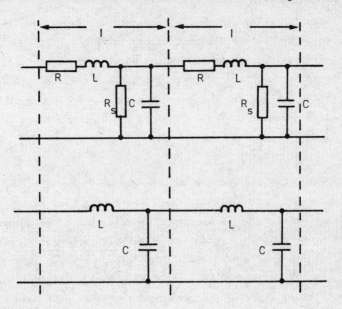

Figure 11.14

The resistance Z_o is called the characteristic impedance of the line. Because the impedance Z_o is determined by the inductance L and the capacitance C of the line, the value is fixed by the physical construction of the cable, i.e., by the dimensions, and by the permittivity and permeability of the insulation. Typical values are 600 ohms for twin parallel open wires, and 75 ohms for small coaxial cable similar to that used with domestic television aerials. Consider an infinitely long length of line of characteristic impedance Z_o. The input impedance of the line is Z_o ohms. If a short length of line is cut off the source end of the line, the input impedance of the short length is no longer equal to Z_o. However, the remaining length of line is still virtually infinitely long; its input impedance is, therefore, still equal to Z_o. If a resistor of value Z_o is used to terminate the short length of line, then conditions in the short length must be identical to those existing when it was part of an infinitely long length. The characteristic impedance Z_o is, therefore, the value of impedance which will correctly terminate a length of transmission line. When correctly terminated, a signal applied to the source end of the line results in the propagation down the line of voltage, and current waves which are correctly related in magnitude by the characteristic impedance Z_o. If the propagating waves reach a discontinuity in the line such as an incorrect terminating load impedance, Z_L, then the load voltage to current ratio

$$\frac{V_L}{I_L} = Z_L \qquad 11.8$$

does not match the propagating voltage and current waves; the load impedance Z_L is unable to accept the propagating signal energy at the correct rate, and some energy is reflected from the mismatched terminating impedance as reflected propagating voltage and current waves. The reflected waves travel back towards the source end of the line. The reflected wave combines with the original forward (incident) wave, altering its phase and magnitude. The resulting wave system on the line is a combination of forward travelling waves carrying energy towards the load, and stationary standing waves giving voltage and current maxima (antinodes) and minima (nodes). If the load impedance is either a short circuit or an open circuit, and can, therefore, accept no energy, all the incident energy is reflected towards the source. The wave structure on the line is then purely standing waves; at the antinodes the voltage or current levels will reach twice the magnitude of the applied signal levels.

The ratio of the maximum value of the standing wave to the minimum value is known as the Standing Wave Ratio, and is used as a measure of the impedance mismatch in a transmission system. Any reflected wave on a line, travelling backwards towards the signal source will again be reflected if an impedance mismatch exists between the line and the signal source.

In any transmission system, therefore, it is desirable that all connecting sections or devices should be matched in impedance. This is an extension of the discussion in chapter 5 regarding power transfer between a signal source and a load, which resulted in the general conclusion that for maximum power transfer from a source to a load, the load resistance should equal the source resistance.

In order to achieve impedance matching between devices or systems of different impedance, impedance transforming circuits must be used. These may take the form of an electronic circuit, such as an emitter follower transistor amplifier, or perhaps the simplest form of a matching transformer.

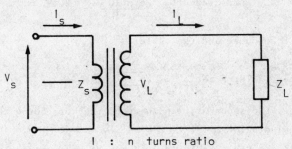

Figure 11.15

Figure 11.15 shows a load impedance, Z_L, supplied via an ideal transformer from a voltage supply V_S. The transformer has a primary to secondary turns ratio of 1:n, and because it is ideal, this results in a secondary load voltage of

$$V_L = nV_S$$

The secondary current is given by

$$I_L = \frac{nV_S}{Z_L}$$

and the primary current by

$$I_S = n \times \frac{nV_S}{Z_L}$$

or

$$I_S = n^2 \frac{V_S}{Z_L}$$

The effective impedance presented to the voltage supply is therefore

$$Z_S = \frac{V_S}{I_S} = \frac{Z_L}{n^2} \qquad 11.9$$

The effect of the transformer is to transform the impedance of the load, Z_L, by a factor equal to the square of the transformer turns ratio.

Example

A transducer, of output resistance 3 ohms is to be connected to an amplifier of input resistance 5 kΩ by a cable of characteristic impedance 600 Ω. Estimate the turn ratios of the required matching transformers.

Between the cable and the amplifier

$$\text{Turns ratio} = \sqrt{\frac{5 \times 10^3}{6 \times 10^2}} = 2.89$$

Between the transducer and the cable

$$\text{Turns ratio} = \sqrt{\frac{6 \times 10^2}{3}} = 14.1$$

In both cases, the secondary winding requires more turns than the primary winding.

Data transmission via a radiated link may be used where the transmission path requires such a link because of the distance involved, or in many transducer applications because of the physical impossibility of making satisfactory connections via wires. Examples are the telemetry of medical data from the human body via a radio link from a swallowed transmitter enclosed in a small 'pill' or the telemetry of temperature data from sensors embedded in the rotating armature of an electric motor.

The use of radio links can enable instrumentation to be applied to a live subject without his knowledge; this is sometimes necessary in behavioural investigations where the attachment of cabling may alter the subject's responses.

The main problem in such data telemetry applications is again one of impedance matching in that energy must be radiated into space in the form of a radio wave. In order to couple efficiently into space the transmitting aerial should have dimensions which are comparable with the wavelength of the radiation; and if directional aerial characteristics are required, the aerial array will need dimensions considerably greater than the wavelength. For this reason, telemetry radio links require frequencies in excess of 10 MHz where the wavelength is 30 metres, with physically small systems using frequencies as high as 80 to 90 MHz, giving a wavelength of 3 to 4 metres. Even so it is often necessary to operate with inefficient aerials which are too small, relying upon short transmission paths for adequate signal to noise ratio.

11.4 Modulation Systems

The process of varying a property of a voltage or current in accordance with a data signal is termed modulation. The data is modulated on to a carrier signal, whose frequency is chosen to suit the characteristics of the transmission link to be used. The carrier waveform may in general be of any waveshape, but is usually of sinusoidal or pulse form.

11.4.1 Amplitude modulation

Amplitude modulation (A.M.) is perhaps the form of modulation most frequently met, many transducer systems inherently producing a signal which is amplitude modulated.

The basic process consists of multiplying the data signal by a constant amplitude, constant frequency carrier signal. If we represent the carier signal in sinusoidal form by

$$v_c = V_c \sin \omega_c t$$

and the data signal, in a simple form, by

$$v_s = V_s \sin \omega_s t$$

then the amplitude modulated carrier is presented by

$$v = (V_c + V_s \sin \omega_s t) \sin \omega_c t \qquad 11.10$$

The form of the amplitude modulated wave of equation 11.10 is shown in Figure 11.16.

The ratio $\qquad k = \dfrac{V_s}{V_c} \qquad\qquad 11.11$

is known as the modulation factor and is usually expressed as a percentage.

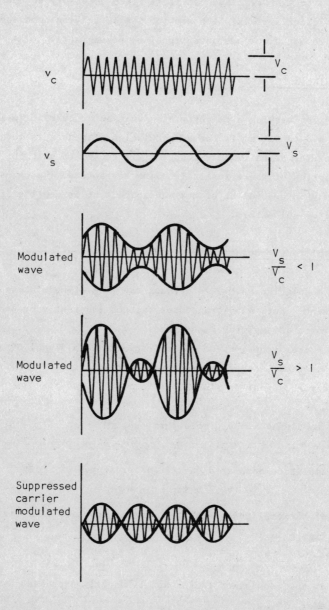

Figure 11.16

The modulated wave equation can be expanded using the identity

$$\sin \alpha \sin \beta = \tfrac{1}{2} \cos(\alpha - \beta) - \tfrac{1}{2} \cos(\alpha + \beta)$$

giving the result

$$v = V_c \sin \omega_c t + \frac{kV_c}{2} \cos(\omega_c - \omega_s)t - \frac{kV_c}{2} \cos(\omega_c + \omega_s)t \qquad 11.12$$

The modulated wave can be seen to consist of three terms:

a term $V_c \sin \omega_c$ which is the unchanged carrier signal,

a term $\dfrac{kV_c}{2} \cos(\omega_c - \omega_s)t$ which is a cosinusoidal term at a frequency given by the difference between the carrier frequency and the modulating frequency.

a term $\dfrac{kV_c}{2} \cos(\omega_c + \omega_s)t$ which is a cosinusoidal term at a frequency given by the sum of the carrier and the modulating frequencies.

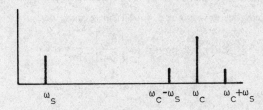

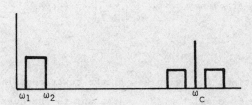

Figure 11.17

In Figure 11.17(a) the frequency spectrum is shown for a sinusoidal carrier, angular frequency ω_c, modulated by a sinusoidal signal of angular frequency ω_s. The two sum and difference frequencies $\omega_c + \omega_s$ and $\omega_c - \omega_s$ are known as the side frequencies. The modulating signal, being of steady sinusoidal waveshape, contains no real information. A real signal will consist of a band of frequencies, from say, ω_1 to ω_2 as shown in the spectrum of Figure 11.17(b).

The side frequencies thus become sidebands of frequencies, and are spoken of as the upper and lower sidebands.

The lower sideband will extend from $\omega_c - \omega_1$ to $\omega_c - \omega_2$ while the upper sideband will extend from $\omega_c + \omega_1$ to $\omega_c + \omega_2$. The bandwidth required for transmission of an amplitude modulated wave is equal to twice the highest frequency component of the modulating signal.

Figure 11.16 shows also the modulated waveform for two modulation depths, one exceeding 100%. In normal modulation systems, it is not desirable for the depth of modulation to exceed 100%, this leading to distortion when the carrier is demodulated to recover the information signal. It is instructive to determine the signal power contained in each of the components of the modulated signal. If the signal voltage was applied across a resistor R, then the power dissipated due to the carrier term would be given by

$$\text{carrier power} = \left[\frac{V_c}{\sqrt{2}}\right]^2 \frac{1}{R} \text{ watts}$$

$$\text{carrier power} = \frac{V_c^2}{2R} \text{ watts} \qquad 11.13$$

Due to each side frequency the power dissipated would be given by

$$\text{side frequency power} = \left[\frac{kV_c}{2\sqrt{2}}\right]^2 \frac{1}{R} \text{ watts}$$

$$= \frac{k^2 V_c^2}{8R} \text{ watts} \qquad 11.14$$

At 100% modulation, i.e. with $k = 1$, the power contained in each side frequency is equal to $\frac{1}{8} \frac{V_c^2}{R}$ watts and is one quarter of the carrier power. Because information is contained only in the side frequencies, the transmission of the carrier signal is a power waste. Systems are in use in which the carrier signal is suppressed at the transmitter, resulting in a power saving. Figure 11.16 shows also a suppressed carrier modulated wave. This technique has been extended to single sideband systems, in which one sideband is also suppressed; this being possible because the total information content of the wave is contained equally in either sideband. This technique also gives a saving in the required frequency bandwidth. In demodulating such a system, however, the signal intelligence can only be recovered from a single sideband radiation if the exact carrier frequency, suppressed at the transmitter, is known at the receiver.

In practice amplitude modulation is achieved by applying the signal and the unmodulated carrier to a non-linear device.

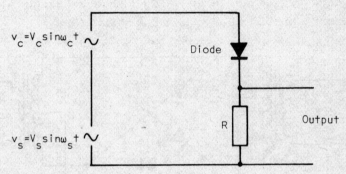

Figure 11.18

Figure 11.18 shows a very simple modulation system in which the two signals are applied to a non-linear circuit made up of a diode and a resistor R in series. The modulated output signal voltage is taken from across the resistor R, whose value is sufficiently low to be negligible in comparison with that of the diode.
A general non-linear characteristic is given by

$$i = a + bv + cv^2 + \ldots\ldots\ldots \quad\quad 11.15$$

If the diode characteristic is represented by the first three terms, then the current may be written

$$i = a + b(V_c \sin\omega_c t + V_s \sin\omega_s t) + c(V_c \sin\omega_c t + V_s \sin\omega_s t)^2 \quad 11.16$$

The first two terms consist of a d.c. component and components at the input frequencies.

The third term may be rewritten as

$$c(V_c \sin\omega_c t + V_s \sin\omega_s t)^2 = cV_c^2 \sin^2\omega_c t + cV_s^2 \sin^2\omega_s t + 2cV_c V_s \sin\omega_c t \sin\omega_s t$$

The $\sin^2$ terms give further d.c. components and also components at twice the input frequency, because of the expansion

$$\sin^2\theta = \tfrac{1}{2}(1 - \cos 2\theta)$$

Term $2cV_cV_s\sin\omega_s t\sin\omega_c t$ represents the modulated wave, expanding to give the side frequencies

$$cV_cV_s\cos(\omega_c - \omega_s)t - cV_cV_s\cos(\omega_c + \omega_s)t \qquad 11.17$$

All the component frequencies will eist in the output signal which can be taken as a voltage signal from across the resistor R. The carrier and side frequencies would be selected by a bandpass filter which would be tuned to the carrier frequency and which would have a bandwidth equal to at least twice the highest modulating frequency. A basic modulator of this sort is not very efficient.

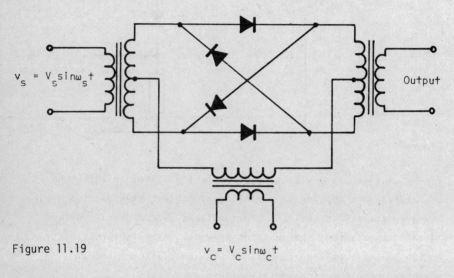

Figure 11.19

Figure 11.19 shows a practical suppressed carrier modulator circuit. This is the double balanced ring modulator consisting of four diodes, which, when driven by a sufficiently large carrier signal, conduct alternately in pairs. The input signal is transmitted unchanged during one half of the carrier signal period, and inverted for the remaining half. The figure also shows the output signal waveform. Because of the symmetry of the circuit, currents at the carrier frequency in the two halves of the output transformer primary winding are equal and opposite, and the carrier output is completely suppressed. Demodulation, or detection of amplitude modulated signals is commonly accomplished using a diode in the circuit of Figure 11.20.

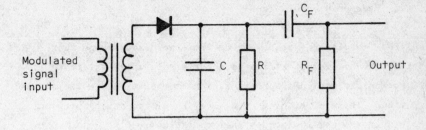

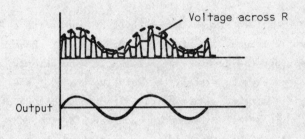

Figure 11.20

The resistor R forms the diode load, and is shunted by a capacitor C, the time constant of the combination being long when compared with the period of the carrier signal, but small when compared with the period of the highest frequency modulating signal. The diode acts as a half wave rectifier, removing one half of the modulated wave. The capacitor C charges during the conducting half cycles almost to the peak value of the applied signal voltage, but, because of the relatively long time constant, discharges only slightly between successive peaks. Because the time constant is relatively short in comparison with the modulating signal period, the output voltage follows normal variations in the peak input voltage. The output voltage consists, therefore of a varying d.c. voltage; the d.c. component can be removed by the capacitor-resistor high pass filter $C_F R_F$ also shown in the figure. The waveforms across the two resistors are drawn in the figure, together with the half sinewave at carrier frequency for comparison. It should be borne in mind when considering these waveforms that the carrier frequency will, in all practical systems, be of much higher frequency than it is possible to represent in such a drawing.

11.4.2 Frequency modulation

Amplitude modulated systems suffer from the disadvantage that much of the noise and extraneous interference which exists in all communication links appears as random variations in the signal amplitude. It is difficult, therefore, to differentiate between the wanted and the unwanted signal variations.
The unmodulated carrier signal may be written

$$v_c = V_c \sin(\omega_c t + \phi)$$

There are two further possible methods of modulating this carrier signal, while maintaining its amplitude constant; these methods both come under the general name of 'angle modulation'.
In a phase modulation system, the relative phase of the carrier signal, ϕ, is altered in accordance with the amplitude of the modulating signal. In a frequency modulation system, the angular frequency of the carrier signal, ω_c, is altered in accordance with the amplitude of the modulating signal.
The two methods are very closely related, and give frequency spectra which are very similar. Frequency modulation systems are more commonly used, and consideration will be restricted to this system.

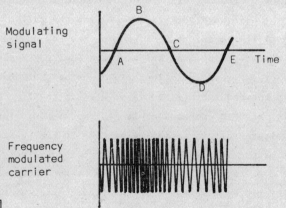

Figure 11.21

Figure 11.21 shows a constant amplitude carrier signal which is frequency modulated by a sinusoidal modulating signal.
At A, C and E the amplitude of the modulating signal is instant-

aneously zero; at these instants the frequency of the carrier is at its normal unmodulated value.

At B, the positive peak value of the modulating signal, the carrier frequency has been increased to a maximum.

At D, the negative peak value of the modulating signal, the carrier frequency has been decreased by the same amount, to a minimum.

The rate at which the carrier frequency is varied is proportional to the frequency of the modulating signal. The magnitude of the carrier frequency variation is known as the 'carrier deviation'. In any system, a maximum carrier deviation is specified which is reached with the maximum allowed amplitude of the modulating signal. In the B.B.C. v.h.f. transmissions, using frequency modulated carriers in the range 88 to 97.6 MHz, the maximum carrier deviation is ±75 kHz.

As with amplitude modulation, frequency modulation of a carrier wave produces side frequencies.

The equation to a frequency modulated wave may be written

$$v = V_c \sin(\omega_c t + m_f \sin\omega_s t) \qquad 11.18$$

where m_f is the modulation index and is defined as

$$m_f = \frac{\text{frequency deviation}}{\text{modulating frequency}}$$

Equation 11.18 can be expanded and yields a centre carrier frequency and an infinite set of side frequencies, each pair separated by an amount equal to the modulating frequency. Although the sidebands extend to infinity, in practice the magnitudes decrease reasonably quickly. It is, therefore, possible to specify the bandwidth necessary for a frequency modulated transmission, a bandwidth of about 200 kHz being required for the v.h.f. broadcast transmissions. This bandwidth requirement is very high in comparison with that required for an equivalent amplitude modulated transmission. With correct design and adequate signal strength, however, frequency modulation has a major advantage over amplitude modulation in noisy signal channels.

11.4.3 Multichannel transmission systems

Many applications occur where transmission is required simultaneously of several data signals. Typical examples are a commercial telephone system where many telephone conversations are being held at the same time over trunk telephone links. In an instrumentation system, the temperature of a particular machine may be measured by thermocouples embedded in the machine at various strategic points, all of which require continuous monitoring.

For such applications there are two basic methods whereby multichannel working may be arranged. One, termed 'frequency division multiplex', involves the positioning of the separate channels side by side in the frequency spectrum. A good example of this technique is given by the use of single sideband modulation to transmit telephone signals. The telephone signals, each with a bandwidth of 300 Hz to 3.4 kHz are modulated on to subcarriers, separated in the frequency spectrum at 4 kHz intervals. Using single sideband techniques, the sidebands produced by the telephone signal fit neatly into the frequency space between adjacent subcarrier signals. The modulated signals are then combined together to give one complex signal which is itself modulated onto a final carrier signal for transmission over a radio or cable link. At the receiving end, the signal is demodulated to produce the complex subcarier system, which is split into separate subcarrier signals by a system of bandpass filters, each of which is tuned to one of the subcarrier frequencies, and with sufficient bandwidth to pass only the subcarrier and its sidebands. After separation into the separate subcarrier channels, each is demodulated to produce the original telephone signal. In this manner, many hundreds of telephone circuits can be simultaneously established over a single transmission link.

The other basic multichannel system is termed 'time division multiplex' and involves the positioning of the separate channels end to end in time. The system is inherently a sampling system and can only give transmission of sampled values of each signal variable.

The principles of sampling, which have already been discussed
in section 11.2, must be applied in determining an adequate sampling
frequency. Each channel is connected in turn to the single data
transmission link for a period t; after all channels have been
so connected, taking a period

$$T = nt$$

where n is the number of channels, the sequence is restarted.
It is necessary for some checking signal to be included to allow
the synchronisation of channels at the transmitter and the receiver.
Each channel, therefore, has available to it a very short pulse
of carrier signal. The pulse characteristics must be varied in
some manner proportional to the signal amplitude at the sampling
instant.

11.4.4 Pulse modulation

Various methods are in common use whereby the characteristics
of a pulse are varied according to a modulating signal.
Pulse-amplitude modulation (PAM). The amplitude of the transmitted
pulse is varied in accordance with the magnitude of the sample
of the modulating signal.
Pulse position modulation (PPM). With a zero modulating signal
the pulses are transmitted in a regular uniform sequence. With
applied modulation, the position of the pulse is varied about
its unmodulated position, with the deviation from the zero position
being proportional to the magnitude of the modulating signal.
Pulse duration modulation (PDM). With applied modulation, the
accomplished by fixing either the leading or trailing edges of
the pulse and varying the position of the other edge. In some
systems, the centre of the pulse is fixed, and both leading and
trailing edges are shifted.
Pulse code modulation (PCM). With this type of modulation the
sample amplitude is encoded, and the code transmitted as a succession
of pulses. The use of this type of code was discussed in section
8.2.

Examples

1. A signal voltage $v_s = 1 \sin 100\,t$ and a carrier voltage $v_c = 2 \sin 1000\,t$ are applied in series to a device whose characteristic is given by

$$i = 10 + 1.5v + 0.06v^2 \quad \text{mA}$$

Determine the magnitudes of the various frequency components of the device current.

(11.5 mA d.c., $\omega = 100$ 1.5 mA peak

$\omega = 200$ .03 mA peak

$\omega = 900$ 0.12 mA peak

$\omega = 1000$ 3 mA peak

$\omega = 1100$ o.12 mA peak

$\omega = 2000$ 0.12 mA peak)

2. For the modulator of Question 1, determine the resulting depth of modulation.

(8%)

CHAPTER TWELVE
Electrical Machines

In 1820 Oersted noticed that a compass needle was deflected when placed near to an electric current. A year later, Faraday invented the first direct current motor and ten years later the first generator. During the following years many different types of motors were invented but because the electrical power source was limited to batteries, they could not compare economically with the steam engine. The early direct current generators employed rotating permanent magnets (Pixii's Generator 1832), and it was not until 1856 that Siemens designed a machine in which the armature was positioned between the poles of permanent magnets. A major step forward in the field of electrical machines was the replacement of permanent magnets by electro magnets in 1866, for which Siemens was again responsible. In 1883 Crompton, in England, and Edison, in America, introduced heavy duty direct current generators. Ferranti, meanwhile, was working on the design of alternating generators and transformers, and in 1887 two-phase and three-phase induction motors were patented by Tesla. Thus the late nineteenth century saw the birth of the electrical machine as we know it today. The function of an electrical machine is to convert mechanical

energy into electrical energy (generator) or electrical energy
into mechanical energy (motor). The method of driving a generator
varies depending on the environmental conditions. The major part
of all the electrical energy generated in this country is produced
by alternators (a.c. generators) driven by high speed steam turbines,
the steam being produced by coal, gas or oil fired boilers or
by nuclear reactors. In North America, however, the major proportion
of the electrical energy is produced by alternators driven by
water wheels (Hydro Electrical Generation). Whatever type of prime
mover is used the important characteristic of an electrical generator
is its regulation curve, i.e. the relation between output voltage
and output current. Electric motors are used in a wide variety
of industrial drives (machine tools, pumps, fans, hoists, transport-
ation etc.). It is usually the mechanical features of a particular
application which determines the choice of electric motor to be
used and therefore the torque-speed characteristic of the machine
is very important.

The characteristics of various types of machines will now be developed
and for purely historical reasons, the direct current machines
will be treated first.

12.1 The d.c. Generator

All conventional electrical machines consist of a stationary member
called the stator separated by an air gap from a rotating member
called the rotor. In direct current machines the stator usually
consists of salient poles with coils wound round them so as to
produce a magnetic field. The rotor is familiarly called the armature
and consists of a series of coils located in slots around its
periphery and connected to a commutator.

Consider the simplest form of d.c. generator shown in Figure 12.1(a)
in which a single coil is rotated in a two pole field. By Faraday's
law the voltage generated in the coil is equal to the rate of
change of the flux linking the coil. When the coil lies in plane
DD' there is maximum flux linking the coil but minimum rate of

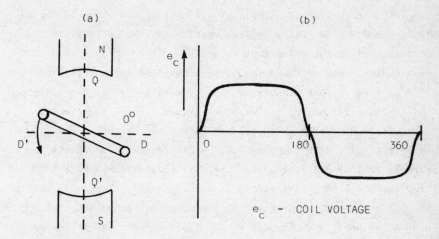

Figure 12.1

change of flux linkage, whereas along plane QQ' there is zero flux linking the coil but the rate of change of flux linkage is a maximum. The variation of the voltage generated in the coil as it moves through 360° is shown in figure 12.1(b). It is seen that the coil voltage is alternately positive and negative, i.e. it is an alternating voltage. To convert this alternating voltage into a direct voltage a commutator is used, as shown in Figure 12.2. The commutator consists of brass segments separated by insula-

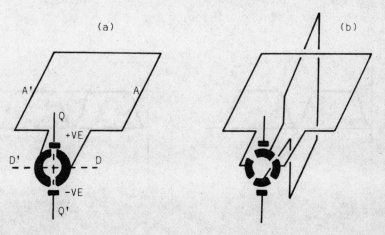

Figure 12.2

ting mica strips. In the simple single coil system shown in Figure 12.2a there are two brass segments and in the two coil system of Figure 12.2b there are four segments.

In a typical machine there may be upwards of 36 coils requiring a 72 or more segment commutator. The commutator is solidly connected to the armature and rotated with it. External connection to the armature is made through stationary carbon brushes which make sliding contact with the commutator as it rotates. Considering Figures 12.1(a) and 12.2(a), as the coil rotates through 180° from the plane DD', coil side A is under a north pole and is connected through its commutator segment to the upper brush, whilst coil side A' is under a south pole and is connected through its commutator segment to the lower brush. As the coil rotates through a further 180° coil side A' is under a north pole and connected to the upper brush whilst coil side A is under a south pole and connected to the lower brush. It is clear then that the coil side under a north pole is always connected to the upper brush whilst the coil side under a south pole is always connected to the lower brush. The voltage waveform across the brushes is as shown in Figure 12.3(a) which is not alternating positive and negative as before, i.e. the commutator has converted the alternating coil voltage into a direct brush voltage. Figure 12.3(b) shows that for the two coil system the d.c. armature output voltage is constant.

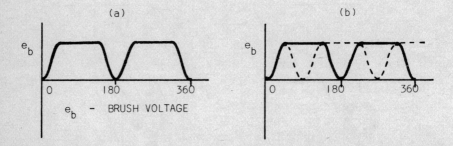

Figure 12.3

12.1.1 Voltage and torque equations of a d.c. machine

Induced armature voltage

Let ϕ be the flux per pole and N be the speed in r.p.s. Consider the single turn armature coil shown in Figure 12.1(a). When the coil moves through 180° from the plane DD', the voltage induced in the coil (by Faraday's law) is

$$E_{coil} = \frac{\text{Flux per pole}}{\text{Time for } \tfrac{1}{2} \text{ revolution}} = \frac{\phi}{1/N.2}$$

$$E_{coil} = 2N\phi$$

For a machine is which there are Z_s armature conductors connected in series ($Z_s/2$ turns), and 2p magnetic poles the total induced armature voltage is,

$$E = 2N\phi \cdot \frac{Z_s}{2} \cdot 2p$$

$$\underline{E = 2N\phi Z_s p \text{ volts}}$$

Torque on the armature

The force on a current carrying conductor (section 1.3.2) is,
$$F = B \ell I$$
The torque on one armature conductor is,
$$T_{cond} = F \cdot R = B_{av} \cdot \ell \cdot I_a \cdot R$$
where R is the radius of the armature

I_a is the current flowing in the armature conductors

ℓ is the axial length of the armature

B_{av} is the average flux density under a pole, and is given by

$$B_{av} = \frac{\phi}{2\pi R \ell / 2p}$$

Therefore the torque per conductor is,

$$T_{cond} = \frac{2p\phi}{2\pi R \ell} \cdot \ell \cdot I_a \cdot R$$

$$T_{cond} = \frac{1}{\pi} p \phi I_a$$

For Z_s armature conductors connected in series the total torque on the armature is

$$T = \frac{1}{\pi} p \phi I_a Z_s \quad \text{Nm}$$

Terminal voltage

If the terminal voltage is V volts and the armature resistance is R_a, then for a generator

$$V = E - I_a R_a$$

and for a motor

$$V = E + I_a R_a$$

12.1.2 Methods of excitation

The schematic representation of a d.c. machine is as shown in Figure 12.4, and the method of excitation depends simply on the interconnection of the field and the armature.

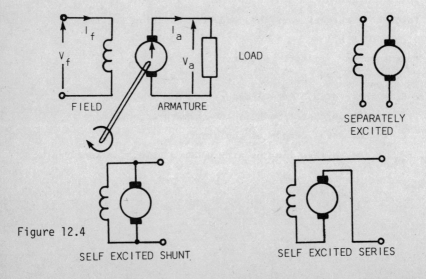

Figure 12.4

12.1.3 Separately excited generator

Open circuit or no-load characteristic (magnetisation curve)

Consider the separately excited generator shown in Figure 12.4 to be driven at constant rated speed. If the field current (and hence the magnetic field) is increased in steps and the terminal voltage is measured at each step, then a plot of terminal voltage versus field current will yield a curve as shown in Figure 12.5. Note that because the armature is open circuited the terminal voltage V_a is equal to the induced voltage E. Assuming that the magnetic circuit of the machine was originally completely demagnetised and that the flux is initially directly proportional to the field current then according to equation

$$E = 2 p \phi Z_s N$$

for constant speed

$$E \propto \phi \propto I_f$$

At higher values of field current the iron begins to saturate and the proportionality between the flux and the field current no longer exists, hence the curve no longer approaches a straight line. Because of magnetic hysteresis, the plot of induced voltage versus field current for decreasing excitation is slightly greater than the increasing excitation curve. The voltage at zero excitation is termed the Residual Voltage, without which self excitation would be impossible.

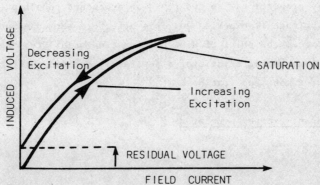

Figure 12.5

476 Basic Electrical Engineering and Instrumentation for Engineers

Load characteristic

Let the machine shown in Figure 12.6 be driven at a constant speed and be supplied with a constant field current I_f. If the load resistance is now varied the plot of terminal voltage versus armature current obtained is shown in Figure 12.7.

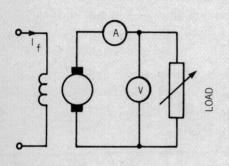

Figure 12.6

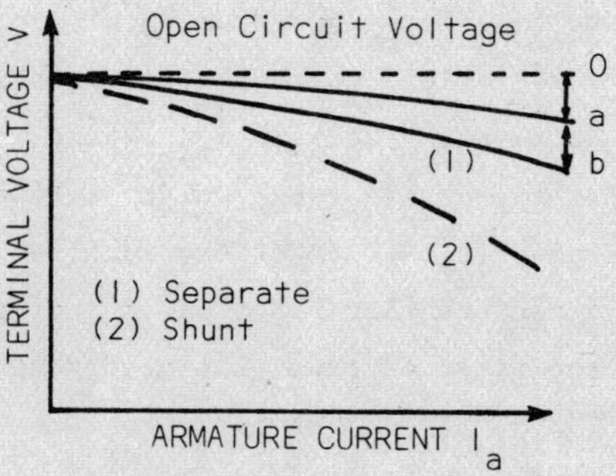

Figure 12.7
oa - armature resistance volt drop
ab - armature reaction volt drop

The drop in terminal voltage as the load current increases is due to (a) the armature (or internal) resistance voltage drop, and (b) the armature reation. When current flows in the armature an armature field is established which causes a nett reduction in the field produced by the exciting winding, and as the induced voltage is proportional to the flux, an attendant reduction in terminal voltage is experienced. This effect is termed Armature Reaction. From the point of view of operating stability, the slightly drooping load characteristic of a separately excited generator is ideal for most applications.

12.1.4 Shunt excited Generator

Voltage build-up on no-load

Consider the self excited shunt generator of Figure 12.4 to be driven at constant speed. If the field is disconnected from the armature the voltage generated across the armature brushes is very small and due entirely to the residual magnetism in the iron. When the field is connected this small residual voltage causes a current to flow in the field winding. If this current is in such a direction as to produce a magnetic flux in the same sense as the residual flux, then the total flux produced by the field winding will gradually build up. The final terminal voltage depends on the total resistance of the field winding and on the magnetisation curve of the machine. The magnetisation curve of a shunt machine is obtained by connecting it as a separately excited generator. Therefore for the successful operation of a shunt generator there must be residual voltage, the field must be connected the right way round and the field resistance must be less than the critical value.

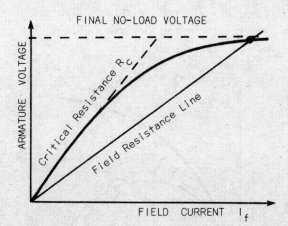

Figure 12.8

Load characteristic

The drop in voltage with increased current is more marked in a shunt generator than in a separately excited generator. The increased drop is due to the fact that as the terminal voltage drops then field current also drops. The voltage reduction with increase in load current is the exact reverse of the voltage build up discussed in the previous section. The load curve for a shunt generator is shown in comparison with a separately excited generator in Figure 12.7.

12.1.5 Series excited generator

The open circuit or magnetisation curve can only be obtained by connecting the machine as a separately excited generator.

Load characteristic

In the series generator shown in Figure 12.4 the armature and field are connected in series, therefore, the armature current determines the flux. Initially, therefore, the voltage increases as the armature (or load) current increases.

$$E \propto \phi \propto I_a$$

However, at large values of load current the combined armature resistance and reaction effects cause the terminal voltage to decrease, as shown in Figure 12.9.

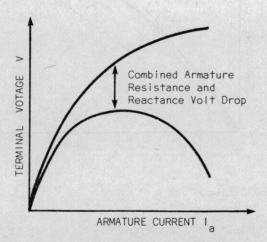

Figure 12.9

12.1.6 Compound generator

Compound generators are produced by combining the effects of shunt and series excitation. Normally a small series field is arranged to assist (cumulative compounding) the main shunt field, the actual shape of the load characteristic depending on the number of series turns, see Figure 12.10.

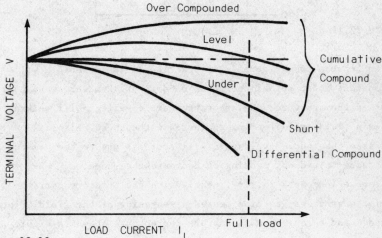

Figure 12.10

If the series field is arranged to oppose the main shunt field (differentially compounded) a fast drooping characteristic is obtained.

12.2.1 Shunt motor

Typical characteristics of a shunt motor can be obtained by considering the three relevant equations derived in section 12.1.1, i.e.

$$E = V - I_a R_a$$
$$N \propto \frac{E}{\phi}$$
$$T \propto I_a \phi$$

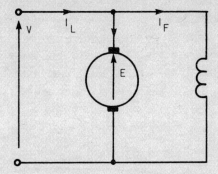

Figure 12.11

It can be seen from Figure 12.11 that the field current I_f will be constant under normal operating conditions. However, when current flows in the armature, the armature reaction effect will weaken the main field, thus tending to increase the speed. Also as I_a increases the induced voltage E will decrease due to the armature resistance volt-drop, tending to decrease the speed.

The torque increases fairly linearly with the armature current until the armature reaction causes a weakening of the field. Plots of speed and torque versus armature current are shown in Figure 12.12, and the torque-speed curve derived from them is shown in Figure 12.13.

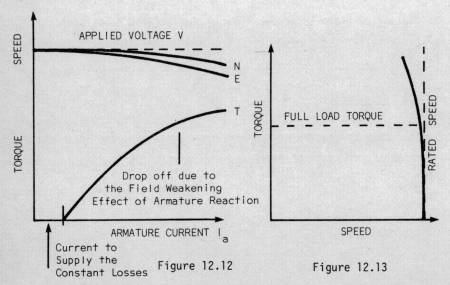

Figure 12.12

Figure 12.13

The torque-speed curve of the shunt motor shows that it could be used for any drive requiring a fairly constant speed from NO-LOAD to FULL-LOAD torque, e.g. machine tools, pumps, compressors, etc.

12.2.2 Series motor

As the load current increases the induced voltage E will decrease due to the armature and field resistance volt drops. Because the field winding is connected in series with the armature, the main flux is directly proportional to the armature current. Armature reaction will, of course, weaken the main field but a fair approximation to the speed versus armature current curve for a series motor would be a rectangular hyperbola, as shown in Figure 12.14. The torque in this case is approximately proportional to the square of the armature current, thus the torque versus armature current curve is approximately parabolic, Figure 12.14. Figure 12.15 shows the dervied torque-speed curve.

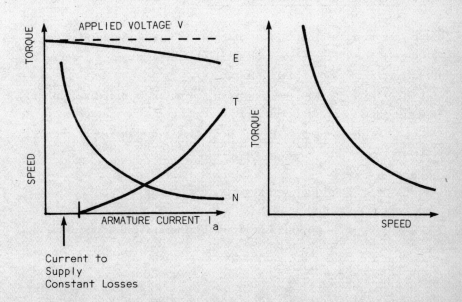

Figure 12.14   Figure 12.15

482 Basic Electrical Engineering and Instrumentation for Engineers

12.2.3 Compound motors

Compound motors can be produced by combining the shunt and series windings giving marginal changes to the characteristics shown above.

12.2.4 Starting a d.c. motor

At standstill the induced voltage of a d.c. motor is zero, therefore the applied voltage is equal to the armature voltage drop, $I_a R_a$. In typical machines the armature resistance is small and hence the armature current at standstill, with rated voltage applied, would be excessive. To limit the starting current, external resistance is connected in series with the armature. As the machine builds up speed, the induced e.m.f. increases and the external resistance is reduced until, at rated speed, all the external resistance has been disconnected.

12.2.5 Stopping or braking a d.c. motor

A d.c. motor will stop if it is disconnected from the supply. The time it takes to reach standstill will depend on its inertia and its friction and windage losses.
If fast braking is required then when the motor is disconnected from the supply, it is immediately reconnected to a resistor. The inertial energy generated is then quickly dissipated in the resistance.

12.2.6 Speed control of d.c. motors

The speed of a d.c. motor is governed by the applied voltage and the flux.

$$N \propto \frac{E}{\phi} \propto \frac{V - I_a R_a}{\phi}$$

It is clear, therefore that the speed can be varied by varying either the applied voltage or the flux.

Variation of Excitation

The main flux of a shunt motor can be weakened by using either a variable rheostat or a potential divider in the field circuit. For a series motor a divertor is used in parallel with the field, see Figure 12.16.

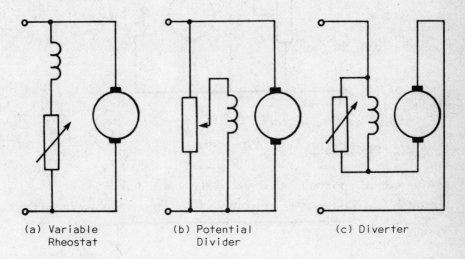

(a) Variable Rheostat   (b) Potential Divider   (c) Diverter

Figure 12.16

In each of the above methods the field can only be weakened so that only increases in speed above the rated speed can be obtained. It should also be noted that too much field weakening will produce a loss in the torque . ($T \propto I_a \phi$)

Variation of armature voltage

The speed can be increased from standstill to rated speed by increasing the armature voltage from zero to the rated value. The main difficulty is in obtaining the variable d.c. voltage.

(a) Potential divider

The potential divider must be rated at the same current level as the motor, consequently this method would only be used for small d.c. motors.

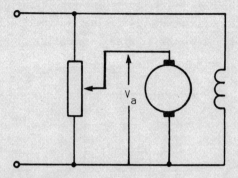

Figure 12.17

(b) <u>Ward Leonard drive</u>

In this case the variable d.c. voltage is obtained from a d.c. generator (variable voltage generator, V.V.G.) driven by an induction

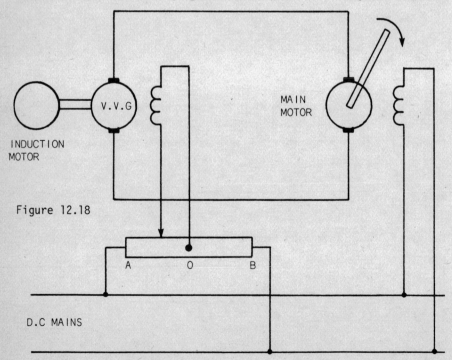

Figure 12.18

motor. The field of the V.V.G. is supplied from a centre tapped potential divider, see Figure 12.18.

When the wiper arm of the field potential divider of the V.V.G. is moved from O to A, the armature voltage of the main motor is increased from zero and the main motor will increase in speed. If the wiper is moved from A through O to B the motor slows down to standstill and then speeds up in the reverse direction.

The advantages of the Ward-Leonard drive are smooth and complete speed control, operation in the forward and reverse directions, and fast regenerative braking.

The disadvantage is the large capital outlay on the induction motor and variable voltage generator.

(c) <u>Chopper control</u>

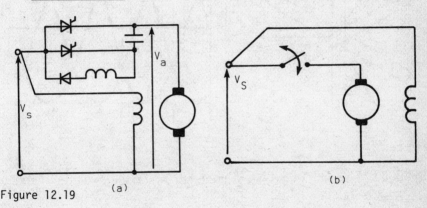

Figure 12.19 (a) (b)

Figure 12.19 shows a thyristor circuit in series with the armature of a d.c. motor, if the thyristor circuit is triggered so that it operates like an on-off switch, Figure 12.19(b), the waveform appearing at the armature terminals is as shown in Figure 12.20.

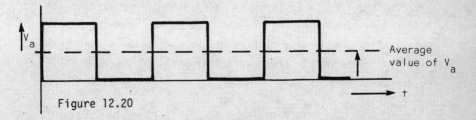

Figure 12.20

486 Basic Electrical Engineering and Instrumentation for Engineers

The mark/space ratio (time on to time off) can be easily changed and so the average armature voltage can be varied from zero (full off) to the full rated value (full on).

The main advantage of this method is its comparative cheapness.

12.3 Three-phase rotating fields

If three coils, physically displaced in space by 120°, are supplied with three-phase currents then a constant magnetic field is produced which rotates at a speed related to the frequency of the currents.

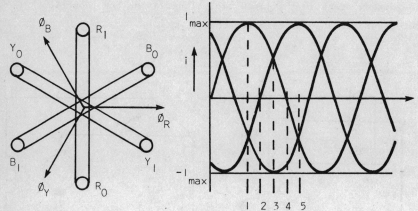

Figure 12.21

Figure 12.21 shows the arrangement of the coils and the time phase relation of the currents.

Consider the position and magnitude of the resultant flux produced by the three coils at the instants in time shown, using the positive direction of the fluxes as shown.

It can be seen that the magnitude of the resulting flux is 3/2 times the magnitude of the phase flux and it rotates at the frequency of the supply currents. If the coils were arranged to give a four-pole system as opposed to the two-pole one shown, then the speed of the rotating field would be halved (i.e. it would rotate through 180° for one cycle of the alternating current). In general, the speed of the field is,

$$N \text{ r.p.m.} = \frac{f \cdot 60}{\text{pole pairs}}$$

Electrical Machines 487

Instant in time	Magnitude of currents	Phasor diagram	Resultant flux $\phi_R + \phi_Y + \phi_B$
1	$i_R = I_{max}$ $i_Y = i_B = \dfrac{-I_{max}}{2}$		$3/2\ \phi_{phase}$
2	$i_R = \dfrac{\sqrt{3}}{2} I_{max}$ $i_B = \dfrac{\sqrt{3}}{2} I_{max}$ $i_Y = 0$		$3/2\ \phi_{phase}$
3	$i_R = i_Y = \dfrac{1}{2} I_{max}$ $i_B = -I_{max}$		$3/2\ \phi_{phase}$
4	$i_R = 0$ $i_Y = \dfrac{\sqrt{3}}{2} I_{max}$ $i_B = \dfrac{\sqrt{3}}{2} I_{max}$		$3/2\ \phi_{phase}$

The speed of the rotating field is normally termed the synchronous speed.

12.4 Three-phase alternators (alternating current generators)

A simple form of three-phase two pole alternator is shown in Figure 12.22. On the stator there are three coils spaced 120° apart. The rotor is a salient pole type and is supplied via slip rings with direct current so producing a uniform magnetic field. If the rotor is now driven by a prime mover voltages will be produced in the coils, and these voltages e_R, e_Y, e_B will be displaced in time phase by 120°. The magnitude of the generated voltages is dependent on the flux produced by the rotor, the number of

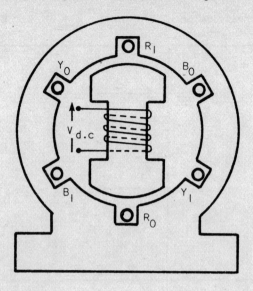

Figure 12.22

turns on the coils and the speed of rotation of the rotor. The rotor speed also determines the frequency of the generated voltages. The no-load voltage versus field current and the no-load voltage versus load current characteristics of an alternator are very similar to those of a d.c. separately excited generator. For constant speed operation the terminal voltage will drop due to armature resistance and armature reaction (the term armature being used in this case to denote the stator winding).

It should be noted that as the load on the alternator is increased then the speed of the prime mover would drop, causing the frequency of the generated voltage to fall. In many applications a small change in frequency could be tolerated but if a constant frequency is required some form of governor system must be used on the prime mover to maintain a constant speed on all loads. When alternators are required to operate in parallel, as in the National Grid, the prime movers are always speed controlled and the output voltage is always automatically regulated at the rated value.

In this country most of the electric power generated is produced by alternators driven by steam turbines. The alternators are usually two-pole machines driven at 3000 r.p.m. so as to produce the rated frequency of 50 Hz. In countries where the natural water resources allow, much of the electric power is produced by alternators driven by water wheels (hydroelectric generation). These alternators are normally slow speed machines with large pole numbers. For example, to produce 60 Hz (standard frequency in America) a 36 pole machine would run at 200 r.p.m.

12.5 Synchronous motors

Synchronous motors are so called because they operate at only one speed, i.e. the speed of the rotating field. They are of exactly the same construction as the alternator discussed in section 12.4 and in fact the term synchronous machine can be used to describe any machine which has its armature connected to the three-phase mains and its field supplied from a d.c. source.

Now if a bar magnet is placed in a strong magnetic field, it will always try to align itself with the field (compass needle in the earth's field). The operation of the synchronous machine is similar in this respect, the constant uniform flux produced by the field (effectively a bar magnet) aligns or "synchronises" with the rotating flux produced by the armature.

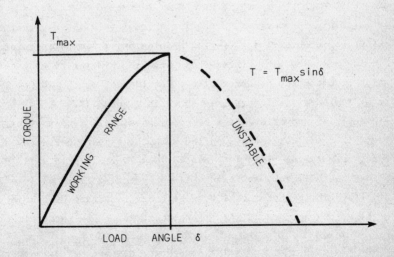

Figure 12.23

When mechanical load is applied to the shaft the uniform field produced by the rotor is pulled out of direct alignment with the rotating field produced by the stator, the angle of misalignment being termed the "load angle". The load characteristic of the synchronous motor is, therefore, the variation of torque with load angle as shown in Figure 12.23.

The machine possesses no starting torque and it is a necessary requirement with a synchronous motor that its rotor can be run up to synchronous speed by some external means.

The Reluctance Motor*, which is a special form of synchronous machine, has found many applications in recent years. These are listed in table 12.1, at the end of the chapter.

12.6 Induction motors

The two basic types of commercial three-phase induction motors are (a) the squirrel cage, and (b) the slip ring. The stators of both these machines are exactly the same as the alternator discussed earlier, i.e. a conventional three-phase winding, in order to produce a rotating field. In the squirrel cage motor the rotor core is laminated and copper (or aluminium) bars are driven through the slots. These bars are brazed on to solid copper (or aluminium) end rings, producing a completely short circuited set of conductors. The slip-ring machine has a laminated rotor housing a conventional three-phase winding, similar to the stator, which is connected to three slip-rings located on the shaft. The rotor lamination silhouette is shown in Figure 12.24.

Figure 12.25 shows three stator coils physically displaced by 120°. If these coils are supplied with three-phase currents, a constant rotating field is produced. Consider a single coil AB placed on the rotor. At standstill the rotating field will induce a voltage in the rotor coil because there is a rate of change of flux linking the coil AB. If the coil AB is now short circuited,

---

* The efficiency of the Reluctance Motor has been greatly increased over the last decade due mainly to the research work of Professor Lawrenson of Leeds University.

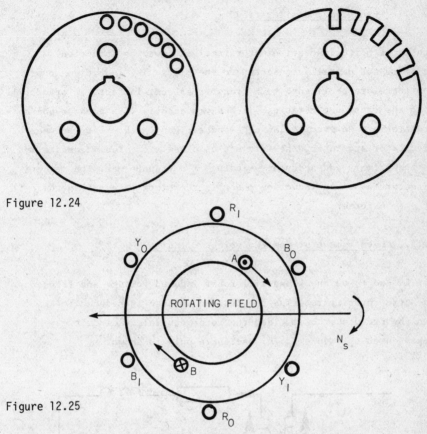

Figure 12.24

Figure 12.25

the induced e.m.f. will cause a current to flow in the direction shown. There will thus be a force produced on the current carrying conductors A and B (Force = $BI\ell$), and the rotor will begin to rotate in the direction of this force. The rotor speed will increase until the torque produced by the machine is equal to the mechanical load torque. The induction motor will never reach synchronous speed because at synchronous speed there would be no relative movement between the rotor and the rotating field, therefore no e.m.f. would be induced in the rotor coils and consequently no torque produced. The ratio of the difference in speed of the rotating field and the rotor to the speed of the rotating field is termed the 'slip'.

$$\text{Slip } s = \frac{N_s - N}{N_s}$$

492 Basic Electrical Engineering and Instrumentation for Engineers

12.6.1 Transformer Similarities

The rotor of the induction motor receives power by induction, as it is not normally connected to the mains. The induction motor can therefore be likened to a transformer and, in fact, at standstill with the rotor not rotating, it behaves exactly like a three-phase transformer. However, because of the air gap which exists between the stator and rotor, they are not as closely coupled magnetically as the primary and secondary winding of a transformer. The magnetising current of an induction motor is therefore larger than that of a transformer.

12.6.2 Development of the Equivalent Circuit

Let E_2 and f_1 be the standstill rotor induced voltage and frequency. When the rotor is rotating the induced voltage is reduced to sE_2 and the frequency to sf_1, and the rotor equivalent circuit may represented by Figure 12.26, one phase only is shown.

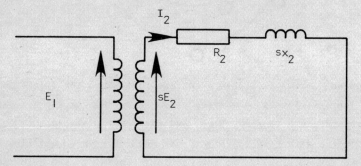

Figure 12.26

The magnitude of the rotor current is $I_2 = \dfrac{sE_2}{\sqrt{R_2^2 + (sX_2)^2}}$

dividing through by s, the rotor current may be written as

$$I_2 = \dfrac{E_2}{\sqrt{\left(\dfrac{R_2}{s}\right)^2 + X_2^2}}$$

and a new rotor equivalent circuit may be drawn in terms of the standstill values plus an extra resistance of $R_2(\frac{1}{s} - 1)$, Figure 12.27.

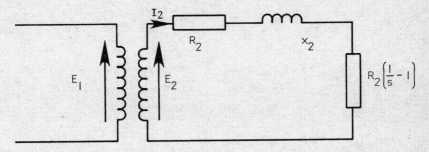

Figure 12.27

Using similar techniques to those explained in Section 7.2 the equivalent circuit of the induction motor referred to the stator, is shown in Figure 12.28.

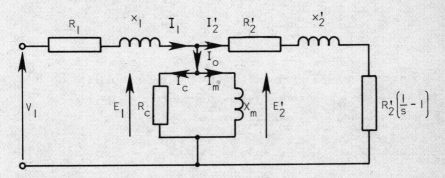

Figure 12.28

The parameters of this equivalent circuit can be measured in a similar way to those of the transformer equivalent circuit, by using (a) Light Running Test (No-Load Test); (b) Locked Rotor Test (Short Circuit Test).

(a) <u>Light Running Test</u>

The machine is operated at rated voltage with no external load connected to the shaft. Its running speed will be nearly synchronous. The stator voltage, current and power are recorded. Assuming negligible voltage drop in the stator resistance and leakage reactance.

$$R_c = \frac{V_{o.c.}}{I_c} \quad \text{and} \quad X_m = \frac{V_{o.c.}}{I_m}$$

$$I_c = I_{o.c.} \cos\theta_{o.c.} \quad \text{and} \quad I_m = I_{o.c.} \sin\theta_{o.c.}$$

Where $\theta_{o.c.}$ is the open circuit power factor angle, and

$$\cos\theta_{o.c.} = \frac{P_{o.c.}}{V_{o.c.} I_{o.c.}} \quad \text{hence} \quad I_c = \frac{P_{o.c.}}{V_{o.c.}}$$

and therefore

$$R_c = \frac{V_{o.c.}^2}{P_{o.c.}}$$

and

$$X_m = \frac{V_{o.c.}}{I_{o.c.} \sin(\cos^{-1}\frac{P_{o.c.}}{V_{o.c.} I_{o.c.}})}$$

(b) Locked Rotor Test

The rotor is locked at standstill, ($s = 1$). Reduced voltage is applied at the stator terminals, and the short circuit stator voltage, current, and power are measured.

The equivalent machine impedance $Z_e = \frac{V_{s.c.}}{I_{s.c.}}$

where $Z_e = \sqrt{(R_1 + R_2')^2 + (x_1 + x_2')^2}$

and $R_1 + R_2' = \frac{P_{s.c.}}{I_{s.c.}^2}$

thus $x_1 + x_2' = \sqrt{Z_e^2 - (R_1 + R_2')^2}$

It is not possible by this method to separate x_1 and x_2' and these are usually taken as equal.

R_1 and R_2' can be taken in the ratio of their d.c. resistances.

12.6.3 Torque and Slip Relationships

Consider the power distribution in the equivalent circuit of Figure 12.28.

The power input to the motor is $V_1 I_1 \cos\theta$ per phase.

The stator and rotor copper losses are $I_1^2 R_1$ and $I_2'^2 R_2'$, and the iron losses are $I_c^2 R_c$.

The power dissipated in the other resistive term corresponds to the only other power sink, i.e. the mechanical load.

The mechanical power output per phase is therefore

$$P_m = I_2'^2 R_2 \left(\frac{1}{s} - 1\right)$$

The power supplied to the rotor, per phase, is

$$P_2 = I_2'^2 \frac{R_2}{s}$$

Thus: $P_m = P_2 - sP_2 = (1-s)P_2$

The torque for an 'm' phase machine is given by:

$$T = \frac{mP_m}{2\pi N} = \frac{mP_2(1-s)}{2\pi N_s(1-s)} = \frac{mP_2}{2\pi N_s} = \frac{m}{2\pi N_s} I_2'^2 \frac{R_2}{s}$$

$$T = \frac{m}{2\pi N_s} \frac{R_2'}{s} \cdot \frac{V_1^2}{(R_1 + \frac{R_2'}{s})^2 + (x_1 + x_2')^2}$$

Maximum torque

Assuming the stator applied voltage, the machine impedances and the frequency to be constant, and the slip 's' to be variable, also write X for $x_1 + x_2'$.

$$T = K \cdot \frac{1}{sR_1^2 + 2R_1 R_2' + \frac{R_2'^2}{s} + sX^2}$$

$dT/ds = 0$ when

$$0 = -R_1^2 + \left(\frac{R_2'}{s}\right)^2 - X^2$$

from which

$$s = \pm \frac{R_2'}{\sqrt{R_1^2 + (x_1 + x_2')^2}} \quad \text{for maximum torque}$$

The negative sign is for supersynchronous speeds giving a negative (generating torque).

Substituting this value in the equation for torque, the maximum value for the torque 'T_{max}' is given by:

$$T_{max} = \frac{m}{2\pi N_s} \cdot \frac{V_1^2}{2(\sqrt{R_1^2 + (x_1 + x_2')^2} \pm R_1)}$$

This expression is independent of R_2' which only determines the position at which the maximum torque occurs.

Starting torque

At standstill $s = 1$, therefore the starting torque T_s is given by:

$$T_s = \frac{m}{2\pi N_s} V_1^2 \frac{R_2'}{(R_1 + R_2')^2 + (x_1 + x_2')^2}$$

If the machine is of the slip ring type, then the starting torque can be increased by adding external rotor resistance via the slip rings.

Torque-slip characteristic

The approximate shape of the torque-slip curve of an induction motor can be obtained by considering the variation of torque when the slip approaches zero, and when the slip approaches unity. From the expression for torque, $s \to 0$.

$$T \simeq \frac{m}{2\pi N_s} \frac{R_2'}{s} \cdot \frac{V_1^2}{\frac{R_2'^2}{s}}$$

hence   T is proportional to s.

Also, when s → 1

$$T \simeq \frac{m}{2\pi N_s} \frac{R_2'}{s} \cdot \frac{V_1^2}{(R_1 + R_2')^2 + (x_1 + x_2')^2}$$

hence   T is proportional to $1/s$.

The complete torque-slip curve is shown in Figure 12.29.

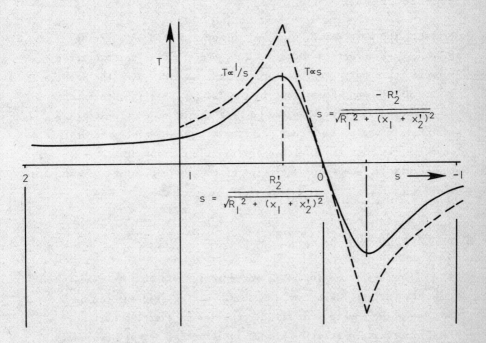

Figure 12.29

12.6.4 Starting

The current drawn from the supply during starting is very large (approximately four or five times the full-load current of the motor). For relatively small machines (less than about 20 h.p.) switching directly on to the supply is permissible. It is usual, however, to obtain the permission of the electricity authorities for the direct on-line starting of larger machines.
There are several ways to limit the current during starting, but they all involve auxiliary gear which is often quite costly.

Star-delta starter

Probably the most widely used and cheapest method of starting high powered induction motors is to connect a star-delta starter between the supply and the stator of the machine. With the machine at standstill and the starter in the 'start' position, the stator is connected in star. When the machine begins to accelerate the switch is moved to the 'run' position which reconnects the stator in delta. By this star-delta operation the current drawn from the supply is reduced to one third of the current during direct on-line starting.

Auto transformer

If the stator of the induction motor is fed via an auto-transformer then the current taken from the supply during starting can be considerably reduced from the direct on-line starting current. Unfortunately the starting torque is also considerably reduced. This technique is costly because the auto-transformer has to be of the same rating as the induction motor.

Rotor resistance

With slip ring machines it is possible to include extra resistance in the rotor circuit. The inclusion of extra rotor resistance not only reduces the starting current but also produces improved starting torque as shown in Figure 12.26.

12.6.5 Braking

Induction motors may be brought to standstill very quickly by either (a) plugging, or (b) dynamic braking.

(a) <u>Plugging</u> - is the term used when the direction of the rotating field is reversed. This is achieved very simply by reversing any two of the three supply leads to the stator. The current taken from the supply during plugging is very large and machines which are to be regularly plugged must be specially rated.

(b) <u>Dynamic braking</u>

In this operation the stator is disconnected from the a.c. supply and re-connected to a d.c. source as shown in Figure 12.30. The direct current in the stator produces a stationary undirectional field and as the rotor will tend to align itself with the stator field, it will therefore come to standstill.

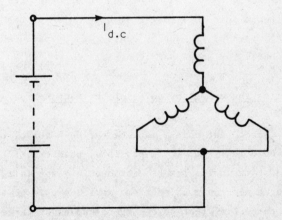

Figure 12.30

12.6.6 Speed control of induction motors

The normal running speed of an induction motor is approximately

98% of the synchronous speed of the field at no-load. At full-load the speed will have dropped to about 94% of the synchronous speed. It is clear, therefore, that to vary the speed of the induction motor the synchronous speed of the rotating field must be varied. Now the synchronous speed is given by

$$N_s = \frac{\text{frequency} \times 60}{\text{pole pairs}} \quad \text{r.p.m.}$$

therefore changes in the synchronous speed must be brought about by either (a) changing the frequency, or (b) changing the number of poles.

(a) <u>Change of frequency</u>

When this type of control was used in the past, the necessary change in frequency was obtained from a specially designed auxiliary machine. With the advent of high power thyristors, static convertors have been designed which will produce a variable frequency. At present these variable frequency convertors are very expensive (many times the cost of the induction motors they are controlling) but it is hoped that this cost will be reduced as the semiconductor technology increases. Frequency convertors do produce a wide range of continuously variable speed control.

(b) <u>Changing the number of poles of the machine</u>

One would expect that once a machine has been built, the number of poles would be fixed. It is possible, however, if the ends of all the stator coils are brought out to a specially designed switch, to change the machine from, say, a 4 pole to a 10 pole. To obtain three different pole numbers and hence three speed ranges, requires very complex switching arrangements. It should be noted that the technique of pole changing does not produce an infinitely variable speed, it only gives discrete speed ranges. For some applications, say, a two speed fan drive, this is often all that is required, in which case pole changing would be the cheapest and most effective method of speed control.

Marginal speed control

Some applications require only a small variation in the machine speed, which can be obtained by either (a) adding external rotor resistance (slip ring machines only), or (b) reducing the stator voltage.

(a) Rotor resistance

It has already been stated that increasing the rotor resistance shifts the speed at which the maximum torque occurs. It is clear, therefore, that the operating speed of the machine can be varied by increasing the rotor resistance. This is shown in Figure 12.31 in that the speed is reduced from N_1 to N_2 and N_3 as the external rotor resistance is increased from R_1 to R_2 and R_3. Although the possible speed variation is small the cost is relatively cheap.

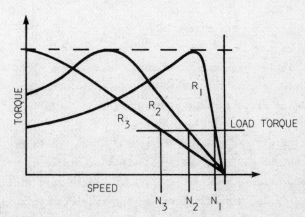

Figure 12.31

(b) Reduced Stator Voltage

If the applied stator voltage is reduced a series of torque-speed curves are obtained as shown in Figure 12.32. It is seen that for a voltage of V_1 the speed for a particular load torque is N_1, if the voltage is reduced to V_2 the speed is reduced to N_2, etc. The main disadvantage of this method is that as the voltage is reduced so too is the torque, ($T \propto V^2$) and thus this technique is only used for obtaining very small changes in speed.

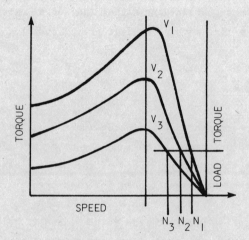

Figure 12.32

12.7 Single Phase Induction Motors

The term single phase induction motor is really a misnomer because the operation of the induction motor depends on the production of a rotating field by the stator winding and this cannot be achieved by a single stator coil alone.

A rotating field can be produced by two stator coils displaced in space by 90° and supplied by currents displaced by 90°. (This is the two-phase equivalent of the three-phase case discussed in section 12.3.) Most single-phase machines attempt to approach the two-phase condition by using some external technique, the most common being (1) the Shaded Pole Motor, and (2) the capacitor motor.

12.7.1 The shaded pole motor

The stator consists of a salient pole single-phase winding and the rotor is of the squirrel cage type. Inset in the pole face is a copper ring as shown in Figure 12.33.

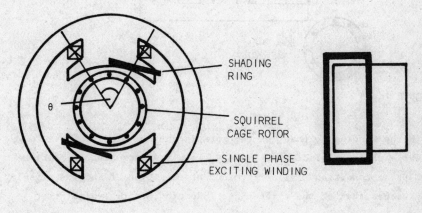

Figure 12.33

When the exciting coil is supplied with alternating current the flux produced induces a current in the shading ring. The angle θ between the magnetic axes is never more than about 60°, and the phase difference between the currents in the exciting winding and the shading ring is very small, therefore the rotating field produced is far from the optimum two-phase condition. The performance of the machine is consequently poor. The efficiency is low due to the continuous losses in the shading rings and the power factor is poor because the ampere-turns for both fluxes are drawn from the single phase supply.

The main advantage of the shaded pole machine is its simplicity and cheapness of manufacture.

12.7.2 Capacitor motor

In the capacitor motor shown in Figure 12.34 the stator has two windings displaced by 90°. A capacitor is connected in series

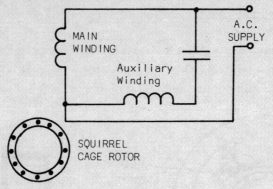

Figure 12.34

with one winding so that the currents in the two windings have a large phase displacement, thereby closely approaching the optimum two-phase condition. The performance of this type of machine closely approaches that of the three-phase induction motor. Typical test curves are shown in Figure 12.35.

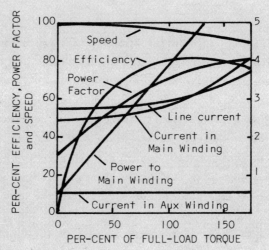

Figure 12.35

12.8 Plain series (Universal) motor

The universal series motor is usually built in the smaller fractional h.p. sizes (up to about $\frac{1}{8}$ h.p.) and is used for mainly domestic applications, see table. The construction is that of a normal d.c. series motor with a totally laminated magnetic circuit so that it can operate on either a.c. or d.c. supplies (hence the term 'universal'). Typical performance curves are shown in Figure 12.36.

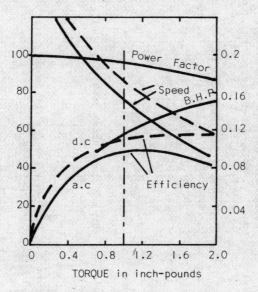

Figure 12.36

TABLE 12.1 CLASSIFICATION OF MOTORS BY SPEED

Speed		Motor	Application
Constant	(a)	Synchronous	Useful for power factor correction.
	(b)	Reluctance and hysteresis	Small power applications, clocks, turntables, recently rolling mill tables, conveyors.
Almost constant	(a)	Induction	Used when a small speed change can be tolerated – fans, blowers, machine tools, pumps, compressors, crushers.
	(b)	D.C. Shunt	
Varying speed	(a)	Series	Cranes, hoists, traction.
	(b)	Universal, repulsion	
Marginal speed control	(a)	Wound rotor induction	Cranes, hoists.
	(b)	D.C. compound with field control	
Variable speed	(a)	D.C. motor with armature voltage control	Used for any drive requiring a wide speed variation. Expensive.
	(b)	Induction motor with variable frequency supply	
Multispeed	(a)	Cascaded induction motors	Used when two discrete speeds are required. Two speed fans.
	(b)	Pole changing induction motors	
		Single-phase induction motors, fractional horsepower universal motors	All domestic applications, cleaners, washers, tumble dryers, central heating pumps.

Examples

1. The open circuit characteristics of a d.c. shunt generator running at 1000 r.p.m. is given below:

E volts	0	40	80	120	160	200	240
I_f amps	0	0.28	0.56	0.86	1.34	2	3.2

   What is the open circuit voltage of the machine with a field resistance of 85 Ω? Estimate the critical field resistance at 1000 r.p.m. What is the approximate critical speed with a field resistance of 100 Ω?

   (225 V, 142 Ω, 698 r.p.m.)

2. The d.c. shunt generator of question 1 is to maintain a constant terminal voltage of 200 V. At full load the speed drops 10% and the armature voltage drop is 10 V. Calculate the required change in field resistance from no-load to full-load.

   (100 Ω to 70 Ω)

3. If the shunt machine of question 1 has an armature resistance of 0.25 Ω, neglecting armature reaction, find the total field resistance required for the machine to run at 1000 r.p.m. as a shunt motor from 200 V d.c. mains if the armature current is 40 A.

   (111 Ω)

4. A series motor with negligible resistance and operating on the unsaturated part of the open circuit characteristic when driving a load takes 40 A at 440 V. If the load torque varies as the square of the speed, find the resistance necessary to reduce the speed by 25%.
   How could the speed be increased?

   (6.42 Ω)

5. Show that two orthogonally placed coils supplied with currents which are 90° out of time phase with each other will produce a rotating filed. What is the magnitude and speed of rotation of this field?

6. A diesel engine drives a 6 pole, three-phase alternator at 1200 r.p.m. The alternator is electrically connected to a 4 pole, three-phase, induction motor running at 4% slip. What is the speed of the induction motor?

(1728 r.p.m.)

CHAPTER THIRTEEN
Measurements

This chapter on measurements will be confined to a description of the instruments commonly used to measure current, voltage, power and energy, and to a few simple bridge techniques to measure resistance, inductance and capacitance. Commercial resistance test sets are mentioned together with a brief outline of one or two digital instruments.

13.1 <u>The basic types of ammeters and voltmeters</u>

The instruments listed below may be used as either ammeters or voltmeters depending on the connection of the external resistance. For example, consider the milliammeter shown in Figure 13.1. The full scale deflection of the meter is 100 mA and its internal resistance is 1 Ω. In Figure 13.1(a) it is used with a diverter resistance to measure currents up to 100 A, whilst in Figure 13.1(b) it is used with a series resistance to measure voltage up to 100 V.

510 Basic Electrical Engineering and Instrumentation for Engineers

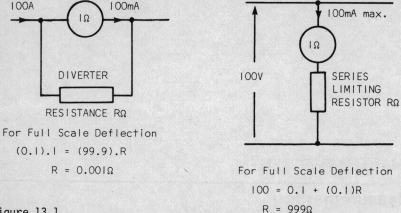

Figure 13.1

The most commonly used instruments are:
- (i) Moving iron,
- (ii) Permanent magnet moving coil,
- (iii) Dynamometer type moving coil,
- (iv) Induction,
- (v) Rectifier type.

13.1.1 <u>Moving iron instruments</u>

There are two types of moving iron instruments in use today, namely the 'Attraction' type and the 'Repulsion' type.

In the attraction type of instrument shown in Figure 13.2 when current flows in the coil a magnetic field is set up. The piece of soft iron tries to align with this field (i.e. is attracted into the coil), causing a deflection of the pointer.

In the repulsion type of instrument shown in Figure 13.3 when current flows in the coil the magnetic field produced magnetises the soft iron rods in the same direction. The free or movable rod is therefore repelled away from the fixed one causing a deflection of the pointer.

To obtain an expression for the torque of the moving iron instrument, we consider the energy balance equation when the pointer

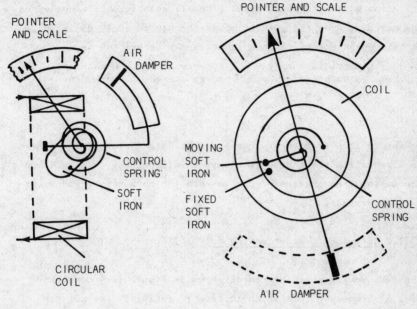

Figure 13.2   Figure 13.3

moves through a small angle dθ due to a current change of di, and a change of inductance of dL.

We have already seen that Faraday's Law can be expressed as

$$e = L\frac{di}{dt}$$

However, if the inductance varies as well as the current, Faraday's Law is written as

$$e = \frac{d(Li)}{dt} = i\frac{dL}{dt} + L\frac{di}{dt}$$

The incremental energy supplied to the coil is e i dt

$$= i^2 dL + iLdi$$

The energy stored in the coil changes from

$$\tfrac{1}{2}Li^2 \text{ to } \tfrac{1}{2}(L + dL)(i + di)^2$$
$$= \tfrac{1}{2}i^2L + iLdi + \tfrac{1}{2}L(di)^2 + \tfrac{1}{2}i^2 dL$$
$$+ i\,di\,dL + \tfrac{1}{2}(di)^2 dL - \tfrac{1}{2}i^2 L$$
$$= i\,L\,di + \tfrac{1}{2}i^2\,dL$$

if second and third order terms of small quantities are neglected. The work done when the pointer moves through an angle $d\theta$ is $Td\theta$, therefore the energy balance equation may be written as,

$$\begin{array}{c} \text{Electrical energy} \\ \text{supplied} \end{array} = \begin{array}{c} \text{Change in} \\ \text{Energy stored} \end{array} + \begin{array}{c} \text{Mechanical} \\ \text{work done} \end{array}$$

$$i^2 \, dL + i \, L \, di = i \, L \, di + \tfrac{1}{2} \, i^2 \, dL + Td\theta$$

$$T = \tfrac{1}{2} \, i^2 \, \frac{dL}{d\theta}$$

It can be seen from the torque expression that even if the current is alternating the torque is always unidirectional ($\propto i^2$). Therefore the moving iron instrument can be used for both direct and alternating currents.

13.1.2 Permanent magnet moving coil

The P.M. moving coil instrument shown in Figure 13.4 consists of a light weight coil which is free to rotate in the air gap formed between the permanent magnet and the soft iron cylinder.

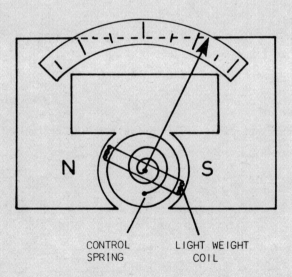

Figure 13.4

When current flows through the coil, a torque is produced according to the equation

$$T = N.B.I.\ell \times 2R$$

where
- N — number of turns on the coil
- B — flux density in the air gap
- I — current in the coil
- ℓ — active length of coil in air gap
- R — radius of coil from centre pivot

It can be seen from the torque expression that if the current is alternating, then the torque is oscillatory, thus the P.M. moving coil instrument can only be used for direct currents.

13.1.3 Dynamometer type moving coil

In the dynamometer instrument shown in Figure 13.5(a) the permanent magnet is replaced by two fixed coils. The flux density produced by the fixed coils is proportional to the current, therefore the torque is proportional to the square of the current, and the instrument can be used with both alternating and direct currents. The instrument finds most application in the measurement of power, as the dynamometer wattmeter, Figure 13.5(b).

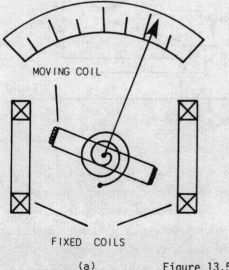
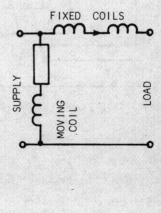

(a) Figure 13.5 (b)

13.1.4 Induction instruments

Induction instruments work on exactly the same principle as the induction motor discussed in Chapter 12 and can therefore only be used with alternating currents. The rotating field produced by the static coils induces currents in the aluminium disc which is free to rotate. The interaction of the rotating field and the field produced by the induced currents in the disc produce a torque which causes the disc to rotate. This type of instrument is commonly used to measure electrical energy (kWhour). The essential details are shown in Figure 13.6.

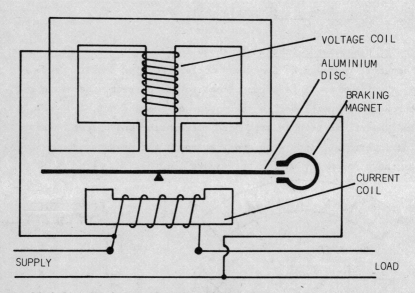

Figure 13.6 Essential Details of a kWhour Meter

Because of the shape of the magnetic circuit of the voltage coil, it is only the leakage flux from this circuit which penetrates the disc. The leakage flux lags the current producing it by nearly 90°, therefore this flux, in conjunction with the flux produced by the current coil, effectively form a two-phase rotating field. The torque produced by an induction meter is proportional to the power supplied to the coils.

$$T \propto V I \cos \phi$$

The retarding torque produced by the braking magnet is proportional to the eddy currents induced in the disc by the magnet and the strength of the magnetic field of the magnet which is effectively constant. The induced currents are proportional to the speed of the disc ($e = B\ell v$), therefore

$$T_{retarding} \propto N$$

For the steady state operation at constant speed the driving torque equals the retarding torque, therefore

$$N \propto V I \cos \phi$$

Integrating both sides w.r.t. time

$$\int N \, dt \propto \int V I \cos \phi \, dt$$

Hence the total number of revolutions is proportional to the energy supplied.

13.1.5 Rectifier instruments

Rectifier instruments are a means of using the permanent magnet moving coil meter with alternating currents. They consist of a moving coil meter in conjunction with a bridge rectifier, as shown in Figure 13.7. On the assumption of sinusoidal current waveshapes, the scales of such instruments are usually marked in terms of 1.11 times the current actually measured to give the r.m.s. value.

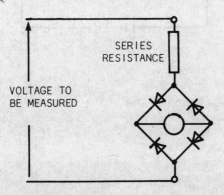

Figure 13.7

516 Basic Electrical Engineering and Instrumentation for Engineers

One of the main disadvantages of rectifier instruments is that they give erroneous readings on non-sinusoidal waveforms. The series resistance is used to swamp the non-linear effect of the resistance of the rectifier.

13.2 Bridge methods to measure resistance, inductance and capacitance

The bridges discussed below are typical examples of a very wide range of bridges available for the measurement of resistance, inductance and capacitance.

13.2.1 Resistance

The most widely used method of measuring resistance is by the Wheatstone Bridge, shown in Figure 13.8. When the bridge is balanced, as indicated by a zero deflection on the galvo, then there is no p.d. between b and d.

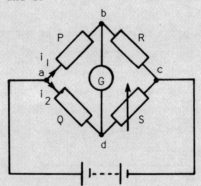

Figure 13.8

Therefore the volt drop across P = the volt drop across Q

$$i_1 P = i_2 Q \qquad 13.1$$

Similarly

$$i_1 R = i_2 S \qquad 13.2$$

By division

$$\frac{P}{R} = \frac{Q}{S}$$

Usually R is the unknown resistance, S is a variable standard resistance, and P and Q are 'ratio arms' which may be varied in multiples of 10.

$$R = \frac{Q}{P} \cdot S \qquad 13.3$$

13.2.2 Inductance

Inductance is a phenomenon associated with alternating currents only, and therefore to measure inductance we use an a.c. bridge, as shown in Figure 13.9.

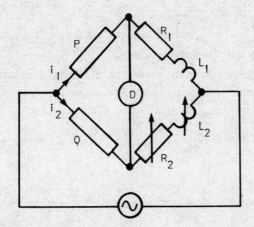

R_1 and L_1 the unknown resistance and inductance

R_2 and L_2 standard variable resistance and inductance

P and Q are resistive ratio arms

D is the detector

Figure 13.9

When the bridge is balanced, i.e. no current in the detector, then the volt drop across P = volt drop across Q

$$i_1 P = i_2 Q \qquad 13.4$$

Similarly

$$i_1(R_1 + j\omega L_1) = i_2(R_2 + j\omega L_2) \qquad 13.5$$

By division

$$\frac{(R_1 + j\omega L_1)}{P} = \frac{(R_2 + j\omega L_2)}{Q} \qquad 13.6$$

518 Basic Electrical Engineering and Instrumentation for Engineers

Equating the real and imaginary terms in equation 13.6 yields

$$R_2 = \frac{Q}{P} \cdot R_1 \qquad 13.7$$

and

$$L_2 = \frac{Q}{P} \cdot L_1 \qquad 13.8$$

13.2.3 Capacitance

One of the simplest a.c. bridges for the measurement of a capacitance which is shunted by a resistance, is Wien's Bridge, as shown in Figure 13.10.

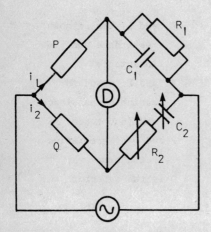

R_1 and C_1 are the unknown resistance and capacitance.

R_2 and C_2 are the variable standards.

P and Q are the resistive ratio arms. D is the detector.

Figure 13.10

At balance

$$i_1 P = i_2 Q \qquad 13.9$$

$$i_1 \frac{R_1}{1 + j\omega C_1 R_1} = i_2 (R_2 - j \frac{1}{\omega C_2}) \qquad 13.10$$

By division

$$\frac{P}{R_1} \cdot (1 + j\omega C_1 R_1) = \frac{Q \cdot j\omega C_2}{(1 + j\omega R_2 C_2)} \qquad 13.11$$

$$1 - \omega^2 R_1 R_2 C_1 C_2 + j\omega R_1 C_1 + j\omega R_2 C_2 = \frac{Q}{P} \cdot R_1 \cdot j\omega C_2 \qquad 13.12$$

Equate real and imaginary terms

$$\omega^2 R_1 R_2 C_1 C_2 = 1 \qquad 13.13$$

$$R_1 = \frac{P \cdot R_2 C_2}{QC_2 - PC_1} \qquad 13.14$$

From which

$$R_1 = \frac{P}{Q} \frac{1 + \omega^2 R_2^2 C_2^2}{\omega^2 R_2 C_2^2} \qquad 13.15$$

$$C_1 = \frac{\frac{Q}{P} \cdot C_2}{1 + \omega^2 R_2^2 C_2^2} \qquad 13.16$$

One of the most important applications of the Wein Bridge is its use as the frequency selective network in R-C oscillators. In which case the bridge would tune (or balance) at various frequencies depending on the values of R_1 and C_1.

13.3 Measurement of small values of resistance

A very useful portable test set for measuring resistances of a few ohms down to one microhm is the DUCTER. The ducter is basically a permanent magnet moving coil meter, as shown in Figure 13.11,

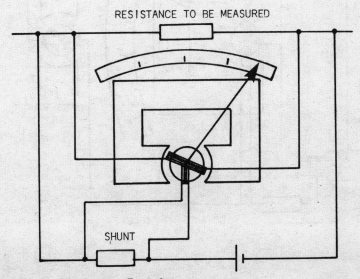

Figure 13.11 The Ducter Test Set

but in this case there are two deflecting coils, one related to the current flowing through, and the other related to the voltage across the resistance to be measured. The position taken up by the pointer is dependent on the ratio of the voltage drop and the current.

13.4 <u>Measurement of large values of resistance</u>

For the measurement of large resistances, particularly the insulation resistance of cables, the MEGGER test set is used. It is very similar in principle to the ducter but in this case the battery is replaced by a hand driven generator capable of producing a voltage up to 2,500 V.

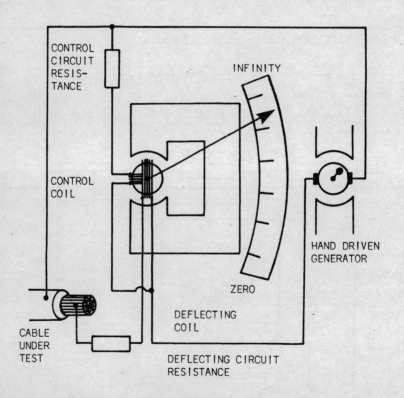

Figure 13.12 The Megger Test Set

The deflecting coil is connected in series with a fixed resistance and the resistance under test. The control coil in series with a fixed resistance is connected across the voltage source.
For large values of resistance, especially the insulation resistance of cables and wires, which is in the order of megohms, the current in the deflecting coil will approach zero. In this case the coil will align along the magnetic axis and the pointer will set at infinity.

13.5 Digital instruments

In general the accuracy obtained with digital instruments is no better than that obtained with the analogue type, even though they cost considerably more. However, they do have certain features which make them very attractive for certain applications. Apart from the ease in reading the meter, digital coded outputs can be used to drive paper tape punches, electric typewriters, etc. or even to feed directly into a digital computer.
It is beyond the scope of this text to cover the circuit details of digital instruments but the fundamental principles of digital voltmeters and frequency meters are outlined below.

13.5.1 Digital voltmeters

Digital voltmeters are basically potentiometers in which the balance condition is achieved automatically. A typical arrangement is shown in Figure 13.13.
An unknown voltage is connected to a variable reference voltage via an amplifier. The variable reference voltage is automatically adjusted until a balance is obtained when the magnitude of the reference voltage (which at balance is also the magnitude of the unknown voltage), is then presented as a digital read out. The reference voltage is usually provided by a compensated zener diode.

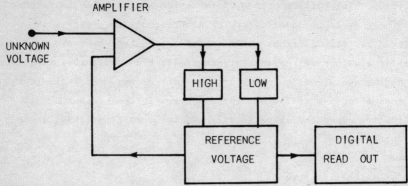

Figure 13.13

13.5.2 Digital frequency meter

Figure 13.14 shows the block diagram of a basic frequency meter.

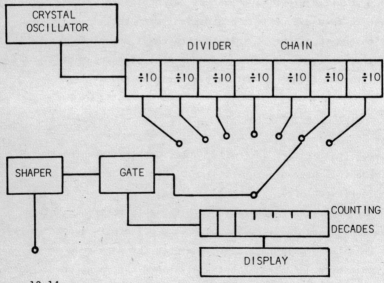

Figure 13.14

The input signal is fed into a shaper which produces standard pulses from almost any type of waveform. The output from the shaper is fed via the gate to the counting decades, which count the number

of pulses. The gate remains open for a certain interval of time
depending on the setting of the divider chain. For most industrial
purposes (not communications) a crystal oscillator of 1 MHz would
be used. The divider chain operates in decades as shown, for example
the sixth decade will produce 1 pulse per second. The counting
decades are, therefore, counting the number of pulses over a certain
time interval which can be directly related to, and displayed
as, the frequency of the input signal. In particular, with the
divider chain on the sixth setting, the counters are counting
the pulses per second which is the direct frequency measurement.

Example 1

One phase of a three-phase cable is found to be short circuited
to the lead sheath. Suggest a method of locating the fault.
Set up a Wheatstone Bridge as shown in Figure 13.8.

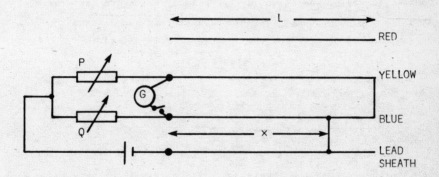

Connect the yellow and blue phases at the remote end.
Let the fault occur at a distance x metres from the near end,
and let the cable resistance be R Ω/metre.
At balance

$$i_1 P = i_2 Q$$

$$i_1 (2L - x) R = i_2 x R$$

By division

$$\frac{(2L - x)}{P} = \frac{x}{Q}$$

Hence $\quad x = 2L \dfrac{Q}{P + Q} \quad$ metres

Example 2

A single core cable has been broken and there is no connection between the core and the sheath. Show how a Wien Bridge may be used to locate the fault.

Let the conductor to sheath capacitance be C μF/metre, and the resistance be R Ω/metre.

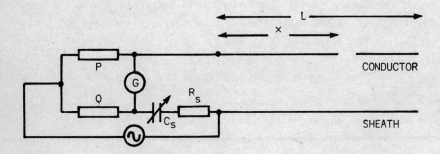

Connect the bridge as shown in Figure 13.10 using standard ratio arms P and Q, a standard variable capacitor C_s and a standard variable resistor R_s. From the equations 13.15 and 13.16

$$R \cdot x = \frac{P}{Q} \frac{1 + \omega^2 R_s^2 C_s^2}{\omega^2 R_s^2 S_s^2}$$

$$C \cdot x = \frac{\frac{Q}{P} C_s}{1 + \omega^2 R_s^2 C_s^2}$$

It is probable that the insulation resistance will be very large in which case R_s will be very small (approaching zero), therefore

$$C \cdot x = \frac{Q}{P} C_s$$

Examples

1. A moving coil instrument gives a full scale deflection with 20 mA flowing. The coil resistance is 1 Ω. What is the value of the shunt required for the instrument to measure 1000 A?

(0.00002 Ω)

2. The coil of a moving iron voltmeter has a resistance of 400 Ω and an inductance of 1 H. The series resistor is 2000 Ω. The meter reads 240 V when connected across 240 d.c. mains. What will it read when connected across 240 V a.c. mains? How could the accuracy of the meter be improved for alternating currents?

(238 V)

3. A dynamometer wattmeter with its voltage coil connected across the load side of the instrument reads 250 W. If the voltage across the load is 240 V and the resistance of the voltage coil is 2400 Ω, what power is being taken by the load?

(226 W)

4. A sinusoidal voltage of 240 V is connected across a series circuit containing a 10 Ω resistor, a silicon diode and a moving coil ammeter. Calculate the reading on the meter.

(10.8 A)

5. A Wien Bridge network is used to measure the capacitance C_1 and the loss resistance R_1 of an imperfect capacitor. The other arms of the bridge are a standard 0.01 μF perfect capacitor in series with a standard 100 Ω resistor, and two perfect resistors each 1000 Ω. The bridge balanced at an angular frequency ω = 10,000. Calculate the values of C_1 and R_1.

(0.01 μF, 1 MΩ)

APPENDIX ONE

Heating, Lighting and Tariffs

Lighting

Electric lighting falls broadly into two categories - (a) tungsten filament lamps, and (b) discharge tubes.

Tungsten filament lamps are the most common form of household lighting and are often used extensively in public and industrial premises. The tungsten filament lamp produces light by heating the filament by passing an electric current through it. Improved performance may be obtained by enclosing the filament in an atmosphere of iodine which reduces the effect of filament loss due to evaporation.

There are three main types of discharge tubes, namely, sodium vapour, mercury vapour, and fluorescent. The first two are used mainly for street lighting whilst fluorescent tubes find wide application in public and industrial premises and to some extent, household premises. All three operate on the same basic principle of producing a strong electric field between the two electrodes situated at each end of a glass tube.

Sodium vapour lamp - Initially the electric field produces a starting discharge in the cold tube containing argon or neon gas at a few millimetres pressure. This starting discharge heats up the metallic sodium which is also in the tube, producing sodium vapour, which gives off the characteristic yellowish light. The warming up process usually takes about fifteen minutes.

Mercury vapour lamp - There are three types of mercury vapour lamp, the cold cathode type, the medium pressure lamp and the high pressure lamp. Discussion will be limited to the medium pressure type which operates at mains voltage and is the most widely used. This type of lamp is made with an inner silica tube in which the discharge takes place and an outer envelope of ordinary glass which maintains a vacuum around the discharge tube, so conserving heat and improving efficiency, and also absorbing the harmful ultra-violet radiation. Inside the silica tube there is argon or neon gas at a few millimetres pressure together with a small amount of mercury. The main discharge is initiated in the inert gas giving off a reddish glow, but when the mercury is vapourised, this changes to the familiar bluey-green colour.

Fluorescent mercury vapour lamps - when ultra-violet light is incident on certain powders they fluoresce brightly, giving off a light which approaches daylight. Fluorescent lamps are manufactured in the form of long tubes of standard lengths 8, 5, 4, 3, 2 and $1\frac{1}{2}$ feet. The glass tubes are internally coated with the fluorescent powder and contain inert gas plus a small amount of mercury. All mercury vapour lamps require a large voltage to establish the discharge and this is usually achieved by causing a rapid change of current through an inductor ($e = L \frac{di}{dt}$). This current change is provided by a starter switch in the form of a bimetal strip, as shown in Figures 1 and 2.

There are two types of starter switch in use, i.e. the glow starter and the thermal starter.

The contacts of the glow starter are housed in a container filled with helium gas and in the cold condition are normally open. When the supply is switched on the full mains is connected across the contacts and a glow discharge takes place in the helium gas. This discharge heats the bimetal strip and the contacts close, thus allowing current to flow through the electrodes in the fluorescent tube. With the glow discharge in the starter now extinct the bimetal strip cools and

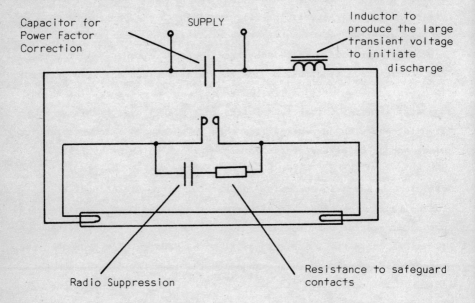

Fluorescent Tube with Glow Starter

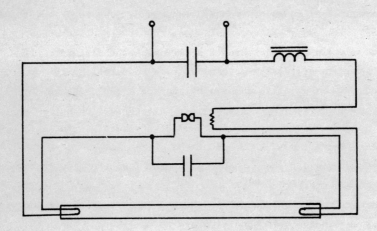

Fluorescent Tube with Thermal Starter

the contacts open. A large voltage surge is produced in the inductor and the main discharge in the tube takes place.
In the thermal starter the contacts are closed when in the off position. When the supply is switched on a heater, which is adjacent to the contacts, heats up the bimetal strip causing the contacts to open. The surge voltage thus produced by the inductor then initiates the discharge in the tube.

Heating

Electric Resistance Heating

By far the most common form of electric heating is resistance heating, which makes use of the fact that when an electric current flows in a piece of wire, the electric power supplied (I^2R) is converted into heat. Examples of this technique are found in every household, e.g. electric fires, irons, kettles, cookers, driers, immersion heaters, central heating systems, etc.
Electric resistance heating is also extensively used in industry. Electrically heated kilns in the pottery industry, heating of billets and strips in the manufacture of steel, glass melting, soldering and brazing, resistance welding, etc.
The main advantages of electric resistance heating are:
1. The conversion of electrical energy into heat energy in a resistance wire is 100%.
2. The electric current which produces the heat can be easily and accurately controlled, often automatically.
3. The maintenance of electro-heat installations is small compared to other forms of heating.
4. With electricity the heat can very often be produced near to where it is required thus reducing the heat losses.
5. Electric heating is probably the cleanest form of heat production.
The disadvantages are mainly restricted to the capital cost of the equipment and the price of electrical energy.

Electric Arc Furnaces

Electric arc furnaces are widely used in the smelting industries especially in the manufacture of steel. Arcs are drawn from three carbon electrodes (fed from the three-phase supply) and the steel charge to be melted. The electric power supplied to the steel is controlled by varying the current supplied to the electrodes by adjusting either the supply voltage or the distance between the electrodes and the charge.

Induction heating

It was shown in section 1.3.7 that when a coil carrying alternating currents is wound round a metallic core, eddy currents are produced in the core, and the electric power supplied to maintain these eddy currents is converted into heat energy. It is clear, therefore, that metallic objects can be indirectly heated by placing them in an alternating magnetic field. It is interesting to note that as the frequency of the alternating magnetic field is increased, the eddy currents produced in the core are concentrated more towards the surface of the material. At very high frequencies ($\simeq$ 2 MHz), the heat generated is at the surface and this technique is used for the surface hardening of castings.

Dielectric heating

This method is used to generate heat in non-conducting materials. The workpiece to be heated is placed between two electrodes which are connected to a high frequency supply. Heat is produced by the work done in re-aligning the molecular structure of the material as the voltage on the electrodes alternates positive and negative. Unlike induction heating there is no "skin" effect as the frequency is increased and the heat is generated uniformly within the material. Frequencies up to 1 MHz and powers in the range of 1 to 25 kW are used in dielectric heating depending on the material and its size.

Microwave heating

Microwaves are electro-magnetic waves occurring in the frequency spectrum from about 300 MHz up to 3 GHz. They are similar to radio waves but of a much smaller wavelength. Microwave heating is sometimes used as an alternative to dielectric heating in applications where it is difficult to constrain the material in between two electrodes. The microwave energy is produced by a magnetron and directed at the material to be heated, the depth of penetration of the microwaves being dependent on their frequency and the properties of the material.

Microwave heating has found wide application in the catering field where microwave ovens are used for the rapid heating of foodstuffs. Industrial microwave ovens operate at a frequency of about 900 MHz using powers up to 25 kW, whilst domestic ovens operate at 2450 MHz at powers of 600 W up to 1 kW.

Tariffs

The cost involved in the generation and transmission of electrical energy can be divided into two categories:
(a) The cost incurred due to the maximum demand on the system, i.e. the capital expenditure of providing sufficient generating plant to supply the maximum needs of all consumers.
(b) The cost incurred in running the system, i.e. fuel costs for the generation of power, maintenance costs, etc.
Therefore all tariffs operate on a two-part basis.
The consumers of electrical energy are divided into classes dependent on their load characteristics, e.g.

      Industrial
      Domestic
      Farm
      Commercial (shops, offices, etc.)

The maximum demand of an industrial user may be either assessed or measured. Whether or not a maximum demand meter is installed usually depends on the size of demand. When the maximum demand is not likely to exceed 50 kVA an assessment is made. The cost of metering is of necessity quite high because of the accuracy required to avoid errors which can amount to large costs to the consumer. Country wide the maximum demand on the grid system occurs at certain times during the day. In an attempt to distribute this demand more evenly, "off peak" tariffs are available. If plant can be used at night, the charge for electrical power may be greatly reduced.

The general domestic tariff is based on a standard charge of 4.969p for the first 72 units per quarter and 1.83p for each unit in excess of 72 units/quarter.

As an incentive to consumers to use electricity during the night period, the electricity authorities introduced two alternative tariffs, (a) the white meter tariff, (b) the "off peak" tariff. The white meter tariff consists of a fixed quarterly charge, a charge for each unit supplied during the day period, and a much lower charge for units used during the night period (usually an eight hour period from about 2300 hours). The "off peak" tariff is designed for equipment which is used solely during specified "off peak" times.

The majority of tariffs now incorporate a fuel adjustment charge, which takes account of the variation of fuel prices.

APPENDIX TWO
The Measurement of Flux Density

The classical method of measuring the flux density in an iron core is to use either a ballistic galvanometer or a fluxmeter. The ballistic galvanometer has a moving coil, would on a non-metallic former, suspended between the poles of a permanent magnet. There is very little damping employed in this type of instrument and it is necessary to short circuit the galvanometer in order to stop the oscillations which result after a current has been passed through the coil.

The fluxmeter is a special form of ballistic galvanometer in which the movement is rendered dead beat by means of electromagnetic damping.

Consider the circuit shown below in which a toroidal coil is connected via a reversing switch to a d.c. supply. When the current in the main coil is reversed an e.m.f. is induced in the search coil which causes a current to flow through the fluxmeter. The deflection on the meter is proportional to the change of flux linkages in the search coil.

If the flux in the core changes from ϕ to $-\phi$ as the current in the main coil changes from I to -I then

534 Basic Electrical Engineering and Instrumentation for Engineers

Change of flux linkage with search coil = $2 \phi N_s$

If θ is the deflection on the fluxmeter and c is the fluxmeter constant in Wb turns per unit of scale deflection

then
$$2 \phi N_s = \theta c$$
$$\phi = \frac{\theta c}{2 N_s}$$

hence
$$B = \frac{\phi}{A}$$

where A is the cross-sectional area of the core

also
$$H = \frac{N_p I}{\ell}$$

where ℓ is the magnetic length of the core

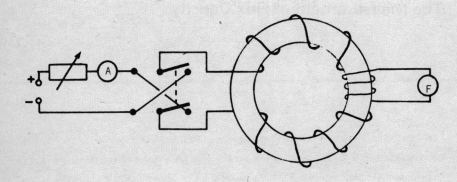

N_p turns on the main coil

N_s turns on the search coil

APPENDIX THREE

A Note on the Manipulation of Complex Numbers

Conjugate

The conjugate of a complex number is obtained by changing the sign of the imaginary term.

Given a complex number $a + jb$ its conjugate is $(a - jb)$. This may be written

$$(a + jb)^* = (a - jb)$$

Addition

The addition of two complex numbers $(a + jb)$ and $(c + jd)$ is performed by adding the real and imaginary parts separately.

$$(a + jb) + (c + jd) = (a + c) + j(b + d)$$

Subtraction

Subtraction is the negative of addition

$$(a + jb) - (c + jd) = (a - c) + j(b - d)$$

Multiplication

To multiply two complex numbers $(a + jb)$ and $(c + jd)$ each term of the first is multiplied by each term of the second.

$$(a + jb)(c + jd) = ac + j\,ad + j\,bc + j^2\,bd$$
$$= ac - bd + j(ad + bc)$$

Division

Division of two complex numbers is performed by 'Rationalising', i.e. multiplying both numerator and denominator by the conjugate of the denominator.

$$\frac{(a + jb)}{(c + jd)} = \frac{(a + jb)(c - jd)}{(c + jd)(c - jd)} = \frac{(ac + bd) + j(bc - ad)}{c^2 + d^2}$$
$$= \frac{(ac + bd)}{(c^2 + d^2)} + j\frac{(bc - ad)}{(c^2 + d^2)}$$

Polar Representation

A complex number may be represented in polar form by a magnitude and an angle, in the same way that cartesian co-ordinates may be changed to polar co-ordinates.

$(a+jb)$ in Polar form is $\sqrt{a^2 + b^2}\ /\underline{\tan^{-1} b/a}$

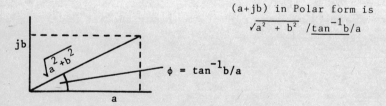

Polar co-ordinates cannot be added or substracted but they are easily multiplied and divided.

Two phasors represented by $A/\underline{\theta}$, and $B/\underline{\phi}$, are multiplied as,

$$A/\underline{\theta} \cdot B/\underline{\phi} = AB/\underline{\theta+\phi}$$

and divided as

$$\frac{A/\underline{\theta}}{B/\underline{\phi}} = \frac{A}{B}/\underline{\theta-\phi}$$

ACSII CHARACTER SET (7-BIT CODE)

L.S. CHAR	M.S. CHAR	0 000	1 001	2 010	3 011	4 100	5 101	6 110	7 111
0	0000	NUL	DLE	SP	0	@	P	`	p
1	0001	SOH	DC1	!	1	A	Q	a	q
2	0010	STX	DC2	"	2	B	R	b	r
3	0011	ETX	DC3	#	3	C	S	c	s
4	0100	EOT	DC4	$	4	D	T	d	t
5	0101	ENQ	NAK	%	5	E	U	e	u
6	0110	ACK	SYN	&	6	F	V	f	v
7	0111	BEL	ETB	'	7	G	W	g	w
8	1000	BS	CAN	(	8	H	X	h	x
9	1001	HT	EM	)	9	I	Y	i	y
A	1010	LF	SUB	*	:	J	Z	j	z
B	1011	VT	ESC	+	;	K	[	k	{
C	1100	FF	FS	,	<	L	\	l	|
D	1101	CR	GS	−	=	M	]	m	}
E	1110	SO	RS	•	>	N	↑	n	~
F	1111	SI	US	/	?	O	← or _	o	DEL

Appendix 4

Index

A.C. resistance 105
Accumulator 356
Address bus 345
Addressing modes 364
Alternators 487
A.L.U. 344
American military specification 290
Ammeters 509
Amplifiers
  adding 195
  bandwidth 161
  current summing 195
  coupling capacitance 166
  effect of series
   capacitance 166
  effect of stray
   capacitance 162
  feedback 172
  gain 141
  integrating 195
  narrowband 147
  operational 192
  stability 182
  tuned 147
  voltage follower 195
  wideband 145

Analogue to digital
  convertors 445
  continuous balance 446
  staircase ramp 447
Aperture time 441
Arithmetic and logic unit 344
Assembler directives 380
Assembler language 364, 377
Astable multivibrator 335
Auto-transformers 238
Auto-transformer starter 498
Avalanche breakdown 107

Bandwidth 62
B-H curve 10
Binary arithmetic 262
Binary code 257
Binary coded decimal 269
Binary coded decimal, digital
  to analogue converter 444
Binary to decimal conversion 260
Binary to octal conversion 261
Binary signal 250
Bipolar transistor 108
Bistable circuit 320
  J.K. 326
  S.R. 321

Index 539

T 326
Bit rate 253
Boole, George 269
Boolean algebra 269
Bottomed operation 117
Braking 499
Byte 348

Capacitance 18
Capacitance transducers 419
Capacitor motor 503
Carrier drift 85
Central processing unit 344
Characteristic impedance
  of line 454
Chopper control 485
Collector-base leakage
  current 119
Common base character-
  istics 111
Common emitter character-
  istics 113
Common mode intereference 438
Commutator 471
Complex numbers 55, 535
Composite characteristic 98
Condition code register 357
Conductivity 2
Constant current method 416
Constant temperature method 416
Contact noise 403
Continuous balance digital
  to analogue convertors 445
Control register 383
Control unit 344
Controlled power recti-
  fication 223
Conversion time 255
Counters 328
Current 2
Current density 3
Current gain - a.c. 185
              - d.c. 111
Current mode logic (CML) 288
Current steering logic
  (CSL) 288
Cut-off operation 117

Data bus 345
Data direction register 383
D.C. generators 470
  compound excited 479

  separately excited 475
  series excited 478
  shunt excited 476
D.C. motors 479
  braking 482
  series 481
  shunt 479
  speed control 482
  starting 482
Decibel 61, 155
Delta connection 67, 69
De Morgan's Theorem 276
Depletion layer 88
Depletion layer capacitance 92
Depletion mode 129
Depth of modulation 460
Development of Equivalent
  circuit 492
Differential mode inter-
  ference 437
Differentiator 34
Diffusion 86
Digital computers 343
Digital signals 245
Digital to analogue
  convertors 441
  B.C.D. 443
  ladder type 444
  weighted resistor 442
Diode
  junction 86
  point contact 93
Diode logic 279
Diode modulator 461
Direct addressing mode 365
Don't care conditions 306
Drain 127
Ducter 519
Dynamic properties of
  transducer 396
Dynamic resistance 105

Earth loop currents 438
Electrodynamic transducers 429
Emitter-coupled logic 286
Encoding 251
Energy 26, 514
Enhancement mode 128
Electrodynamic transducers 429
EPROM 352
Equivalent circuit for diode 101

Erasable programmable read-
  only memory 352
Excitation 474
Exclusive OR function 314
Extended addressing mode 365

Fan out 285
Faraday's law 6, 8, 9, 11
Feedback
  application to amplifiers 172
  effect on distortion and
    noise 178
  effect on gain 173
  effect on input resistance 175
  effect on output resist-
    ance 176
  positive 182
Field effect transistor 123
First order system 397
Flip-flop 321
Flux 2, 4, 18
Flux density 5
Force 8
Fourier series 449
Frequency 50
Frequency division multi-
  plex 466
Frequency meter 522
Frequency response 60
Full-wave rectifier 201

Gate 127
Gate symbols 290
Gauge factor 406
Gauss's law 19
Gray binary 266

Half-wave rectifier 199
Harmonics 449
Hexadecimal code 259
Hot wire measurements 415
Hysteresis 11
Hysteresis loop 13, 15

Immediate addressing mode 364
Impedance 56
  input 147
  matched 151
  output 147
Incremental resistance 105
Index register 356
Indexed addressing mode 369

Induction motors 490
  braking 499
  single phase 502
  speed control 499
  starting 498
Inductive transducers 425
Inherent addressing mode 364
Instruments 510
  digital 521
  dynamometer 513
  induction 514
  moving coil 512
  moving iron 510
  rectifier 515
Insulated gate field effect
  transistor 126
Integrated circuits 193
Integrator 35
Interference
  common mode 438
  differential mode 437
Inverters 500
Inverting logic 283
Iron losses 231

"j" notation 55, 535
J.K. Bistable 326
Johnson noise 436, 404
Junction diode 86
Junction field effect
  transistor 123
Junction transistor 108

Karnaugh map 295
  use for minimisation 300
Kilobyte 348
Kirchoff's laws 35

Label field 377
Ladder type digital to
  analogue converter 443
Large scale integration 340
Least significant bit 257
Lenz's law 6
Light running test 493
Linear variable differential
  transformer (LVDT) 427
Load angle 490
Load regulation 217
Locked rotor test 494
Logarithmetic units - the
  decibel 61, 155

L.S.I. 340

Machine code 359
Magnetic cores 9
Magnetic inductance 230
Magnetic intensity 5
Magneto-motive-force 4
Majority carriers 85
Matched impedances 151
Maximum power transfer 151
Mean free path 107
Measurement of resistance
   516, 519
   capacitance 518
   inductance 517
Medium scale integration 340
Megger 520
Memory 344
Memory Map 350
Mesh analysis 39
Microprocessor 343
   data accuracy 348
   memory organisation 348
Minority carriers 85
Modulation 457
   amplitude 457
   angle 464
   frequency 464
   phase 464
   pulse 467
Monitor programme 375
Motorola 6800 micro-
   processor 355
Monostable multivibrator 336
M.O.S. logic 288
Most significant bit 257
M.S.I. 340
Multichannel transmission
   system 466
Multiplying out 274
Multivibrator 335

Nand gates 307
Negative logic 278
Nibble 348
Nodal analysis 41
Noise 435
NOR gates 307
Norton's theorem 45

Open circuit test 234
Operational amplifier 192

Parallel signals 251
Peak inverse voltage 200
Permeability 4, 10
Permeance 5
Permittivity 18, 20
Phase angle 51
Phase difference 51
Phase sensitive detector 428
Phasor diagrams 52
Photo diodes 418
Piezoelectric transducer 431
Point contact diode 93
Pole changing 500
Polyphase rectifiers 205
Positive logic 278
Potential difference 2
Potential divider 38
Potential gradient 2
Potentiometer transducer 403
Power 26
   in a.c. circuits 58
   in three-phase circuits 70
   measurement in three phase
     circuits 70
Power gain of amplifier 145
Power supply regulation 217
Programme counter 357
Programmable read only memory 352
PROM 352
Propagation delay 286
Pulse amplitude modulation
   (PAM) 467
Pulse coded modulation (PCM) 467
Pulse duration modulation
   (PDM) 467
Pulse position modulation
   (PPM) 467

Q factor 62
Quantisation 250

Random access memory 350
RAM 350
Read only memory 351
ROM 351
Rectifier
   hardwave 199
   fullwave 201
Referred values 232
Reflected waves 454
Registers 328
Regulated power supplies 218

Regulation 217
Relative addressing mode 366
Reluctance 5
Reservoir capacitor 208
Resistance 3, 36
Resistivity 3
Retarding torque 515
Ring main 76
Ring modulator 462
Ripple factor 211
Ripple filters 214
Ripple reduction 208
R.M.S. values 54
Rotating field 486, 491

Sample and hold circuit 440
Saturated operation 117
Second order system 401
Self inductance 8, 17
Semiconductors 80
  doped 94
  intrinsic 82
Sequential networks 319
Serial signals 251
Series regulator 219
Settling time 399, 400
Shaded pole motor 503
Shaft position encoder 267
Shift register 329
Short circuit test 233
Shot noise 436
Shunt regulator 219
Sidebands 460
Signal bandwidth 451
Signal to noise ratio 435
Silicon controlled rectifier 131
Single sideband operation 460
S.L.S.I. 340
Slip 491
Small scale integration 340
Source 127
S.R. bistable 321
S.S.I. 340
Stabilised power supplies 218
Stack pointer 357
Staircase ramp analogue to digital converter 447
Standing waves 454
Star connection 67, 69
Star delta starter 498
State table 324

Strain gauges
  semiconductor 407
  wire 406
Super large scale integration 340
Super position theorem 45
Suppressed carrier operation 461
Synchronous motor 489
Synchronous speed 487

Transistor
  bipolar 108
  field effect 123
  junction 108
  metal oxide semi-conductor (MOST) 126
  switching 129
  unijunction 135
Transmission lines 452
Travelling waves 454
Triac 135
Trigger (T) bistable 326
Truth table 274

Unijunction transistor 135
Unipolar transistor 123
Universal motor 504

Very large scale integration 340
V.L.S.I. 340
Volatile memory 350
Voltage doubler 216
Voltage multiplying rectifier 216
Voltmeters 509
  digital 521

Ward Leonard drive 484
Weighted codes 266
Weighted resistor digital to analogue converter 442
White noise 404

Zenor diode 106
Zero order system 396

Computer, electronics and electrical
engineering books from Granada:

PROGRAMMING WITH FORTRAN 77
J Ashcroft, R H Eldridge,
R W Paulson and G A Wilson
0 246 11573 4

PROGRAMMING MICROCOMPUTERS WITH PASCAL
Martin D Beer
0 246 11619 6 In preparation

ELECTRONIC TEST EQUIPMENT
Edited by A M Rudkin
0 246 11478 9 In preparation

SIXTEEN-BIT MICROPROCESSORS
I R Whitworth
0 246 11572 6 In preparation

MICROPROCESSOR DATA BOOK
S A Money
0 246 11531 9 In preparation

MICROPROCESSOR DEVELOPMENT AND DEVELOPMENT SYSTEMS
Edited by Vincent Tseng
0 246 11490 8 In preparation

ELECTRIC CABLES HANDBOOK
Edited by D McAllister
0 246 11467 3 In preparation

HANDBOOK OF FIBER OPTICS
Edited by H F Wolf
0 246 11535 1

BASIC ELECTRICAL ENGINEERING
E C Bell and R W Whitehead
0 246 11477 0